NEW
MATHS
IN ACTION

S4³

Members of the
Mathematics in Action Group
associated with this book:

D. Brown
M. Brown
R.D. Howat
G. Meikle
E.C.K. Mullan
K. Nisbet

Nelson Thornes
a Wolters Kluwer business

Published in 2005 by:
Nelson Thornes Ltd
Delta Place
27 Bath Road
CHELTENHAM
GL53 7TH
United Kingdom

07 08 09 / 10 9 8 7 6 5 4 3

A catalogue record of this book is available from the British Library

ISBN 978 0 7487 9045 6

Cover llustration by The Studio Dog, Photodisc 26B (NT)
Illustrations by Oxford Designers and Illustrators
Page make-up by Tech-Set Ltd

Printed and bound in Croatia by Zrinski

Acknowledgements
Digital Vision 4 (NT): 8; Corel 497 (NT): 18; Jean Ryder (DHD Multimedia Gallery): 24;
Photodisc 76 (NT): 50, 56; Jeff Miller: 77; Photodisc 43 (NT): 95; Digital Stock (NT): 113;
Corel 412 (NT): 115; Ancient Art and Architecture Collection: 156; Corel 626 (NT): 165;
Corel 100 (NT): 175; Alfred Wegener/Heinz Nixdorf Museum: 192; Stockpix 3 (NT): 193;
Digital Stock 7 (NT): 193; Corel 174: 198; Photodisc 10 (NT): 203; Photodisc 4 (NT): 216;
Photodisc 54: 243; Photodisc 66 (NT): 287; Corel 626: 289; Photodisc 38 (NT): 291.

The publishers have made every effort to contact copyright holders but apologise if any
have been overlooked.

Contents

Introduction

This book has been specifically written to address the needs of the candidate following the Standard Grade mathematics course at Credit level. It is the second part of the course, completing the syllabus and complementing the coverage in Book S3[3].

The content has been organised to ensure that the running order of the topics is consistent with companion volumes S4[1] and S4[2] aimed at the Foundation and General level candidate respectively. This will permit flexibility of dual use and facilitate changing sections during the course if necessary.

Following the same model as in S3[3], each chapter follows a similar structure.

- A review section at the start ensures the knowledge required for the rest of the chapter has been revised.
- Necessary learning outcomes are demonstrated and exercises are provided to consolidate the new knowledge and skills. The ideas are developed and further exercises provide an opportunity to integrate knowledge and skills in various problem solving contexts
- Challenges, brainstormers and investigations appear throughout to provide an opportunity for some investigative work for the more curious.
- Each chapter ends with a recap of the learning outcomes and a revision exercise which tests whether or not the required knowledge and skills addressed by the chapter have been picked up.

The final two chapters in the book are devoted to revision.

- Chapter 13 contains revision for each chapter, revisiting each of the twelve chapters in this book. This complements the equivalent chapter in Book S3[3], together providing a topic by topic revision of the whole course.
- Chapter 14 provides an opportunity to prepare for assessment. The revision exercises here give the student a chance to select strategies. Where, before, questions on Pythagoras were encountered in the Pythagoras chapter and trigonometry could not be confused with it, being in the trigonometry chapter, the student now encounters mixed revision in the form of six two-part tests. Part 1 of each test contains non-calculator and calculator neutral questions; part 2 provides the questions where a calculator may be used.

A Teacher Resource CD-ROM provides additional material such as further practice and homework exercises and a preparation for assessment exercise for each chapter.

To assist with final revision, in association with Chapter 13 a revision checklist is included, and in association with Chapter 14 there is a 'prelim exam' laid out in a similar fashion to the final exam. A marking scheme is also included.

1 Area and volume

The 'surface area to volume' ratio is a very important one.

Animals with a high 'surface area to volume' ratio lose heat more rapidly.

Manufactured objects with a high 'surface area to volume' ratio cost more to produce.

1 Review

◀◀ **Exercise 1.1**

1 a Which of these units would be used to measure area?

  b Which of these units would be used to measure volume?

    i m^3 **ii**   litres   **iii** cm^2 **iv** mm^2 **v** cm^3

    vi hectare   **vii** ml   **viii** mm^3 **ix** km^2 **x** m^2

2 Calculate the area of each square using the formula $A = l^2$.

3 cm

7 cm

45 mm

3 Calculate the area of each rectangle using the formula $A = lb$.

2 cm

6 cm

51 mm

32 mm

3·2 m

7 m

4 **a** Write each length in centimetres.
      i 50 mm   **ii** 4 mm   **iii** 3 m   **iv** 12·74 m
   b Convert each length to millimetres.
      i 48 cm   **ii** 62·9 cm   **iii** 2 m   **iv** 28·88 m
   c Write these lengths in metres.
      i 900 cm   **ii** 385 cm   **iii** 6000 mm   **iv** 55 mm
   d How many metres are there in:
      i 7 km   **ii** 4·5 km   **iii** 2·632 km   **iv** 82·47 km?
   e Write each distance in kilometres.
      i 5000 m   **ii** 8500 m   **iii** 8427 m   **iv** 63 m

5 Each of these containers holds 1 litre when full.
   How much water is in each one? Give your answers in millilitres.

a    **b** 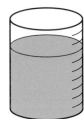   **c**

6 Convert these volumes to litres (1 litre = 1000 ml).
   a 7000 ml   **b** 3500 ml   **c** 4250 ml   **d** 750 ml

7 Convert these volumes to millilitres.
   a 2 litres   **b** 4·750 litres   **c** 0·25 litre   **d** 9·5 litres

8 Calculate   **i** the volume   **ii** the surface area of these cuboids.

a    **b** 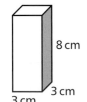   **c**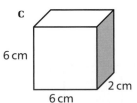

9 Convert each weight to kilograms (1 kg = 1000 g).
   a 6000 g   **b** 8425 g   **c** 7250 g

10 Convert each of these weights to grams.
   a 7 kg   **b** 3·576 kg   **c** 63 kg

11 Calculate the area of each right-angled triangle.

a 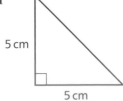   **b** 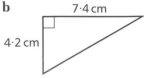   **c**

12 These are the nets of solids. Name the solids.

a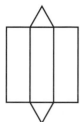
b
c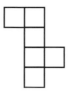
d

2 The area of a triangle

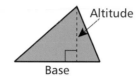

On this triangle the base and the altitude (height) are marked.

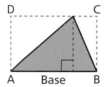

Area of rectangle ABCD = AB × BC

$\qquad$ = altitude × base

$\Rightarrow$ Area of $\triangle = \frac{1}{2}$ of altitude × base

In a right-angled triangle the altitude is one of the sides.
Area of $\triangle = \frac{1}{2}$ of altitude × base

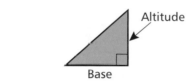

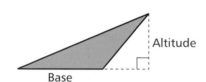

Here the altitude lies outside the triangle.

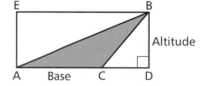

$\triangle ABC = \triangle ABD - \triangle BCD$

$\qquad = \frac{1}{2} AD \times BD - \frac{1}{2} CD \times BD$

$\qquad = \frac{1}{2} BD(AD - CD)$

$\qquad = \frac{1}{2}$ of altitude × base

The area formula still works.

The formula for the area of any triangle is: $A = \frac{1}{2} ab$

Example   Calculate the area of this triangle.

$A = \frac{1}{2} ab$

$\Rightarrow A = \frac{1}{2} \times 8 \times 6$

$\Rightarrow A = 24$

So the area of the triangle is 24 cm².

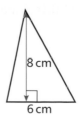

Exercise 2.1

1 Calculate the area of each triangle.

a
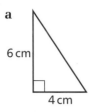
6 cm
4 cm

b
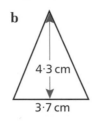
4·3 cm
3·7 cm

c
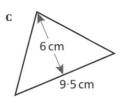
6 cm
9·5 cm

d

71 mm
23 mm

e
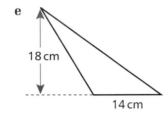
18 cm
14 cm

f
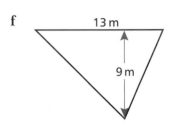
13 m
9 m

2 This is the flag for the eighth
hole at Shanklee Golf Club.
Calculate the area of one
side of the flag.

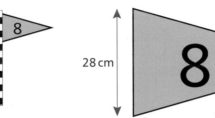
28 cm
51 cm

3 This sign is mounted on a wall in a city centre.
Calculate the area of the sign.

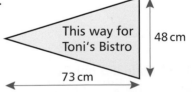
This way for
Toni's Bistro
48 cm
73 cm

4 The surface of this candle stand is an isosceles
right-angled triangle.
Calculate its area.

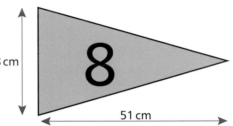
20 cm
20 cm

5 Calculate the area of this symmetrical road sign.

52 cm
GIVE
WAY
60 cm

6 A wallpaper border is made up of a line of congruent triangles.
Calculate the area of one triangle.

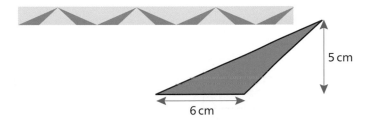

5 cm

6 cm

Exercise 2.2

1 Calculate the area of each triangle.

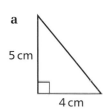

a
5 cm
4 cm

b
6 cm
38 mm

c
395 cm
4·2 m

d
14 cm
12 cm

2 Calculate:
  i the altitude (hint: Pythagoras)
  ii the area
of each isosceles triangle.

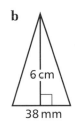

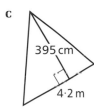

a
18 cm
10 cm

b 9 cm
10 cm

3 A flower bed in a park is an isosceles triangle.
Two sides of the flower bed are each 5 metres long.
The third side is 3 metres long.
Calculate the area of the flower bed.

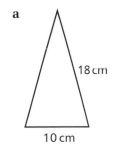

5 m
5 m
3 m

4 The diagram shows a ramp for disabled access to a village hall.
  a How much does the ramp rise over the 2 m?
  b Calculate the area of the side of the ramp.

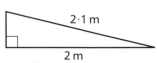

2·1 m
2 m

5 This triangular piece of cloth is just right for Mrs Rural's patchwork.
  a Calculate the length of side AB.
  b Calculate the area of the triangle.

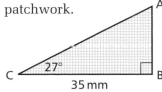

A
C 27°
35 mm
B

3 The area of a parallelogram

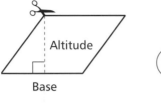

Diagram 1

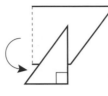

Diagram 2

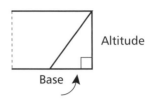

Diagram 3

Clearly since diagram 1 and diagram 3 are made up of the same parts, they will have the same area. Diagram 3 is a rectangle so area = base × altitude.

Therefore the area of a parallelogram = base × altitude
$A = ab$

Example
Calculate the area of this parallelogram.

$A = ab$

$\Rightarrow \quad A = 6 \times 14$

$\Rightarrow \quad A = 84$

So the area of the parallelogram is 84 cm².

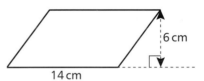

Exercise 3.1

1 Calculate the area of each parallelogram.

a

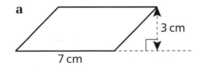

b

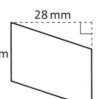

c

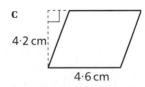

2 The side of an escalator is a parallelogram. Calculate its area.

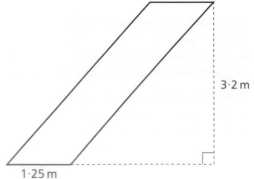

3 A path connects two paved areas in a park.

The path is in the shape of a parallelogram.

The paving slabs are squares of side 60 cm.

The path is 3 slabs wide.

Calculate the area of the path.

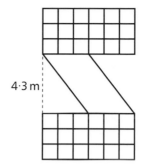

4·3 m

---

4 The rhombus and kite

A kite is a quadrilateral with one axis of symmetry passing through a pair of vertices.

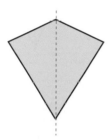 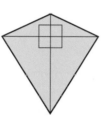 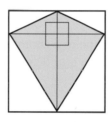 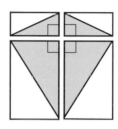

The axis means …

… diagonals intersect at right angles …

… so the kite can be surrounded by a rectangle …

… and we see the kite is half of this rectangle.

> The area of a kite = half of the area of the surrounding rectangle

In the example above we see that:

i the breadth of the rectangle is equal to the short diagonal

ii the length of the rectangle is equal to the long diagonal.

Area of kite $= \frac{1}{2} \times$ area of rectangle $= \frac{1}{2} \times$ length $\times$ breadth $= \frac{1}{2} \times$ diagonal$_1 \times$ diagonal$_2$

> The area of a kite = half of the product of the diagonals
> $$A = \tfrac{1}{2} d_1 \times d_2$$

A rhombus is just a special kite. It has a second axis of symmetry.

So the same formula will work for it.

Exercise 4.1

1 Calculate the area of each kite.

a

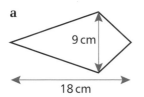

9 cm

18 cm

b

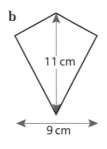

11 cm

9 cm

c

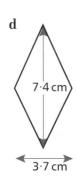

34 mm

15 mm

d

7·4 cm

3·7 cm

2 Charlie is buying a new kite and he decides that he wants the one with the biggest area.
Which of these three kites will he choose?

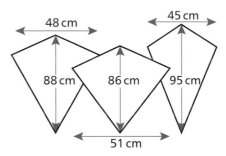

48 cm

88 cm

86 cm

45 cm

95 cm

51 cm

3 A fire-side rug is in the shape of a rhombus.
Its diagonals measure 155 cm and 62 cm.
Calculate its area.

4 A plantation of pine trees has a perimeter in the shape of a kite. The plantation is known locally as the Kite Wood.
What area does it cover?
Give your answer in:
a m²
b hectares.

125 m

450 m

5 Calculate the area of this sign on the front of a jeweller's shop.
It is a rhombus with diagonals of length 2·2 m and 1·6 m.

A Girl's Best Friend

6 This shape is known as a V-kite.
a Calculate its area by thinking of it as a big isosceles triangle with a small isosceles triangle cut out of it.
b Calculate its area by using the formula for the area of a kite.

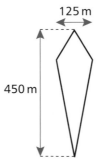

2 cm

5 cm

5 cm

5 The trapezium

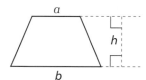

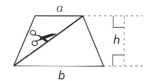

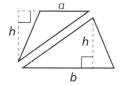

A trapezium ...   ... cut along one   ... gives two triangles.
diagonal ...

Area of the trapezium $= $ sum of the areas of the triangles

$$= \tfrac{1}{2}ah + \tfrac{1}{2}bh$$
$$= \tfrac{1}{2}h(a + b)$$

$$\boxed{\text{For a trapezium } A = \tfrac{1}{2}h(a + b)}$$

This is often considered as
half the sum of the parallel sides $\times$ the distance between them.

Example   Work out the area of this trapezium.

$$A = \tfrac{1}{2}h(a + b)$$
$$= \tfrac{1}{2} \times 8 \times (5 + 9)$$
$$= 4 \times 14$$
$$= 56$$

So the area of the trapezium is 56 cm².

Exercise 5.1

1 Calculate the area of each trapezium.

a

6 cm
7 cm
10 cm

b
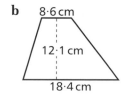
8·6 cm
12·1 cm
18·4 cm

c
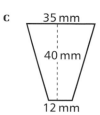
35 mm
40 mm
12 mm

2 Calculate the area of the side of this garden shed.
It is trapezium-shaped.

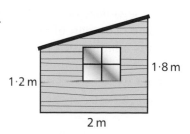

1·8 m

1·2 m

2 m

3 Work out the area of the side of this dust-pan.
It is trapezium-shaped.

4 The diagram shows a container for road salt.
The council have placed it beside a steep hill in case
cars get stuck in wintry conditions.
Its end is a trapezium.
Calculate the area of this end.

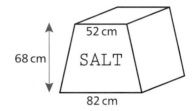

Exercise 5.2

1 Calculate the area of each quadrilateral.

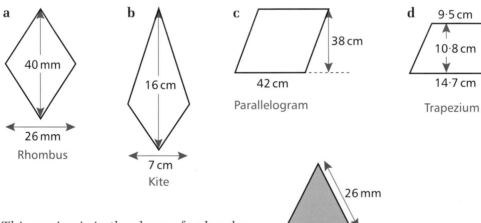

a
40 mm
26 mm
Rhombus

b
16 cm
7 cm
Kite

c
38 cm
42 cm
Parallelogram

d
9·5 cm
10·8 cm
14·7 cm
Trapezium

2 This earring is in the shape of a rhombus.
a Calculate the length of the other diagonal.
b Calculate the area of the earring.

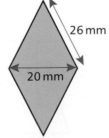

26 mm
20 mm

3 The lid of a jewellery box is made from two types
of wood. The main part of the lid is made from
ash which is inlaid with a parallelogram of walnut
as shown. The lid measures 15 cm by 12 cm.
a Calculate x.
b Work out the area of the parallelogram.

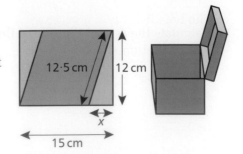

12·5 cm
12 cm
x
15 cm

4 The blade of this bricklayer's trowel
   is kite-shaped.
   a Calculate the length of a.
   b The blade's perimeter is 46·2 cm.
     Calculate the length of b.
   c Work out the area of the trowel's blade.

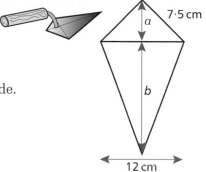

5 The faces of the shade for this table lamp are
   congruent symmetrical trapezia. One of the faces is
   shown in the diagram.
   a Use trigonometry to calculate y.
   b Work out the length of z.
   c Calculate the area of the face.

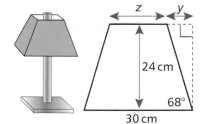

---

6 The area of composite shapes

Composite shapes are a combination of other basic shapes.

Example
Calculate the area of this road sign.
It is a composite of a rectangle and an isosceles
triangle.

Area of rectangle $= lb = 30 \times 110 = 3300$ cm^2
Area of triangle $= \frac{1}{2}ab = \frac{1}{2} \times 30 \times 15 = 225$ cm^2
So the total area $= 3300 + 225 = 3525$ cm^2

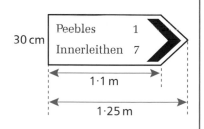

Exercise 6.1

1 Calculate the area of this road sign.

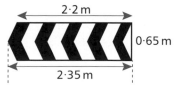

2 When laid flat, a poncho is a rhombus with a
   circular hole in the middle.
   The hole has a radius of 20 cm.
   Calculate the area of the material in the poncho.

   (Hint: area of circle $= \pi r^2$)

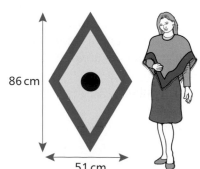

3 Sharon has made some 'bats' to help decorate
her house for Halloween.
The bats are made from three congruent kites,
each of which has diagonals of 18 cm and 10 cm.
Calculate the area of one bat.

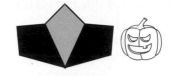

4 Meggie has bought a set of circular coasters.
Each coaster has a radius of 4·5 cm and has a
design based on six parallelograms.
  a Calculate the area of one parallelogram.
  b Calculate the total area of all the
  parallelograms.
  c Calculate the area of the background.

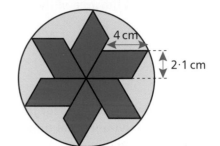

5 a Calculate the area of this sign which is suspended
above an aisle in a DIY superstore.

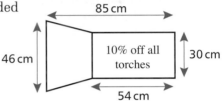

  b Another sign in the DIY superstore is made
  up of a rectangle and two congruent
  parallelograms.
  Calculate its area.

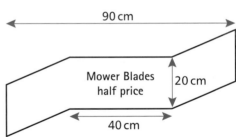

Exercise 6.2

1 This is the side view of the rubbish bin in a kitchen.
Calculate its area.

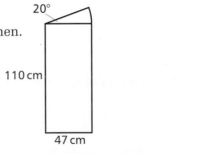

2 The diagram shows a field that can be split into a
rectangle and a right-angled triangle.
Calculate its area in:
  a m²
  b hectares.

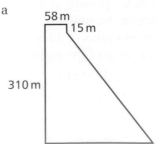

3 By first calculating the altitude of the isosceles triangle, work out the area of this sign.

(Hint: triangle and rectangle.)

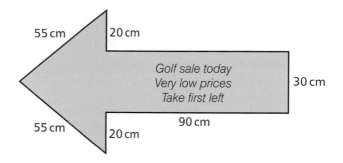

55 cm 20 cm

Golf sale today
Very low prices
Take first left

30 cm

55 cm 20 cm

90 cm

4 This drawing of a Christmas cracker is made of a rectangle and two congruent trapezia. Its total area is 2740 cm². Calculate the total length of the cracker.

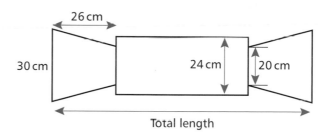

26 cm

30 cm 24 cm 20 cm

Total length

7 The volume and surface area of a prism

A prism is a solid that has a constant cross-section.
Wherever you cut it, parallel to the base, you get a shape congruent to the base.

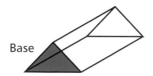
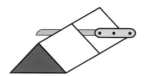
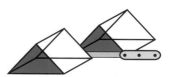

Base

This diagram shows a triangular prism but there are many types of prisms.
Each prism takes its name from the shape of the base.

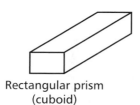

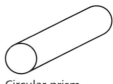

Rectangular prism
(cuboid)

Octagonal prism

Circular prism
(cylinder)

For every type of prism $V = Ah$
where A is the area of the base (or cross-section)
and h is the height (or distance between ends).

Example
Calculate the volume and the surface area of this triangular prism.

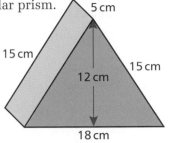

Volume
Area of base $= \frac{1}{2}ab = \frac{1}{2} \times 12 \times 18 = 108$ cm²
$$V = Ah$$
$\Rightarrow \qquad V = 108 \times 5$
$\Rightarrow \qquad V = 540$ cm³
So the volume is 540 cm³.

Surface area
Consider the net.

Area of triangular face C $= \frac{1}{2}ab = \frac{1}{2} \times 18 \times 12 = 108$ cm²
Area of sloping face A $\quad = 5 \times 15 = 75$ cm²
Area of bottom face B $\quad = 18 \times 5 = 90$ cm²
Total surface area $\qquad = 2 \times 108 + 2 \times 75 + 90$
$\qquad\qquad\qquad\qquad = 456$ cm²

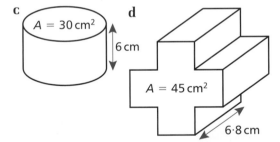

All lengths are in centimetres.

Exercise 7.1

1 Calculate the volume of each prism.

a
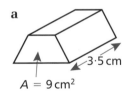
$A = 9$ cm²
3·5 cm

b
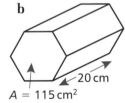
$A = 115$ cm²
20 cm

c
$A = 30$ cm²
6 cm

d
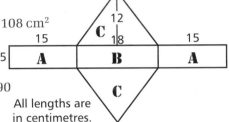
$A = 45$ cm²
6·8 cm

2 A cuboid is a rectangular prism.
  a Use the formula $V = lbh$ to work out the volume of this cuboid.
  b Treat the cuboid as a prism and use the formula $V = Ah$ to calculate the volume of the cuboid. Do this three times using a different face as the base each time.

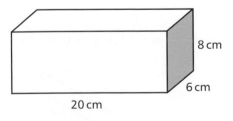
8 cm
6 cm
20 cm

3 Calculate the volume and surface area of each triangular prism.

a
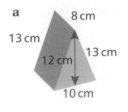
8 cm
13 cm
13 cm
12 cm
10 cm

b
15 cm 15 cm
40 cm   12 cm
18 cm

c
8 cm
21 mm
24 cm   26 cm 25 mm
10 cm

d
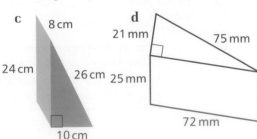
75 mm
72 mm

4 Mr Reed's doorstop is a triangular prism.
Its cross-section is a right-angled triangle.
Calculate its volume and surface area.

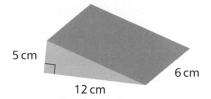

5 cm
12 cm
6 cm

5 This window box is a triangular prism.
It is made from plywood.
Its cross-section is a right-angled triangle.
  a What area of plywood is required to make the
    window box? (It has no top.)
  b What is the volume of the window box?
  c Compost is sold in 65 litre bags.
    How many bags are required to fill seven of these
    window boxes?

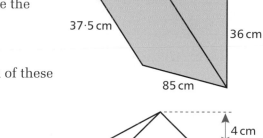

18 cm
37·5 cm
36 cm
85 cm

6 Stevie gets a pen for his birthday. It
comes in a presentation box which is a
triangular prism. The cross-section is
an isosceles triangle.
Calculate the volume and surface area
of the presentation box.

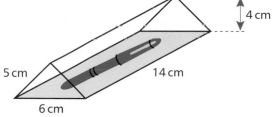

4 cm
5 cm
14 cm
6 cm

Exercise 7.2

1 A chocolate bar comes in a box which is a
triangular prism.
The cross-section is an equilateral triangle.
Calculate:
  a the altitude of the triangle
  b the total surface area of the box
  c the volume of the box.

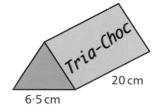

Tria-Choc
20 cm
6·5 cm

2 This is the net of a triangular prism.
  a Calculate the altitude of the triangular face
    where the base is 8 cm.
  b Calculate the surface area of the prism.
  c Calculate its volume.

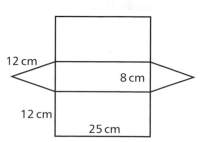

12 cm
8 cm
12 cm
25 cm

3 This tent is a triangular prism.
  a Calculate its volume.
  b Work out the area of the waterproof groundsheet
    (the floor).
  c Calculate the length of sloping edge a.
  d Work out the area of the canvas (everything but
    the floor).

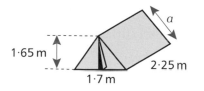

1·65 m
a
2·25 m
1·7 m

4 This vase is a triangular prism.
Its base is an equilateral triangle.
  a Calculate the altitude of the triangular base.
  b Calculate the area of the base.
  c The vase contains 1·3 litres of water.
     Calculate (to 1 decimal place) the depth of water in the vase.
  d Calculate the total surface area of the vase.
  e Each square centimetre of this glass (at this thickness) weighs 0·52 g.
     Each cubic centimetre of water weighs a gram.
     Calculate the total weight of vase and water.

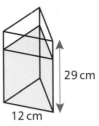

29 cm

12 cm

8 The surface area of a cylinder

Consider the net of this cylinder.

  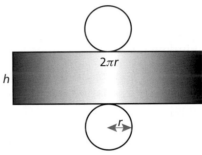

$2\pi r$

h

The curved surface of the cylinder opens out to be a rectangle.
The breadth of this rectangle is the same as the height of the cylinder.
The length of the rectangle is the same as the circumference of the lid.

> Area of rectangle = lb
> $\Rightarrow$ curved surface area of a cylinder = $2\pi rh$

Example

Find:
a the curved surface area to the nearest cm^2
b the total surface area of this cylinder.

4 cm

12 cm

a Curved surface area = $2\pi rh$
$$= 2 \times \pi \times 4 \times 12$$
$$= 301\cdot59 \ldots cm^2$$
$$= 302\ cm^2 \text{ (to the nearest } cm^2)$$

b Area of top = πr^2
$$= \pi \times 4^2$$
$$= 50\cdot265\ldots cm^2$$

Total surface area of cylinder = curved surface + top + bottom
$$= 301\cdot59\ldots + 2 \times 50\cdot265\ldots$$
$$= 402\cdot12\ldots$$

So the total surface area is 402 cm^2 (to the nearest cm^2).

Exercise 8.1

1 Calculate the curved surface area of each of these cylinders to the nearest whole number. Use the $\boxed{\pi}$ button on your calculator.

a

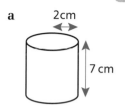

2 cm

7 cm

b

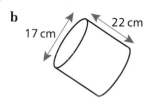

22 cm

17 cm

c

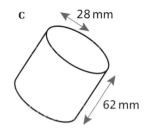

28 mm

62 mm

d

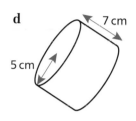

7 cm

5 cm

e

150 cm

2·2 m

2 Work out the area of the label on a tin of beans which is cylindrical with a diameter of 7·5 cm and a height of 11·5 cm.

3 The rollercoaster at Brightriver Vale Theme Park goes through this pipe.
The outside of the pipe needs repainting.
The pipe is 5·4 metres long and has a radius of 2·2 metres.
Calculate the area that needs repainting.

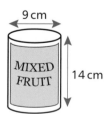

4 Calculate the total surface area of these cylinders to the nearest whole number.
  a diameter 8 cm, height 10 cm
  b radius 7·2 cm, height 9·5 cm
  c radius 52 mm, height 78 mm
  d radius 32 mm, height 6·7 cm
  e diameter 85 cm, height 1·25 m

5 Calculate the total surface area of this tin.

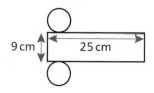

9 cm

MIXED FRUIT

14 cm

6 Calculate the total surface area of this mug (no top).
Its radius is 4 cm and its height is 10 cm.

7 Here is the net of a cylinder.
Work out:
  a its diameter
  b its total surface area.

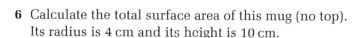

9 cm

25 cm

9 The volume of a cylinder

A cylinder is a circular prism

So $V = Ah$

$\Rightarrow \quad V = \pi r^2 h$

Example Calculate the volume of this cylinder.

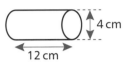

$r = 4 \div 2 = 2$ cm

$V = \pi r^2 h$

$\Rightarrow \quad V = \pi \times 2^2 \times 12$

$\Rightarrow \quad V = 150\cdot796\ldots$

$\Rightarrow \quad V = 151$ to 3 significant figures

So the volume is 151 cm³ to 3 significant figures.

Exercise 9.1

1 Calculate the volume of these cylinders (to 3 s.f.):
  a radius 8 cm, height 17 cm
  b radius 3 cm, height 7 cm
  c radius 8·2 cm, height 9·8 cm
  d diameter 145 cm, height 2·15 m

2 The picture shows cylindrical storage tanks at an oil refinery.
The tanks are 9 metres high and 2·5 metres in diameter.
Calculate the volume of one tank, correct to 2 s.f.

3 These slim cylindrical glass tumblers have a radius of
3·5 cm and a height of 18 cm.
Calculate the capacity of one tumbler (to 2 s.f.).

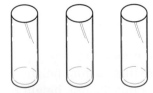

4 Calculate the volume of this cylindrical laundry basket.
It has a height of 86 cm and a diameter of 65 cm.

5 Elaine has three candles.
The greater the volume, the longer the candle burns.
Which one will burn for the longest period of time?

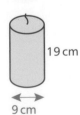

Exercise 9.2

Give your answers to 3 s.f. where appropriate.

1 This cylindrical rubbish bin has a diameter of 75 cm and a height of 1·15 m.
   Calculate its volume.

2 Karen buys her salt in a cylindrical container like the one shown here.
   She then puts it into her salt cellar.
   How many times can she fill her salt cellar from one container?

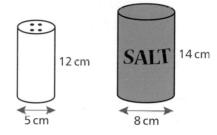

3 The guttering pictured here has a semi-circular cross-section.
   The guttering is therefore half a cylinder with a diameter of 15 cm.
   Calculate how much water can be held by a 5 metre length of this guttering.

4 Both of these cylinders have a volume of 1018 cm³.
   Calculate the missing dimension in each cylinder.

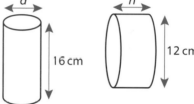

10 Composite solids

Exercise 10.1

1 The picture shows Mr Wood's garden shed.
   a Calculate the area of the trapezium that is the side of the shed.
   b Calculate the volume of the shed.
   c We can view the shed as a prism on top of a cuboid.

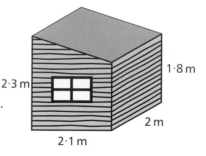

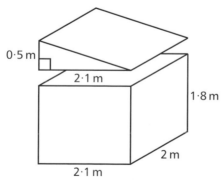

   Calculate: i the volume of the cuboid
               ii the volume of the triangular prism
               iii the total volume of the shed.

2 a Use Pythagoras' theorem to calculate the length of the sloping roof of Mr Wood's shed on page 19.
b Calculate the total surface area of the shed (including the floor).

3 a Calculate the volume of Mrs Anderson's garage.
b She intends to paint the outside of the walls with a wood preservative and re-felt the roof. What area requires painting?
c What area of felt will she need?

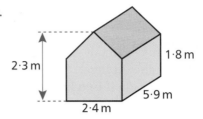

2·3 m 1·8 m 5·9 m 2·4 m

4 Myra's jewellery box is pictured on the right. Its cross-section is also shown.
The box is 27 cm long.
Calculate:
a its volume
b its surface area.

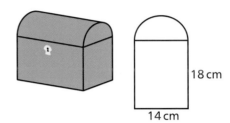

18 cm 14 cm

5 Calculate the volume and surface area (excluding the floor) of this greenhouse if its length is 2·4 m.

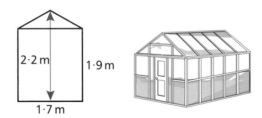

2·2 m 1·9 m 1·7 m

6 This plastic pipe has an inside diameter of 10·5 cm and an outside diameter of 11 cm.
For a 4·8 metre length of pipe, calculate:
a the volume of water that it can hold
b the volume of plastic required to make the pipe
c the curved surface area of the outside of the pipe.

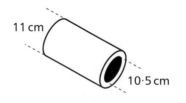

11 cm 10·5 cm

7 The upright for a standard lamp is made from three cylinders.
Each cylinder is 46 cm tall and they have diameters of 6 cm, 4 cm and 3 cm.
Calculate the total volume of the upright.

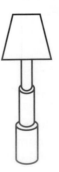

◀◀ **RECAP**

Area of a triangle

$$A = \tfrac{1}{2}ab$$

where a is the altitude and b is the base of the triangle.

Area of a parallelogram

$$A = ab$$

where a is the altitude and b is the base of the parallelogram.

Area of a kite or rhombus

$$A = \tfrac{1}{2}d_1 \times d_2$$

where d_1 and d_2 are the diagonals of the kite or rhombus.

Area of a trapezium

$$A = \tfrac{1}{2}h(a + b)$$

where a and b are the parallel sides of the trapezium and h is the perpendicular distance between them.

Volume of a prism

$V = $ Area of the base $\times$ height

for a *triangular prism*

$$V = \tfrac{1}{2}abh$$

where a is the altitude of the triangle, b is the base of the triangle and h is the distance between the ends, measured in the same units

for a *cylinder*

$$V = \pi r^2 h$$

where r is the radius of the base and h is the distance between the ends, measured in the same units.

Surface area of a triangular prism

$S = 2 \times$ area of triangular base $+$ areas of the three rectangular faces

Curved surface area of a cylinder

$$S = 2\pi rh$$

where r is the radius of the base and h is the distance between the ends, measured in the same units.

Volume and surface area of composite shapes

The volume and surface area of a composite shape can be calculated by splitting the shape into its simpler component parts.

1 Calculate the area of each of these quadrilaterals.

a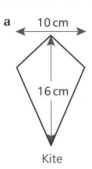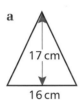
10 cm
16 cm
Kite

b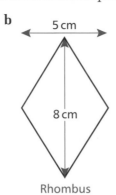
5 cm
8 cm
Rhombus

c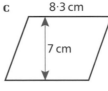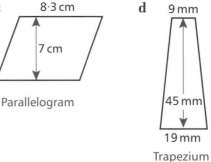
8·3 cm
7 cm
Parallelogram

d
9 mm
45 mm
19 mm
Trapezium

2 Calculate the area of each triangle.

a
17 cm
16 cm

b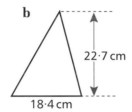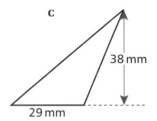
22·7 cm
18·4 cm

c
38 mm
29 mm

3 This shop sign is made up from a parallelogram and a triangle.
Calculate its area.

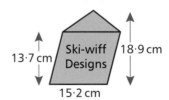
13·7 cm
Ski-wiff
Designs
18·9 cm
15·2 cm

4 **a** Work out the length of the slope, *l*, on this ramp (to the nearest centimetre).
b Calculate the volume of the ramp.
c Calculate the surface area.

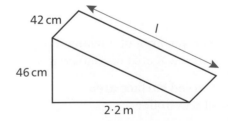
42 cm
l
46 cm
2·2 m

5 This rolling-pin is made from three cylinders.
The middle section is 32 cm long and has a diameter of 6 cm.
The two end-sections are 9 cm long and have a diameter of 2 cm.

a Calculate the total volume of the rolling-pin.
b What area does the pin cover when rolled once (curved surface area of middle section)?

REVISE

6 a Calculate the volume of this loaf of bread.
b Calculate its surface area.

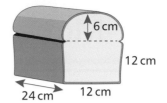

7 The front of a lawnmower is shaped
like a triangular prism sitting on
top of a cuboid.
Calculate its volume.

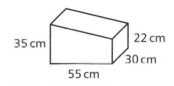

2 Money

In 1799, Prime Minister William Pitt the Younger introduced income tax to fund the war against Napoleon. The rate was 10% of a person's total income above £60 per year.

In 2003–04 the average wage for all full-time employees was £476 per week. The average credit card debt was £1100 for each adult in the UK.

1 Review

◀◀ Exercise 1.1

1 a Round each of these amounts to the nearest 10p:
 i 85p **ii** £2·83 **iii** £34·07 **iv** £63·99

 b Round each of these amounts to the nearest 1p:
 i 27·3p **ii** £4·333… **iii** £25·975 **iv** £59·999

2 Don works seven and three-quarter hours each day.
 How many hours does he work in a 5 day week?

3 Find:
 a $\frac{1}{2}$ of £8·50 **b** $\frac{1}{3}$ of £11·76 **c** $\frac{1}{4}$ of £6·24 **d** $\frac{3}{4}$ of £7·96

4 Calculate:
 a 1% of £3000 **b** 4% of £700 **c** 10% of £39 **d** 5% of £274
 e 25% of £17 **f** 75% of £826 **g** 15% of £8500 **h** 8% of £164

5 a Giles saves a regular amount each week for one year.
  At the end of the year he has saved £1248.
  How much does he save each week?
 b Walter repays a loan by making 12 monthly payments of £175.
  Calculate his total repayments.
 c Gill's flat costs her a total rent of £3336 for a year. Calculate the rent for one month.
 d Jo's weekly newspaper bill comes to £5·35. Calculate the amount she spends on newspapers in a year.

6 **a** Write these as decimal fractions:

    i 25%   **ii** $17\frac{1}{2}\%$ **iii** 7%   **iv** 250%

 b Write these as percentages:

    i 0·03   **ii** 0·625   **iii** $\frac{2}{5}$ **iv** $\frac{1}{8}$

7 Calculate:

  a 12% of £9360   **b** 2·5% of £64·73   **c** 17·5% of £238·20

8 **a** Last year the population of Blackloch was 1580. It has risen by 5% since then. What is the population now?

 b The population of Redburn is 798.

   Since last year the population has fallen by 5%.

   What was the population last year?

---

2 Money calculations

Example 1   Estimate £48·89 × 72.

              £48·89 is approximately £50 and 72 is approximately 70.

              Estimate is £50 × 70 = £3500.

Example 2   Calculate £5·83 × 80.

              £5·83 × 80 = £5·83 × 10 × 8 = £58·30 × 8 = £466·40

Example 3   Calculate £5·65 ÷ 0·05.

              £5·65 ÷ 0·05 = £565 ÷ 5 = £113

Example 4   Calculate 17·5% of £54·80.

              10% of £54·80 = £5·48

               5% of £54·80 = £2·74   (half of 10%)

             2·5% of £54·80 = £1·37   (half of 5%)

            17·5% of £54·80 = £9·59   (10% + 5% + 2·5%)

Exercise 2.1

1 Estimate:

  a £195·99 + £324·49 + £283·79   **b** £595·80 − £213·45

  c £78·49 × 62   **d** £3999 ÷ 83   **e** £6·85 × 365   **f** £38 875 ÷ 52

2 Calculate:

  a £17·99 + £25·99   **b** £9·65 + £28·99 + £36·35   **c** £2000 − £78·62

  d £58·75 × 400   **e** £568 ÷ 25   **f** 63·26 × 45 + £63·26 × 55

  g £4705 ÷ 500   **h** £67·72 ÷ 0·8

3 Calculate:

  a the total cost of the TV and video

  b the difference between the prices.

VIDEO
£479·99

Flat Screen TV
£1125·45

4 **a** Paul pays £9·99 per month for his internet connection.
How much does it cost him over 18 months?

b Pat pays £17·99 per month for broadband.
Calculate her costs for 12 months' connection.

5 Calculate 17·5% of:

a £48   **b** £5860   **c** £7·60   **d** £0·80

6 Mr MacDonald pays £8539 for an extension.
He takes out a loan which costs him an extra 20%.
He makes 24 equal monthly repayments.
Calculate:

a the total amount he has to repay

b the monthly repayment.

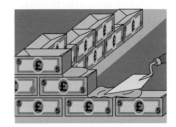

7 Copy and complete the table.

Fraction	$\frac{4}{5}$			$\frac{2}{3}$			$2\frac{3}{4}$		
Decimal		0·96			0·625			3·166…	
Percentage			45%			2·5%			462%

8 Calculate the missing entries in the table.

Item	Cost price	Selling price	% profit/loss
Vacuum cleaner	£180·00	**a**	8% profit
Washing machine	£360·00	£480·00	**b**
Dish washer	**c**	£305·20	9% profit
Spin dryer	£156·00	£148·20	**d**

9 In one week Petra fills her car's tank three times.
The table shows the amount of fuel and the price.
Calculate the average (mean) price of a litre of fuel,
correct to 1 decimal place of a penny.

Volume	Price per litre
37 litres	82·9p
42 litres	84·9p
45 litres	79·9p

10 **a** Which is the better buy?

TOP BREW TEA
£5·99 for 360 bags

TOP BREW TEA
£3·99 for 240 bags

b Can you see a quick, non-calculator, method of comparing prices?

3 Wages and salaries

Example 1   How much will the successful applicant earn in a month?

> SYSTEMS ANALYST
> Annual salary £30K

Monthly salary = £30 000 ÷ 12 = £2500

Example 2   Alice works a 35 hour week. Her hourly rate is £7·85 per hour. Calculate her:
    a weekly wage   **b** total earnings in a year.

    a Weekly wage = £7·85 × 35 = £274·75
    b Earnings in a year = £274·75 × 52 = £14 287

Exercise 3.1

1 Hillary's weekly wage is £367·65.
How much does she earn in:
    a 4 weeks
    b 20 weeks?

2 Jim is paid £24 750 a year as a lecturer. He works 200 days in a year.
How much does he earn per day?

3 How much will the successful applicant earn in a year?

> **EVENING BUGLE**
> **JOURNALIST REQUIRED**
> Salary £3190 per month

4 Calculate:
    a the minimum
    b the maximum
monthly salary for this post.

> CIVIL ENGINEER
> Car Included
> £27 120 – £32 700 p.a.

5 Martin gets the job.
How much will he earn in:
    a a week
    b a year?

> WANTED
> **MACHINE OPERATOR**
> £6·74 per hour for 37 hour week

6 The managing director of Beta Bio-chemicals is paid an annual salary of £1·5 million.
He claims to work a 50 hour week.
How much does he earn per hour, correct to the nearest £10?

7 Delia sees two jobs for chefs.

How many hours per week would she have to work to earn more at the Riverside Restaurant than the Kosy Kitchen?

8 Mr Smith is a technician. He works a 37·5 hour week at an hourly rate of £8·45. His wife, a personal assistant, is paid a monthly salary of £1596.
  a Who earns more in a year? By how much?
  b Calculate their total annual income.

Example 3   Glenn's annual salary is £18 640. He receives a pay rise of 4·5%. Calculate his salary after the rise.

Increase = 4·5% of £18 640 = 4·5 ÷ 100 × £18 640 = £838·80
New salary = £18 640 + £838·80 = £19 478·80

Example 4   Josie's new weekly wage, after a 5% pay rise, is £363·93. Calculate her wage before the rise.

Previous weekly wage is 100%.
Wage after the rise is 105%.
105% of original = £363·93
100% of original = £363·93 ÷ 105 × 100 = £346·60

Exercise 3.2

1 Mr Thompson is paid £7·84 per hour. He works 37·5 hours per week.
  a Calculate his weekly pay.
  b He is given a pay rise of 38p per hour.
    Calculate his new weekly wage.

2 Mrs Davies starts as a manager at Giant Gas Plc on an annual salary of £28 600.
  a Calculate her monthly salary.
  b She receives annual increments of £1240.
    Calculate her
    i annual   **ii** monthly salary after 2 years.

3 Ms Rutherford is appointed.
  a Calculate her annual salary.
  b After 1 year she is promised a 5% pay rise.
    Calculate her
    i monthly   **ii** annual salary after the pay rise.

INTERNATIONAL CHEMICALS
EXPERIENCED CHEMIST
monthly salary £2975

4 Last year Mr Wood earned £18 600 as a carpenter. He receives a 3·5% pay rise.
  a Calculate his new annual wage.
  b He works a 36 hour week.
    Calculate his new   **i** weekly wage   **ii** hourly rate of pay.

5 Patrick works at a DIY store for 36 hours a week.
  His hourly rate of pay is increased from £6·50 to £6·76.
  a By how much has his weekly pay risen?
  b What is his new
    i weekly
    ii annual wage?
  c Calculate his pay rise as a percentage of his old rate.

6 Ron and Ruth both work for Eureka Electronics.
  Ron earns £1250 per month and Ruth £2180 per month.
  The company awards everyone an 8% pay rise.
  a Calculate their monthly salaries after the rise.
  b How much more is Ruth's pay rise worth compared to Ron's?

7 Since this advert was placed the pay rates have
  increased by 2%.
  a What should the new annual pay scale be?
  b Asha gets the job.
    Calculate her monthly salary, including the pay rise,
    if she is employed at the   **i** bottom   **ii** top of the pay scale.

> Creative Media
> Assistant
> *£16 975 – £20 418 p.a.*

8 After a pay rise of 3·5%, Matt's annual salary rose to £27 324.
  Calculate his salary before the increase.

---

4 Time-sheets and overtime

Example 1

TIME-SHEET					
Name: Leah Jackson		Employee No. 65			Week No. 12

	In	Out	In	Out	No. of hours worked
Mon	08 00	12 00	13 00	17 00	8
Tue	08 00	12 00	13 00	17 30	8·5
Wed	08 30	12 30	13 15	17 00	7·75
Thu	08 45	13 15	14 30	17 15	7·25
Fri	08 45	12 30	13 15	16 00	6·5
				Total	38 hours

Look at Friday. Leah worked from 08 45 to 12 30 before lunch.

That's 3 hours 45 minutes (3·75 hours).

After lunch she worked from 13 15 to 16 00.

That's a further 2 hours 45 minutes (2·75 hours).

Total number of hours worked on Friday = 3·75 + 2·75 = 6·5 hours.

Exercise 4.1

1 Hillary works a 5 day week from 8.45 am until 5 pm each day.
She has 45 minutes for her lunch break.
How many hours does she work:
a each day
b in a week?

2 Ernie works night shifts. Each night he starts at 20 30 and works through to 08 15.
He has a 1 hour break and works 4 nights a week.
How many hours does he work:
a each night
b in a week?

3 Saeed works mornings, from 7.30 am until 12.15 pm,
from Monday to Saturday. He earns £8·24 per hour.
a Calculate his total hours for a week.
b Calculate his weekly pay.

4 This is Ann Farmer's time-sheet.

TIME-SHEET

Name: Ms A Farmer   Employee No. 61   Week No. 8

	In	Out	In	Out	No. of hours worked
Mon	08 15	13 00	13 45	17 00	…
Tue	08 30	12 45	13 30	17 15	…
Wed	08 00	13 15	14 00	17 00	…
Thu	08 45	13 00	13 45	17 15	…
Fri	08 30	13 15	14 15	17 45	…
					Total … hours

a Write down the number of hours she worked each day.
b Find her total hours for the week.
c Her hourly rate of pay is £8·42. Calculate her wage for this week.

5 This is Bert Sharp's time-sheet.

TIME-SHEET

Name: Mr B Sharp   Employee No. 99   Week No. 52

	In	Out	In	Out	No. of hours worked
Mon	09 00	13 00	14 00	**a**	8
Tue	08 30	13 00	**b**	17 30	8
Wed	08 15	**c**	14 30	17 00	7·5
Thu	**d**	12 45	13 30	17 15	7·75
Fri	08 45	13 15	14 00	**e**	…
					Total 38 hours

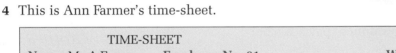

Find the entries **a–e**.

Example 2

When John works overtime he is paid at time and a half. His basic rate is £6·20 per hour. Calculate his pay for **a** 1 hour of overtime **b** 2 hours 30 minutes of overtime.

a For 1 hour of overtime his pay = £6·20 × 1·5 = £9·30

b For 2 hours 30 minutes of overtime his pay = £9·30 × 2·5 = £23·25

Exercise 4.2

1 Calculate the overtime pay.

Basic pay per hour	Overtime rate	Overtime pay per hour
£6·40	Time and a half	
£9·75	Double time	
£7·85	Time and a third	
£8·24	Time and a quarter	
£10·92	Time and three-quarters	

2 At the factory where Carly works, overtime is paid at time and a half.
The basic rate is £8·60.
Calculate Carly's pay for:
    a 1 hour of overtime   **b** 4 hours 30 minutes of overtime.

3 On Saturday Laura works 4 hours at the basic rate and does
3 hours' overtime. The basic rate is £7·20 per hour.
The overtime rate is time and a third.
    a How much does she earn at the basic rate?
    b How much is she paid for the overtime?
    c What is her total pay for the day?

4 Michael's basic week is 35 hours. His basic rate of pay is £9·20.
Overtime is paid at time and a quarter.
For a week when he works 40 hours calculate his:
    a basic pay for the 35 hours   **b** overtime pay   **c** total wage for the week.

5 Freddie Lee's basic week is 37·5 hours. Any time above this is paid at time and three-quarters. The basic rate is £7·84.

TIME-SHEET				
Name: Mr F Lee		Employee No. 21		Week No. 27

	In	Out	In	Out	No. of hours worked
Mon	08 30	13 00	13 45	17 15	...
Tue	08 30	12 30	13 30	17 15	...
Wed	08 00	13 00	13 45	17 00	...
Thu	08 15	13 00	13 30	17 00	...
Fri	08 45	13 00	13 45	17 00	...
				Total ... hours	

Fill in his time-sheet and calculate his total wage for the week.

5 Piecework and commission

Not everyone is paid by the hour.
Some people are paid for the number of items made or tasks completed.
This is called **piecework**.
People in selling jobs are often paid according to the value of the goods sold.
This is called **commission**.

Example 1   Stella does private tuition for pupils who want extra help for their
maths exams.
She charges £12·50 a lesson.
a How much does she earn in a week when she gives 9 lessons?
b What would she charge for a course of 20 lessons?

a Pay for 9 lessons = £12·50 × 9 = £112·50
b Pay for 20 lessons = £250

Example 2   Sales staff at Premier PCs are paid a commission of 3% on their sales.
Calculate the commission earned on a sale of value £3500.
Commission = 3% of £3500 = 3 ÷ 100 × 3500 = £105

Exercise 5.1

1 Mandy charges £9·99 for a manicure. How much does she earn in:
a a day when she does 8 manicures
b a month when she does 200 manicures?

2 Countrywide Estate Agents pay their staff a commission of 0·5% on their sales.
Calculate the commission on the sale of a property for:
a £35 000
b £225 000.

3 Meera produces special headed notepaper
for individuals and companies.
What does she charge:
a Mrs Wright for 40 sheets
b Atom Advertising for 5000 sheets?

Headed Notepaper
85p per 10
55p per 100

4 The sales staff at Country Caravans are paid a basic salary
plus commission of 2% of their sales. Gary's basic salary
is £1280 per month.
a How much commission does he earn in a month
when his sales total £45 000?
b How much, in total, should he expect to earn in a year
if this is a typical month?

5 One day Colin repairs 12 pairs of shoes
that need only their heels replacing and
5 pairs that need both heels and soles.
Calculate his total earnings.

SHOE REPAIRS
Heels £5·50
Heel and Sole £11·80

6 Charlotte works at a stockbrokers, selling shares.
She earns a commission of 0·75% on her sales.
   a Calculate her commission on shares worth £25 000.
   b On one deal she earns £450. What is the value of the shares she has sold?

7 Mr and Mrs Holmes run a campsite.
   a Calculate the total fees they collect on a night
   when they have 15 caravans and 20 tents.
   b On another night the total revenue is £247.
   If the number of caravans is 17, how many tents
   are there?

COSY CAMPING
CARAVANS
£12·50 PER NIGHT
TENTS £5·75 PER NIGHT

8 Sanjeev sells office furniture.
He is paid a basic salary of £1700 per month plus
commission of 3% on the value of his sales.
In January he sells goods worth £15 000.
   a Calculate his total earnings in January.
   b His total earnings for the year actually add up to £25 440.
   Calculate the total value of his sales for the year.

6 Payslips

Workers receive payslips which show all their earnings for the latest week or
month.

It also shows all the deductions.

Earnings include basic wage, overtime, commission and bonuses.

Deductions include income tax, National Insurance Contributions (NIC),
superannuation (pension) fund and union dues.

A superannuation fund provides an income when a worker is retired.

The following terms should be understood:

● gross pay – total earnings before deductions are made, often referred to as 'the top
   line'
● deductions – money paid out of your earnings
● net pay (take-home pay) – total earnings after deductions, often referred to as 'the
   bottom line'.

Net pay = Gross pay − Deductions

Example 1   This is Ian Wilde's payslip.

	Name I. Wilde	Employee number 135		NI number YM135678A	Week number 20
	Basic pay £473·50	Overtime £48·40	Commission —	Bonus £25	Gross pay
	Income tax £90·37	NI £42·08	Superannuation £24·54	Other deductions —	Total deductions
					Net pay

Personal information → (Name row)
Basic pay and extras → (Basic pay row)
Deductions → (Income tax row)
Take-home pay → (Net pay row)

Calculate his: **a** gross pay   **b** total deductions   **c** take-home pay.

Gross pay = £473·50 + £48·40 + £25 = £546·90
Total deductions = £90·37 + £42·08 + £24·54 = £156·99
Net pay = Gross pay − Total deductions = £389·91

Exercise 6.1

1 For Nisa Parker calculate her:
  a gross pay   **b** total deductions   **c** take-home pay.

Name N Parker	Employee number 246		NI number YK560489B	Week number 33
Basic pay £502·65	Overtime —	Commission £64·75	Bonus £20	Gross pay
Income tax £104·13	NI £45·28	Superannuation £27·28	Other deductions —	Total deductions
				Net pay

2 For Rob Steele calculate his:
  a gross pay   **b** total deductions   **c** take-home pay.

Name R Steele	Employee number 39		NI number XY741596A	Week number 5
Basic pay £2648·76	Overtime £134·63	Commission —	Bonus £50	Gross pay
Income tax £526·33	NI £255·28	Superannuation £138·59	Other deductions £14·85	Total deductions
				Net pay

3 Make out a payslip for J Chang; employee number: 205; NI number CD543294A; Month 8; Basic pay: £1953·70; Commission: 3% of £2500; Income tax: £324·60; Superannuation: 6% of gross pay; NIC: £157·38; Other deductions: £17·90.

4 Find the missing entries in this payslip.

Name C Ryder	Employee number 329		NI number MN423087B	Week number 27
Basic pay	**Overtime** £48·05	**Commission** —	**Bonus** £24	**Gross pay**
Income tax	**NI** £43·02	**Superannuation** £28·53	**Other deductions** £9·32	**Total deductions** £174·34
				Net pay £385·13

7 National Insurance Contributions (NIC)

This is a government tax on earnings, intended to contribute towards unemployment, ill-health and retirement payments.
The table gives the rates for 2004–5.

Rate	Weekly	Annual
0%	<£91	<£4745
11%	£91–£610	£4745–£31 720
1%	>£610	>£31 720

Example 1
Calculate the amount these workers pay in National Insurance Contributions each week.
a Fred, wage £70 per week
b Mavis, wage £121 per week
c Darren, earnings £750 per week

a NIC for Fred = 0% of £70 = £0
b NIC for Mavis = 11% of (£121 − £91) = 11% of £30 = £3·30
c NIC for Darren = 11% of (£610 − £91) + 1% of (£750 − £610)
$\qquad$ = £57·09 + £1·40 = £58·49

Exercise 7.1

1 **a** Mac earns £91 per week. Does he have to pay any National Insurance Contributions?
b Scott earns £92. Does he have to pay any National Insurance Contributions? If so, how much?

2 Mrs Potts' weekly wage is £351.
   a On how much of this wage does she pay National Insurance Contributions?
   b How much does she pay?

3 Ms Webb's weekly pay is £800.
   a How much of this is charged at **i** 11% **ii** 1%?
   b Calculate her total contribution.

4 Karen's total earnings for the year are £18 445.
   Calculate her NIC for the year.

5 Mrs Brodie's annual salary is £45 000.
   How much does she pay in National Insurance Contributions
   a in the year   **b** per month?

6 Brian's weekly NIC is £32·45. Find his weekly wage.

7 Patsy earns £592 per week. She receives a 6% pay rise.
   By how much does her National Insurance Contribution rise?

8 Robbie, a football star, pays £10 150·05 in
   National Insurance Contributions.
   Calculate his earnings for the year.

---

8 Income tax

If your income in a tax year is below a certain amount you do not pay tax.
This amount is called your tax allowance.
The tax allowance is made up from a personal allowance plus any other special allowances.
(Other tax allowances include expenses such as special clothing or equipment needed for work and membership of professional bodies.)

In the tax year 2004–5 the personal allowance was £4745.
Income above the tax allowance is called taxable income.

Income tax is paid on your taxable income.
The table shows the income tax rates for 2004–5

Taxable income	Rate of tax
£0–£2020	10%
£2020–£31 400	22%
over £31 400	40%

Example 1 Rebecca's total annual income is £27 000. Her personal allowance is £4745.
Calculate: **a** her taxable income
              b the tax payable at   **i** 10%   **ii** 22%
              c the total tax payable.

a Taxable income = £27 000 − £4745 = £22 255
b i   tax at 10% = 10% of £2020 = £202
   ii tax at 22% = 22% of (£22 255 − £2020) = 22% of £20 235 = £4451·70
c Total tax payable = £202 + £4451·70 = £4653·70

Exercise 8.1

1 Alison's total income for the year is £4600. Her sister, Amy, has an income of £5000. They both have a tax allowance of £4745. Do either of them have to pay tax?

2 Calculate the taxable income for these people:

Name	Income in tax year	Tax allowance	Taxable income
Ms Hawke	£12 000	£4745	
Mr Swift	£4900	£5000	
Mrs Partridge	£34 043	£5105	
Mr Martin	£17 103	£4995	

3 Mr Patel's taxable income is £1500.
   a At what rate is he taxed?
   b How much tax does he pay?

4 Mrs Parry's taxable income is £2520.
   a How much of her income is taxed at   **i** 10%   **ii** 22%?
   b What is the tax payable at   **i** 10%   **ii** 22%?
   c What is the total tax payable?

5 Elaine has a tax allowance of £5000. Her total income is £6000.
   a On how much of her income is she taxed?
   b How much tax does she pay?

6 Ron's total income is £21 545. His tax allowance is £4745.
   Calculate:
   a his taxable income
   b the tax payable at   **i** 10%   **ii** 22%
   c the total tax payable.

7 Copy and complete the table for these people. (Tax allowance is £4745.)

Name	Total yearly income	Taxable income	Amount payable at 10%	Amount payable at 22%	Total payable
Mr Hawthorn	£5000				
Ms Greenwood	£10 000				
Mrs Ash	£20 000				
Mr Appleyard	£30 000				

8 Gordon is paid £10 per hour for a 36 hour week. His tax allowance is £4745.
   Calculate:
   a his total annual income
   b the amount of income tax he pays
   c his annual income after paying income tax.

Example 2 Paula, a successful business woman, has a total income of £70 000.
Her tax allowance is £4745.

Calculate: **a** the total tax payable **b** her income after tax is deducted.

a Taxable income = £70 000 − £4745 = £65 255
Tax payable at 10% = 10% of £2020 = £202
Tax payable at 22% = 22% of (£31 400 − £2020) = £6463·60
Tax payable at 40% = 40% of (£65 255 − £31 400) = £13 542
Total tax payable = £202 + £6463·60 + £13 542 = £20 207·60
b Income after tax = £70 000 − £20 207·60 = £49 792·40

Exercise 8.2

1 Marvin pays tax at the 40% rate on £6700 of his salary.
Calculate the tax he pays on this part of his income.

2 How much can a person earn in a tax year before they pay tax at the
a 22% rate **b** 40% rate? (Assume a tax allowance of £4745.)

3 Mrs Crawley, a head teacher, has an income of £52 000. Her tax allowance is £5750.
Copy and complete the following to calculate her tax demand.

Taxable income =

Tax payable at 10% = 10% of … =

Tax payable at 22% = 22% of (… − …) =

Tax payable at 40% = 40% of (… − …) =

Total tax payable = …

4 Repeat question **3** for Mr Smart, a media personality, with an income of £120 000 and a tax allowance of £7450.

5 During the tax year Sam Swinger's earnings as a professional golfer are £260 000.
His tax allowance is £8500.
a How much income tax does he have to pay?
b How much will he have left after tax?

6 Vicky, a barrister, is paid a monthly salary of £6280. Her tax allowance is £6345.
Calculate:
a the tax due for the year
b the tax due each month (assume the same amount is paid each month)
c her monthly salary after tax.

7 a Calculate the tax payable on an annual income of £80 000 (tax allowance £6000).
b Express the income tax paid on £80 000 as a percentage of the total income.

8 Louise's total earnings for the year are £24 500. Her tax allowance is £5000.
She pays 6% of her salary into her superannuation fund.
a Calculate these total deductions for the year:
i income tax **ii** National Insurance Contributions **iii** superannuation.
b What is her total net pay for the year?
c Express her net pay as a percentage of her gross pay.

Brainstormer

Vince gets a tax statement.

Calculate his taxable income.

**INLAND REVENUE
Amount to pay £13 834**

Investigation

Find out:

a when the financial year starts and ends
b who gives the 'Budget' speech
c when and where the speech is made
d what the present allowances and rates of income tax are.

9 Bank and building society accounts

Bank and building society accounts help us to manage our money.

Accounts may have the following facilities:

- cheque book – to pay for goods or services
- credit and debit card – to pay for goods or services
- cash card – to withdraw money from cash machines
- bank statement – to record transactions and give an up-to-date balance.

Here is an example of a statement.

STATEMENT			
Date	**Paid out**	**Paid in**	**Balance**
1 Feb			125·00
5 Feb	150·00		25·00 DR
12 Feb		136·75	111·75
24 Feb	120·40		8·65 DR

On 5 February there had been £125 in the account but £150 was withdrawn.
£125 − £150 = −£25 (overdrawn).
On 12 February £136·75 was paid in.
£136·75 + (−£25) = £111·75.
On 24 February £120·40 was withdrawn but only £111·75 was in the account.
£111·75 − £120·40 = −£8·65 (overdrawn).

Exercise 9.1

1 One of Mr P Kirk's cheques is shown below.

  a Who is the cheque made out to?

  b How much is it for?

  c What is his account number?

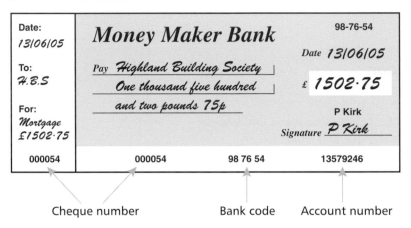

Cheque number   Bank code   Account number

2 Copy these statements and complete the balance columns.

a

STATEMENT			
Date	Paid out	Paid in	Balance
1 Apr			280·00
8 Apr	136·00		
18 Apr		75·00	
26 Apr		179·00	

b

STATEMENT			
Date	Paid out	Paid in	Balance
1 Sep			124·70
7 Sep	88·90		
17 Sep	64·30		
24 Sep		146·50	

c

STATEMENT			
Date	Paid out	Paid in	Balance
1 Oct			50·00 DR
14 Oct	25·35		
20 Oct		158·00	
31 Oct	95·99		

d

STATEMENT			
Date	Paid out	Paid in	Balance
1 Dec			153·73 DR
8 Dec		214·76	
19 Dec	94·49		
23 Dec		624·08	

3 Calculate the balances **a–d** for this statement.

STATEMENT				
Date	Details	Paid out	Paid in	Balance
1 Jan	Balance brought forward			1254·86
8 Jan	Direct Debit Highland Building Soc.	1275·48		**a**
11 Jan	Paid in at MM Bank Glasgow		580·00	**b**
20 Jan	Cheque 000055	467·99		**c**
23 Jan	Cash Edinburgh	200·00		**d**

4 Calculate the balances **a–e** for this statement.

STATEMENT				
Date	Details	Paid out	Paid in	Balance
1 May	Balance carried forward			137·42 DR
5 May	Titan Textiles		1859·63	a
12 May	Direct Debit Eversafe BS	1543·87		b
19 May	Cheque 000654	632·96		c
20 May	Paid in at MM Bank Inverness		500·00	d
22 May	Gold Credit Card	85·59		e

10 Savings and interest

When money is placed in a bank or building society account it normally earns interest. The interest paid is a percentage of the amount deposited.

Example 1   Joe invests £5000 in a savings account. It earns 4·5% p.a. interest (p.a. means *per annum* or *per year*).
a How much interest is he paid at the end of the year?
b He leaves the £5000 in the account for 4 years.
He withdraws the interest at the end of each year.
How much interest is paid in total?

a Interest = 4·5% of £5000 = 4·5 ÷ 100 × £5000 = £225
b Total interest over 4 years = £225 × 4 = £900

Example 2   £300 is invested at 6% p.a. How much interest will it earn in 5 months?
1 year   ...   6% of £300 = £18
1 month   ...   £18 ÷ 12 = £1·50
5 months   ...   £1·50 × 5 = £7·50

Exercise 10.1

1 Jack invests £200 in the Mushroom Building Society.
Jill invests £470.
How much interest does
a Jack
b Jill earn in 1 year?

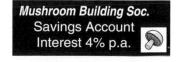

Mushroom Building Soc.
Savings Account
Interest 4% p.a.

2 Mr B Spender invests £7000 in an internet account that pays 7·5% p.a. interest.
a How much interest does he earn after 1 year?
b He leaves the £7000 in for 5 years.
He withdraws the interest at the end of each year.
Calculate the total interest earned.

3 Janis invests £3000 in an account that pays 5% p.a. interest.
Calculate the interest earned in
a 1 year   **b** 1 month   **c** 10 months.

4 Calculate the interest earned.

Amount invested	Interest rate	Time invested	Interest earned
£200	3% p.a.	6 months	
£720	5·5% p.a.	3 months	
£4000	4·2% p.a.	8 months	
£60 000	6·25% p.a.	9 months	

5 Petra invests £800 in a savings account. After 1 year she receives £24 interest.
Calculate the annual rate of interest.

6 Ms Goldsmith deposits £30 000 in Invincible Investments.
At the end of the year there is £30 630 in her account.
What annual rate of interest has been paid?

7 Jumal received a windfall 12 months ago and invested it in an account which paid
5·8% p.a. interest. There is now £7829·20 in the account.
How much did he deposit in the account 1 year ago?

Investigation

Look at the interest rates offered by different financial institutions. Find out about
fixed interest rates, special offers when opening accounts, internet banking, etc.

Compound interest

Example 3
Mrs Lamb deposits £6000 in an account that pays interest at the rate of 4% p.a.
She leaves it in the account for 3 years. She does not withdraw the interest at the
end of each year. How much is in her account at the end of the third year?

Year 1
Interest = 4% of £6000 = £240
At end of year 1: balance = £6000 + £240 = £6240

Year 2
Interest = 4% of £6240 = £249·60
At end of year 2: balance = £6240 + £249·60 = £6489·60

Year 3
Interest = 4% of £6489·60 = £259·58 (to nearest penny)
At end of year 3: balance = £6489·60 + £259·58 = £6749·18

Notes:
- When the interest becomes part of the account each year, it is called **compound** interest.
- When the interest does not become part of the account each year, it is called **simple** interest.
- Sometimes the interest is calculated on 'complete pounds only', e.g. if £524·67 is in the account, interest is only received for the £524.

There is a quick method of calculating compound interest.

Example 4
Do *Example 3* on your calculator.
Enter £6000 and multiply by 1·04 (100% + 4%) to get the total in the account after 1 year.
Multiply the answer by 1·04 to get the second year total.
Multiply the second year total by 1·04 to get the third year total.

Total amount in bank after 3 years
$$= £6000 \times 1·04 \times 1·04 \times 1·04 = £6000 \times 1·04^3 = £6749·18.$$
So the compound interest is £6749·18 − £6000 = £749·18 (to the nearest penny).

Example 5
Similarly, if you want to find out how much money is in the bank when £750 is put in for 6 years at 3% p.a., perform the calculation: $£750 \times 1·03^6 = £895·54$ (to the nearest penny).

The interest received can then be found by subtraction: £895·54 − £750 = £145·54

Exercise 10.2

1 Mr Fox invests £4000 at 6% p.a. for 2 years. He leaves the interest in the account at the end of the first year. Copy and complete:

Year 1
Interest = 6% of £4000 = £…
At end of year 1: balance = £…

Year 2
Interest = 6% of £… = £…
At end of year 2: balance = £…

2 Ms Archer deposits £600 with the Moneymaker Bank at a rate of 3·5%.
She makes no withdrawals.
Calculate the total amount in her account at the end of:
a 1 year
b 2 year
c 3 years.

3 Glynn invests £1250 at 3·2% p.a.
He leaves the money and the interest in the account for 3 years.
How much will be in his account after 3 years?

4 Mrs Rich puts £6500 in an account which pays 4·8% p.a. interest.
She leaves the money in the account for 3 years.
How much interest will she earn in the 3 years if she:
a withdraws the interest at the end of each year (simple interest)
b leaves the interest in the account (compound interest)?

5 Use the quick method described on page 43 to calculate compound interest for each
of the following:
a £500, for 3 years at 6% p.a. (Hint: $500 \times 1·06^3$)
b £1000 for 5 years at 2·5% p.a. (Hint: $1000 \times 1·025^5$)
c £950 for 4 years at 10% p.a.
d £340 for 10 years at 4·25% p.a.

6 Six years ago Willie was left an inheritance of £10 000.
The money has been left to grow in an account at an interest rate of 7% p.a.
Calculate the total amount in the account after 6 years.

7 Mel deposits £1000 at 10% p.a. interest. She makes no withdrawals.
How many years will it take to double her money?

8 Mr Moore's savings account has grown to £3931·29 at a fixed rate of 4·5% p.a. over
2 years.
How much did he deposit in the account 2 years ago?

Investigation

Investigate these formulae:

Simple interest

$$I = \frac{PRT}{100}$$

Compound interest

$$I = P\left(1 + \frac{R}{100}\right)^T - P$$

where I is the interest; P is the principal (amount invested);
R is the rate as a percentage; T is the number of years.

Brainstormer

Wayne discovered that his grandfather had put £10 000 in the Van Winkle Bank
100 years ago at a fixed interest rate of 7·25% per annum. When Wayne went to take
the money home in a case he had a problem. How big was the problem?

11 Appreciation and depreciation

Houses tend to **appreciate**, or increase in value, year by year;
cars usually **depreciate**, or decrease in value, year by year.

Example 1   Jude and Jenny bought a house for £120 000. In the following 3 years
its value appreciated by 10%, 5% and 2%. Calculate its value each year.

First year
Value = £120 000
Appreciation = 10% of £120 000 = £12 000

Second year
Value = £120 000 + £12 000 = £132 000
Appreciation = 5% of £132 000 = £6600

Third year
Value = £132 000 + £6600 = £138 600
Appreciation = 2% of £138 600 = £2772
Value after 3 years = £138 600 + £2772 = £141 372

Exercise 11.1

1 Gary bought £500 worth of shares in a chemical company.
In the last year they have appreciated by 7%. What is the value of the shares now?

2 The value of Ellie's car has depreciated by 30% since she bought it.
She paid £800 for it.
How much is it worth now?

3 Mr and Mrs Castle bought a flat for £80 000. In the first year it appreciated by 15%
and in the second year by 8%.
Calculate its value after **a** the first **b** the second year.

4 Jock's car depreciated badly. It cost him £6000.

   a In the first year it depreciated by 25% of its value.
   What was it worth then?
   b In the second year it lost a further 20%.
   Calculate its value then.
   c In the third year it went down by another 15%.
   What was it worth after the 3 years?

5 Clive's pottery collection cost him £1500. He reckons it is now worth £2475.
By what percentage of the original cost has the collection appreciated?

6 Penny's mountain bike has fallen in value from £500 to £300.
Express the depreciation as a percentage of the original value.

7 Insight Investments buy a painting for £240 000 and a sculpture for £180 000.
One year later the painting has risen in value to £258 000.
The sculpture is worth £195 300.
   a Express each rise as a percentage of the price Insight Investments paid.
   b Which has been the better investment?

8 For insurance purposes the Clarkes have their house valued each year.
Calculate the rate of appreciation or depreciation as a percentage for:
a 2002 to 2003
b 2003 to 2004
c 2004 to 2005
d 2002 to 2005.

Year	Value
2002	£160 000
2003	£198 400
2004	£248 000
2005	£235 600

Example 2   Mr Mole sells his technology shares for £2000.
They have fallen in value by 60% since he bought them
How much did he pay for them?

Original value is 100%
Current value = 100% − 60% = 40%
40% of original cost is £2000
⇒   1% of original cost is £2000 ÷ 40
⇒   100% of original cost is £2000 ÷ 40 × 100 = £5000
He paid £5000 for the shares.

Exercise 11.2

1 Sharon's caravan cost her £8000.
It has depreciated by 10% for each of the last 3 years.
Calculate its value now, correct to the nearest £100.

2 The Khans' house cost £175 000 in 1998. By 2002 its value had depreciated by 20%.
By 2005, however, it was worth 12% more than in 2002.
Calculate:
a its value in 2005   b the percentage change in value from 1998 to 2005.

3 The table shows some of the values and appreciation rates of
23 Railway Cuttings between 2002–4.
At the beginning of 2002 it was valued at £100 000.
Find the missing figures.

Year	End of year value	Appreciation rate
2002	a	5%
2003	b	10%
2004	£138 600	c

4 The value of Trisha's coin collection fell by 8% to £460.
Calculate its previous value.

5 The Vulcan Vintage Car Company sold an old Rolls Royce for £125 000.
They made a profit of 150%.
How much did they pay for the car?

6 The Rollercoaster Real Estate Company speculated on a land deal.
They paid £500 000 for a plot of land 10 years ago.
Since then the value has fallen by 5% each year.
What is the land's current value, to the nearest £10 000?

7 The value of the Great Strike Goldmine has appreciated by 50% each year for the last 3 years. It is now worth £27 million pounds.
What was its value 3 years ago?

8 Mrs Kelly's share portfolio is now worth £18 900. The shares fell by 12·5% in the first year, then rose by 12·5% in the second year.
  a Calculate the value of the shares 2 years ago.
  b Find the percentage change in the value, correct to 2 d.p., over the 2 years.

◀◀ RECAP

Money calculations
You should be able to carry out calculations involving money, including finding percentages of amounts. You should also be able to round amounts of money to a suitable degree of accuracy and estimate answers to calculations.

Wages and salaries
You should be able to do calculations involving wages, salaries, piecework, commission, overtime, time-sheets and pay rises.

Payslips and deductions
You should be able to do calculations involving gross pay, net pay, income tax, National Insurance Contributions, bonuses and superannuation funds.

> Net pay = gross pay − deductions
> Taxable income = total income − personal allowance

Savings
You should be able to understand cheques, bank statements, savings accounts and interest. You should be able to calculate percentages without a calculator, for example: 1%, 2%, 4%, 5%, 8%, 10%, 15%, 20%, 25%, 30% 35%, 50%, 75%, etc.

You should be able to calculate compound interest and know how to use a quick method for working out a rate which is repeated over several years.

Example   For a rate of 6% for 4 years:

> Final amount = amount invested $\times 1\cdot06^4$

Appreciation and depreciation
You should be able to work with values that appreciate and depreciate over several years.

Note: you should be able to express a rise or fall as a percentage and be able to calculate the original quantity given the percentage change and the resulting quantity.

Example   After a rise of 5% Jasmine's salary is £462.
           Original salary is 100%. New salary is 105%.
           105% of the original salary = £462
           ⇒ 1% of the original salary = £462 ÷ 105
           ⇒ 100% of the original salary = £462 ÷ 105 × 100 = £440

1 Helen works in the catering industry. Last year her total earnings came to £13 923.
  a Calculate her weekly wage.
  b She works 35 hours per week. Calculate her hourly rate of pay.

2 Ron earns £2748 per month.
  a Calculate his annual salary.
  b He receives a 7·5% pay rise.
    Calculate his new
    i   monthly
    ii annual salary.

3 Rosie's basic rate of pay for a 36 hour week is £9·72 per hour.
  Overtime during the week is paid at time and a third.
  At weekends overtime is paid at time and three-quarters.
  On Monday to Friday she works from 8.30 am to 5 pm with 45 minutes for lunch.
  On Saturday she works from 8.45 am to 12.15 pm.
  Calculate her total wage for the week.

4 **a** Mrs Baker is a freelance hairdresser.
    How much does she earn in a month
    when she cuts and dries hair for
    80 ladies and 50 men?

  b Harry, a car salesman, is paid
    commission of 2% on his sales.
    i How much commission is he paid for sales worth £17 000?
    ii What value of sales does he need to make to earn £500 commission?

> **Ladies' Cut/blow dry £22·50**
>
> **Men's Cut/blow dry £17·50**

5 Calculate Diane Simpson's:
  a gross pay
  b total deductions
  c take-home pay.

Name D Simpson	Employee number 246		NI number AB100079B	Week number 43
Basic pay £496·58	Overtime —	Commission £73·07	Bonus £30	Gross pay
Income tax £85·64	NI £44·61	Superannuation £27·31	Other deductions —	Total deductions
				Net pay

6 Income tax rates

Taxable income	Rate of tax
£0–£2020	10%
£2020–£31 400	22%
over £31 400	40%

National Insurance Contributions

Rate	Annual
0%	<£4745
11%	£4745–£31 720
1%	>£31 720

Calculate the annual amounts paid in   **i** NICs   **ii** income tax for:
  a Mr Long, wages £100 per week, tax allowance £4745
  b Mrs Short, annual salary £28 500, tax allowance £5000
  c Ms Middleton, annual salary £46 000, tax allowance £5445.

7 Copy this statement and complete the balance column.

STATEMENT			
Date	Paid out	Paid in	Balance
1 Jun			62·73
9 Jun	103·75		
13 Jun		85·38	
25 Jun	67·39		

8 Ms Spencer invests £12 000 at 5·5% p.a.
She leaves the money and the interest in the account for 3 years.
How much will be in her account after 3 years?

9 Since Mr and Mrs Singh bought their house it has risen in value by 8% to £156 600. Calculate the amount they paid.

10 Mark paid £28 000 for his mobile home. In the first year it depreciated in value by 15%, in the second by 12% and in the third by 8%. Calculate its value now.

3 Similarity

When we look in a mirror we expect to see an image **similar** to ourselves.
At the funfair, the hall of mirrors shows that this is not always the case.
We can appear 'squashed' or 'stretched'.

Some objects look the same on any scale no matter how many times you magnify them.
They are **self-similar**.

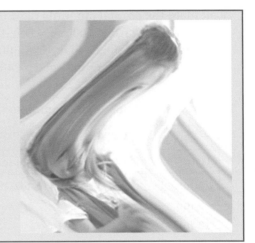

1 Review

◄◄ Exercise 1.1

1 Find the actual distances.

Map distance	Scale	Actual distance
2 cm	1 cm represents 6 m	
5 cm	1 cm represents 50 m	
6·5 cm	1 cm represents 500 m	
11 cm	1 cm represents 6·5 km	

2 Murray walked west for 480 metres and then south for a further 360 metres.
   a Letting 1 cm represent 100 metres, make an accurate drawing of Murray's route.
   b How far will Murray have to walk to go directly back to the start?

3 This is a plan of Jade's bedroom.
   1 cm represents 0·5 m.
   a What is the actual length and breadth of
      i the bed   ii the bookcase?
   b Jade also has a desk in the room.
      It measures 75 cm by 1·5 m.
      What size will it appear on the plan?

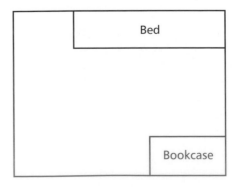

4 a Draw these shapes double their original size.

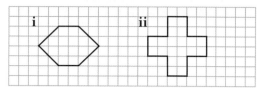

b Draw these shapes half their original size.

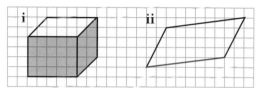

2 Representative fractions

A scale can be given as the **ratio** measurement on map : measurement on ground, sometimes called the **representative fraction**.

Example   1 cm represents 10 m

$$= 1 \, cm : 1000 \, cm$$

$$= 1 : 1000$$

Note that, once they are the same in the ratio, the units can be dropped.
The above scale would be read as 'one to one thousand' and we can think of it as '1 unit represents 1000 units'.

Example
What length will represent a distance of 4 km on a map with a scale of 1 : 50 000?

50 000 units on the ground is represented by 1 unit on the map

$\Rightarrow$ 1 unit on the ground is represented by $\frac{1}{50\,000}$ unit on the map

$\Rightarrow$ 4 km on the ground is represented by $\frac{4}{50\,000}$ km on the map = 0·000 08 km

0·000 08 km = 0·08 m = 8 cm

A length of 8 cm represents 4 km on this map.

Exercise 2.1

1 Copy and complete this table.

	Scale	Representative fraction
a	1 cm represents 40 m	
b	1 cm represents 150 m	
c		1 : 6000
d		1 : 200 000

2 Copy and complete the table.

	Scale	Representative fraction	Map distance	Actual distance
a	1 cm represents 10 m		4 cm	
b	1 cm represents 200 m		5 cm	
c	1 cm represents 500 m		2·5 cm	
d	1 cm represents 800 m		3 cm	

3 Copy and complete the table.

	Scale	Representative fraction	Map distance	Actual distance
a	1 cm represents 5 m			80 m
b	1 cm represents 40 m			240 m
c	1 cm represents 2 km			5 km
d	1 cm represents 1·5 km			9 km

4 On a map, a ship's route was marked as 6 cm east and then 8 cm south.
The scale of the map is 1 : 500 000.
  a Work out the actual distances sailed by the ship.
  b Calculate the direct distance between the ship and its starting point.

5 Amy ran in an orienteering race.
The table gives the distances on
the map between checkpoints.
The scale of the map was 1 : 15 000.
Work out the actual distance of
each leg.

Leg	Distance on map
Start to checkpoint 1	3 cm
Checkpoint 1 to checkpoint 2	7 cm
Checkpoint 2 to checkpoint 3	8 cm
Checkpoint 3 to finish	5·5 cm

6 This is the layout of the field for the
Dunban Highland Games.
  a Calculate the distance between the
    Secretary's tent and
    i the Highland dancing
    ii the funfair entrance
    iii the Piobroch.
  b The running track has two straight
    sides and two semi-circular ends.
    Calculate the actual distance round the
    track.
  c A caber is 6 metres in length. Using
    the same scale, how long would it
    appear on the map?

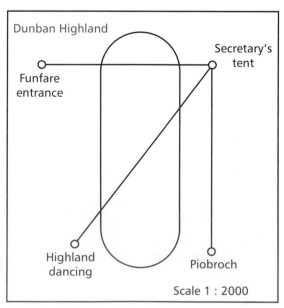

7 The Water of Leith is 35 km in length.
In a visitor centre, there is a large map showing the watercourse on the wall.
The scale of the map is 1 : 8000.
How long is the Water of Leith on the map?

---

3 Scale drawings

Important points to remember:

- choose a suitable scale, large enough to fit the available space (increases the accuracy)
- make sure all measurements are accurate
- measure all bearings from north in a clockwise direction.

Example	A ship sailed on a bearing of 040° for 15 kilometres. It then turned and sailed for a further 18 kilometres on a bearing of 170°. Make an accurate drawing of the ship's route using the scale 1 : 300 000.
Step 1	Calculate scaled lengths. 15 km is represented by $15 \div 300\,000 \times 100\,000$ cm = 5 cm. 18 km is represented by $18 \div 300\,000 \times 100\,000$ cm = 6 cm.
Step 2	Pick a starting point and define north.
Step 3	Draw a bearing of 040°.
Step 4	Extend the line for 5 cm.
Step 5	Add a new north line, parallel to the first, and draw a bearing of 170°.
Step 6	Extend the line for 6 cm.

Exercise 3.1

1 A plane left Newcastle on a bearing of 160° and flew for 500 km.
Turning to a bearing of 200° it then flew for 600 km.
Finally, following a bearing of 225°, it flew for 2000 km to Funchal.

  a Make an accurate scale drawing of the route the plane took.
  b On the return flight, the plane headed straight to Newcastle.
    How far did it fly this time?

2 Two planes flew from Gatwick heading in different directions but at the same altitude.
   The first plane headed on a bearing of 210° and flew for 2700 km.
   The other plane headed on a bearing of 075° for 2400 km.
   a On the same diagram, make an accurate scale drawing of their routes.
   b How far apart were the planes at the end of their journeys?
   c At the end of their journeys, what was the bearing of the first plane from the second?

3 A plane flies for 2 hours on a bearing of 040° at an average speed of 400 mph.
   From the same airport, another plane flies on a bearing of 110° for 3 hours at 450 mph.
   a Make an accurate scale drawing of their routes.
   b How far apart are they at the end of their journeys?
   c What is the bearing of the first plane from the second at the end of their journeys?

4 Two friends walk for 1·5 hours on a bearing of 045°, averaging a speed of 6 km/h.
   They then walk for a further 45 minutes, this time at 4 km/h on a bearing of 150°.
   a Make an accurate scale drawing of their route.
   b How far away are the friends from their starting point?
   c What is the bearing of their starting point from their end point?
   d On the return journey, following a direct route, they walk at 4 km/h.
      How long will it take them to get back to their starting point?

5 From port, Fergus and Becky sailed for half a day at a steady
   15 miles per hour on a bearing of 235°. After tacking, they
   followed a bearing of 200° for 6 hours, this time at a speed of
   20 miles per hour.
   a Make an accurate scale drawing of their route.
   b At the end, how far away were they from the port?
   c If they were to head straight back to port, what bearing would
      they need to follow?

6 A captain of a ship spies a lighthouse on a bearing of 145°.
   After sailing south for 30 miles he sees the same lighthouse, this time on a bearing of
   045°.
   a Make an accurate scale drawing of the situation.
   b How far was the ship away from the lighthouse when each of the observations was
      made?
   c What was the closest the ship got to the lighthouse?

7 A runner maintained a steady pace on a straight course.
   An observer first saw the runner 1·5 kilometres away at a bearing of 190°.
   20 minutes later, at a distance of 1 kilometre, the observer saw him at a bearing of 160°.
   a Make a scale drawing of the situation showing both the observer and the runner.
   b What bearing was the runner following?
   c How far had he run in the 20 minutes?
   d At what speed was he running?

8 In one leg of an orienteering race, Paul and Shona headed off from the start at the same time.

Paul ran at a speed of 5 km/h on a bearing of 069° for 15 minutes.

Shona ran for 20 minutes at a speed of 4 km/h on a bearing of 147°.

The first checkpoint was actually 2 km from the start on a bearing of 110°.

a Make a scale drawing of the situation including the checkpoint.

b What bearings must each now follow to reach the checkpoint?

c If they maintain their original speeds, who will arrive at the checkpoint first and by how many minutes will they beat the other? Round your answer to the nearest minute.

---

4 Working out scales

The representative fraction of a map or diagram can be worked out provided you have one distance from the map and the corresponding actual distance.

Example

The map distance between BeinnTarsuinn and Beinn Bhreac is 2 cm.
The actual distance is 5 km.

The representative fraction is 2 cm : 5 km

$$= 2\,cm : 500\,000\,cm$$
$$= 1 : 250\,000$$

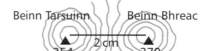

Exercise 4.1

1 The actual length of the side of a swimming pool is 30 m.
On the plan it measures 5 cm.
What is the scale of the diagram:
a expressed in words
b as a representative fraction?

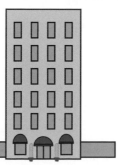

2 The diagram shows the front of a new hotel.
The actual height of the hotel is 18 m.
a What is the scale of the diagram in words?
b What is the representative fraction?

3 On a map of Tenerife, a road marked on the map is actually 58 km long.
On the map it measures 7·25 cm.
What is the scale of the map expressed as a representative fraction?

4 One giant redwood tree is 303 metres tall.
In a photograph, the same tree is only 6 cm high.
Express the scale of the photograph as a representative fraction.

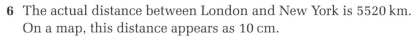

5 The diameter of a table in the plan of a restaurant
measures 1·5 cm.
The actual diameter of the table is 1·20 m.
  a Work out the representative fraction of the plan.
  b Each table requires a space of 3·20 m to allow for chairs.
    On the same plan, the rectangular restaurant measures
    12·5 cm by 16·3 cm.
    What is the maximum number of tables that can be fitted
    round the outside of the room?

6 The actual distance between London and New York is 5520 km.
  On a map, this distance appears as 10 cm.
  a What is the scale of the map given as a representative fraction?
  b The distance between New York and Chattanooga is 1240 km.
    On the same map, what would the distance between the cities be?

Challenge

  a Measure your height.
    Find a recent photograph of yourself.
    Measure your height from the photograph and work out the scale of the photograph.
  b Use a digital camera to photograph a friend next to your school.
    i Use your friend's height to give you a scale for the photo.
    ii Use this scale to find the height of the school.

5 Similar triangles

Mathematically, shapes are said to be similar if one is just a scaled version of the
other. For this to be the case, all measurements in one shape must be a fixed factor
of the corresponding shapes in the other.

If not, distortion will happen, like in this example from the Hall of Mirrors at a funfair.

If there is no distortion, then the sizes of the angles will be preserved in the enlargement.

If we know shapes are similar then deductions can be made about the lengths of sides.

If we know triangles are similar, and we know which sides correspond with which,
we can work out missing sides.

Corresponding sides will be opposite the equal angles.
So first we must identify the equal angles.

Example 1   These two triangles are similar. Equal angles are marked. Find the missing length, x cm.

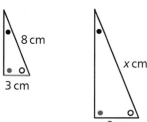

Note that the side marked x cm is an enlargement of the side marked 8 cm, ... and that the side marked 9 cm is an enlargement of the side marked 3 cm.

The enlargement scale factor $= \dfrac{\text{measurement from larger}}{\text{measurement from smaller}} = \dfrac{9}{3} = 3$

i.e. the large triangle is 3 times bigger than the smaller.

$$\begin{aligned} \text{So } x &= \textbf{scale factor} \times 8 \\ &= 3 \times 8 \\ &= 24 \text{ cm} \end{aligned}$$

Example 2   These two triangles are similar. Equal angles are marked.

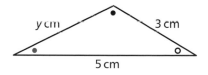 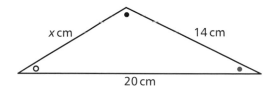

     a Calculate the size of the side marked x cm.
     b Calculate the value of y.

The equal angles allow us to pair off the sides:

3 cm with x cm (both opposite the angles marked with orange dots)

14 cm with y cm (both opposite the angles marked with white circles)

5 cm with 20 cm (both opposite the angles marked with black dots)

The third pairing, in this case, can be used to find the scale factors.

a We find x by enlarging 3. The enlargement factor is $\frac{20}{5} = 4$.
  $x = 3 \times 4 = 12$ The side is 12 cm long.

b We find y by reducing 14.

  The reduction scale factor $= \dfrac{\text{measurement from smaller}}{\text{measurement from larger}} = \dfrac{5}{20} = 0 \cdot 25$

$$y = 14 \times 0 \cdot 25 = 3 \cdot 5$$

Exercise 5.1

1 In each case, the two triangles are similar. Equal angles are marked.
   Find:
   i the relevant scale factor (enlargement or reduction)
   ii the missing lengths.

a
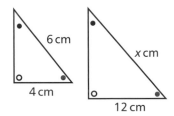

b
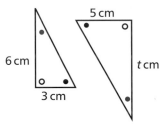

c
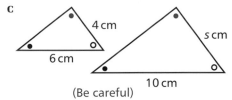
(Be careful)

d

e
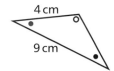

f
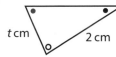

2 Each pair of triangles is similar.
   i Calculate the enlargement factor.
   ii Identify pairs of equal angles.
   All lengths are in centimetres.

a
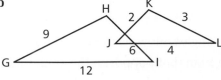

b

c
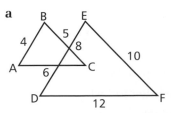

d
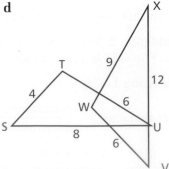
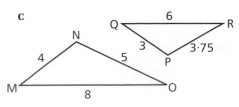

3 Each pair of triangles is similar. Equal angles are marked.
All lengths are in centimetres.
Calculate the length of the labelled side.

a

b

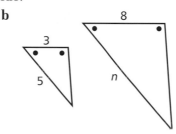

c

d

4 Find the size of each labelled side, given that the triangles in each pair are similar.
All lengths are in centimetres.

a

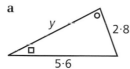 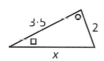

b

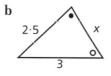 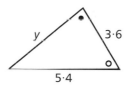

Example 3
In the diagram, AB is parallel to ED.

ACD and BCE are straight lines.
Triangles ABC and CDE are similar.
Calculate the size of the labelled sides.

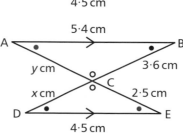

Mark equal angles. This is a guide to
which side of one triangle matches, or
corresponds, with which side of the other.

y is an enlargement of 2·5.

Enlargement factor $= \dfrac{5\cdot4}{4\cdot5} = \dfrac{6}{5} = 1\cdot2$

$\Rightarrow \quad y = 1\cdot2 \times 2\cdot5 = 3$ cm

x is a reduction of 3·6.

Reduction factor $= \dfrac{4\cdot5}{5\cdot4} = \dfrac{5}{6}$

$\Rightarrow \quad \dfrac{5}{6} \times 3\cdot6 = 3$ cm

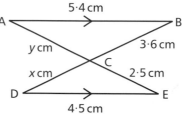

Example 4

In the diagram, BC is parallel to DE.
Triangles ABC and ADE are similar.
Calculate the size of the labelled sides.

Mark equal angles to identify the corresponding sides.

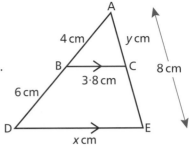

y is a reduction of 8.

$$\text{Reduction factor} = \frac{AB}{AD} = \frac{4}{10}$$

$$\Rightarrow \quad y = 8 \times 0{\cdot}4 = 3{\cdot}2 \text{ cm}$$

x is an enlargement of 3·8.

$$\text{Enlargement factor} = \frac{AD}{AB} = \frac{10}{4}$$

$$\Rightarrow \quad x = 3{\cdot}8 \times 2{\cdot}5 = 9{\cdot}5 \text{ cm}$$

Exercise 5.2

All lengths, unless otherwise stated, are in centimetres.

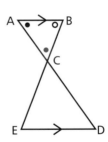

1 In the diagram, AB is parallel to ED.
   ACD and BCE are straight lines.
   a Copy the diagram and mark an angle equal to:
     i ∠CAB
     ii ∠ABC
     iii ∠CBA.
   b Triangles ABC and CDE are similar.
   Name the side in triangle CDE which corresponds to
     i AB
     ii BC
     iii CA.

2 In the following diagrams, two straight lines and two parallel lines intersect to form a
   pair of similar triangles.
   In each case:
   i sketch the diagram
   ii identify the equal angles
   iii calculate the enlargement factor
   iv find the labelled side.

a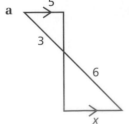

b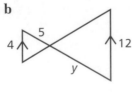

c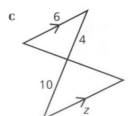

d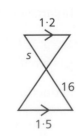

3 The surface of an ironing board is parallel to the ground.
Its legs cross as shown, with the distance AB being adjustable to allow for a change in height. All distances are in metres.
In all positions, the two triangles formed by the legs are similar.

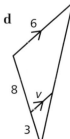

a How far apart are the feet, E and D, when AB has a length of
  i 0·6 m   **ii** 1 m   **iii** 70 cm?
b What is the length of AB when the distance between E and D is
  i 1·2 m   **ii** 2·1 m   **iii** 1·8 m?

4 The diagonals of any trapezium form two similar triangles with the parallel sides always forming a pair of corresponding sides.
Calculate the value of the letter in each case.

a
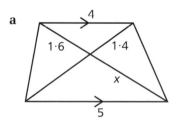

b
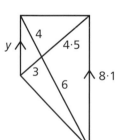

c
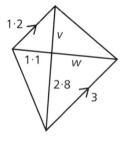

5 Each diagram below contains a pair of similar triangles.
  i Identify the equal angles.
  ii Calculate the labelled lengths.

a
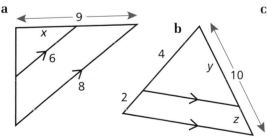

b

c
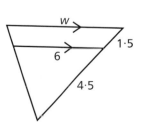

d
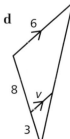

6 Whenever a line is drawn across a triangle, parallel to a side, similar triangles are formed. Use this fact to help you find the unknown lengths in each diagram.

a

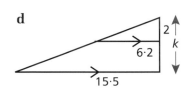

b

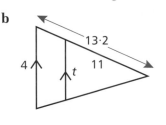

c

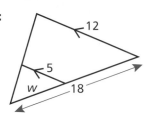

d

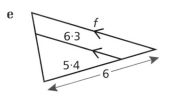

e

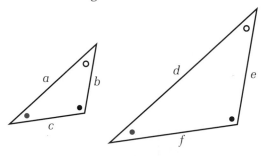

f

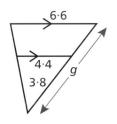

---

6 More about similar triangles

In a pair of similar triangles the ratio of corresponding sides is a constant, always producing the same enlargement or reduction factor.

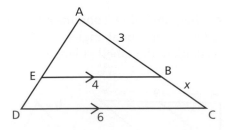

$$\frac{a}{d} = \frac{b}{e} = \frac{c}{f}$$

We say that the corresponding sides are **in proportion**.
This can be useful when solving problems.

Example 1   Find the value of x given that the triangles are similar.

The corresponding sides are in

proportion so $\dfrac{AB}{AC} = \dfrac{EB}{DC}$

$\Rightarrow \quad \dfrac{3}{3 + x} = \dfrac{4}{6}$

$\Rightarrow \quad 3 \times 6 = 4(3 + x)$

$\Rightarrow \quad 18 = 12 + 4x$

$\Rightarrow \quad 4x = 6$

$\Rightarrow \quad x = 1 \cdot 5$

Example 2 Find the value of x where the two triangles are similar.

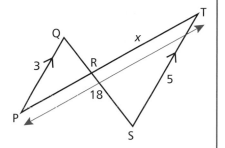

The corresponding sides are in proportion so $\dfrac{PQ}{ST} = \dfrac{PR}{RT}$

$$\Rightarrow \qquad \frac{3}{5} = \frac{18 - x}{x}$$

$$\Rightarrow \qquad 3 \times x = 5(18 - x)$$

$$\Rightarrow \qquad 3x = 90 - 5x$$

$$\Rightarrow \qquad 8x = 90$$

$$\Rightarrow \qquad x = 11{\cdot}25$$

Exercise 6.1

Unless otherwise stated, all measurements are in centimetres.

1 Given that the pairs of triangles are similar, calculate the value of x in each case.

a

b
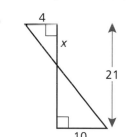

c
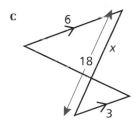

2 The diagram tries to show how a camera lens works.
Two similar triangles are formed.

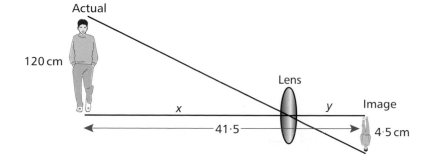

The actual boy is 120 cm tall. His image in the camera is 4·5 cm tall.
Calculate:
a the distance of the boy from the camera, x cm
b the distance of the image from the lens.

3 A farmer is surveying his land. A tree is growing in one field.
Because of the presence of a bull, he can't get into the field.
He makes the following measurements from outside the field.

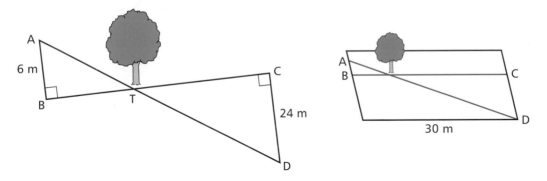

Triangles ABT and CDT are similar.
How far is the tree from B?

4 In each diagram there is a pair of similar triangles.
Calculate the value of *x* in each.

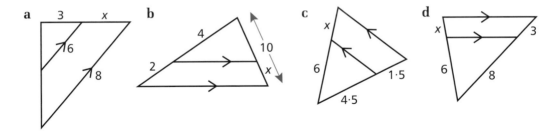

5 The two triangles formed in the stepladder are similar.

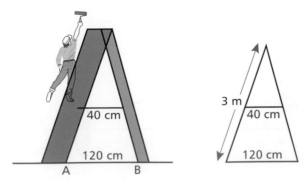

The rope is 40 cm long. When fully extended, the feet of the ladder are 120 cm apart
and the rope is parallel to the ground.
The length of the ladder is 3 metres.
What is the distance from A to the rope?

6 The legs of a sun bed form two similar triangles.

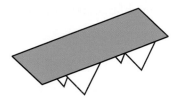 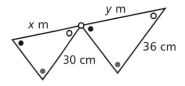

The overall length of the sun bed is 2·2 metres.
Equal angles are marked.
Calculate the length of:
a x　　　**b** y.

7 The front of a balcony is made up of similar isosceles triangles.
The bases of three triangles are 1·4 m, 80 cm and 60 cm as shown.

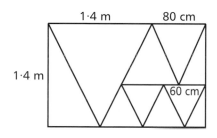 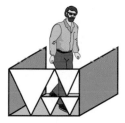

The height of the balcony is 1·4 metres.
a Calculate the length of the sloping edge of the largest triangle.
b Calculate the lengths of the sides of the triangle with
　　i an 80 cm base　　　**ii** a 60 cm base.

---

7 Other similar figures

Any shapes that are similar, like triangles, will have the ratios of their corresponding sides constant.
Again, corresponding sides can be identified by considering the equal angles of the shapes.

Example
These quadrilaterals are
similar and so their
corresponding sides are
in proportion.
a Find the scale factor.
b Find the value of x.

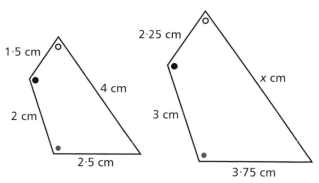

a Having marked up the equal angles, the corresponding sides can be spotted as lying between the equal angles. So in this case the enlargement factor can be worked out in one of three ways, namely:

$$\frac{2 \cdot 25}{1 \cdot 5} = \frac{3}{2} = \frac{3 \cdot 75}{2 \cdot 5} = 1 \cdot 5$$

The enlargement factor is $1 \cdot 5$.

b x is an enlargement of $4 \Rightarrow x = 1 \cdot 5 \times 4 = 6$ cm.

Exercise 7.1

1 Each pair of quadrilaterals is similar. For each
   i find the reduction/enlargement factor
   ii check it is constant by trying more than one pair of corresponding sides
   iii find the length of the labelled side.

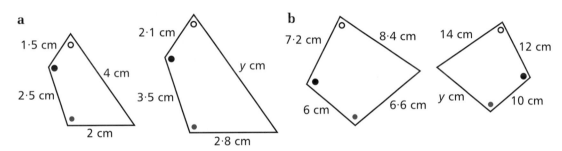

a

b

2 This pair of quadrilaterals are similar.
   a Calculate
      i the enlargement
      ii the reduction factor.
   b Use this information to find the value of
      i x
      ii y
      iii z.

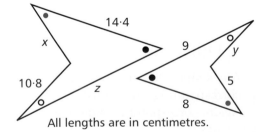

All lengths are in centimetres.

3 The picture shows one shape drawn in another, sharing a common angle.
The shapes are similar. Equal angles have been marked.

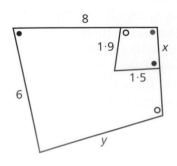

   a There appears to be only one angle marked with a coloured dot. Explain this.
   b Calculate the size of
      i x
      ii y.

4 The bases of the showers at the side of a pool come in two different sizes.
The shapes of the bases are similar.

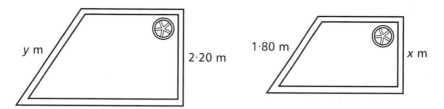

The right-hand edge of the larger base is 2·20 m.
The smaller base is 60 cm shorter.
The sloping edge of the smaller base is 1·80 m.

a What is the value of x?

b Calculate the relevant scale factor.

c Calculate the length of the sloping edge of the larger base, y m.

5 Two kites are similar in shape. The enlargement factor is 1·75.

a The length of the larger kite's longer diagonal is 63 cm.
How long is the longer diagonal of the smaller kite?

b The length of the other diagonal in the smaller kite is 30 cm.
What is the corresponding length in the larger kite?

6 Two similar rectangular flags fly outside a school.
The first one has a diagonal of length 1·75 m.
The breadth of the rectangle is 1·05 m.
The second flag has a length of 80 cm.

a Use Pythagoras to find the length of the larger flag.

b Calculate the relevant scale factor.

c What is the breadth of the second flag?

7 In the middle of the turning circle outside a hotel is a large
pagoda. It consisted of two similar regular octagons.
The length of the diagonal of the larger octagon is 20 m.
The reduction factor is $\frac{3}{5}$.

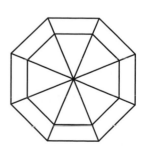

a Calculate the length of one side of the larger octagon to the
nearest centimetre. (Hint: use trigonometry.)

b Calculate the length of one side of the smaller octagon by
scaling, to the nearest centimetre.

c What length of wood is required to construct the smaller
octagon and its diagonals?

8 Ratios of areas

These two triangles are similar.   The enlargement factor is k.

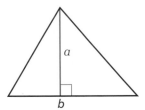   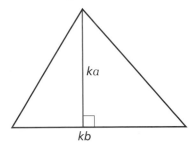

The area of the smaller triangle $= \frac{1}{2}ab$
The area of the larger triangle $= \frac{1}{2}kakb = \frac{1}{2}abk^2$

The larger area is found by multiplying the smaller by k^2.

We can look at the area formula for any of the shapes we know and we will get the same result each time.

> The scale factor for area is the square of the scale factor for length.

Example   Two sauce bottles are similar with heights as shown.
The larger bottle has a label of area 40 cm².

What is the area of the label on the smaller bottle?
Reduction factor (length) $= 18 \div 24 = 0.75$
$\Rightarrow$ reduction factor for area $= 0.75^2 = 0.5625$
$\Rightarrow$ area of smaller label $= 40 \times 0.5625 = 22.5$ cm²

Note that it doesn't matter what shape the labels are, as long as they are similar.

Exercise 8.1

1 Two rectangles are similar. One is 2 cm high by 5 cm wide. The other is 8 cm high.
  a Write down the enlargement factor for
     i length   ii area.
  b What is the area of the smaller rectangle?
  c What is the area of the larger rectangle?

2 Two triangles are similar.
  One has a base of 2 cm and the other a corresponding base of 7 cm.
  The area of the smaller triangle is 6 cm².
  Calculate the area of the larger triangle.

3 A company produces two sizes of parasols.
  The shapes used in the production of the parasols are similar.
  The larger parasol is cut from a circle of radius 1 metre.
  The smaller parasol is $\frac{1}{2}$ the radius of the larger one.
  a Calculate the area of the larger parasol in
     square metres, giving your answer correct
     to 2 decimal places.
  b Calculate the area of the smaller parasol.

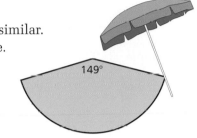

149°

4 To tile a particular area, a tiler uses 50 rectangular tiles.
  Each tile is of length 40 cm.
  How many tiles similar in shape but of length 10 cm would she need to tile the
  same area?

5 A regular pentagonal slab has a surface area of 2150 cm².
  The area of a smaller, similar slab is $\frac{1}{25}$ of the larger slab.
  a Calculate the length of one side of the large slab using trigonometry.
  b Calculate the length of one of the sides of the smaller slab.

6 The areas of two similar displays are 4·5 m² and 1800 cm².
  The length of the larger display is 3 metres.
  Calculate the length of the smaller display.

7 Mr and Mrs Mohammed spent £140 on their bedroom carpet.
  The cost is proportional to the area of the floor.
  They want to carpet their living room using the same carpet.
  Both the bedroom and the living room floors are of similar shape.
  Each length of the living room is 1·7 times larger than the corresponding length in
  the bedroom.
  How much will it cost to carpet the living room?

8 Alistair painted two similar rooms.
  One tin of paint can cover up to 18 m².
  The smaller of the two rooms needed exactly one tin of paint.
  The larger room is 1·75 times larger in length than the smaller.
  How many tins of paint did he need for the larger room?

Paint

9 Surface areas of similar solids

The same rule applies to 3-D shapes as it does to 2-D shapes.

> If two solids are similar, then the scale factor for surface area is the square of the scale factor for length.

Example   These two cuboids are similar. Calculate the surface area of the larger cuboid.

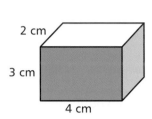
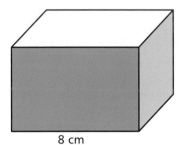

The smaller cuboid has edges of length 2 cm, 3 cm and 4 cm.
Its surface area $= 2 \times (2 \times 3) + 2 \times (2 \times 4) + 2 \times (3 \times 4)$
$\qquad\qquad\quad = 52 \text{ cm}^2$

The enlargement factor (length) $= \frac{8}{4} = 2$

Thus the enlargement factor for area $= 2^2 = 4$

The surface area of the larger cuboid $= 4 \times 52 = 208 \text{ cm}^2$

Exercise 9.1

1 A display box is a cuboid with length $= 3$ cm, height $= 4$ cm and depth $= 5$ cm. Another similar display box has a height of 12 cm.

   a Calculate the surface area of the smaller box.

   b Calculate the surface area of the larger box.

2 These two Greek columns are similar in shape.
   The smaller of the two has a radius of 50 cm and a height of 3 metres.
   The other column has a radius of 75 cm.

   a Calculate the curved surface area of the smaller column (ignoring the decorations).

   b Hence, calculate the curved surface area of the larger column.

3 A model of an aircraft hanger is a cuboid
20 cm long, 10 cm high and 10 cm broad,
on which sits a half-cylinder for a roof.
The actual hanger is 30 metres in length.
a Calculate the curved surface area of the
model's roof.
b Calculate the curved surface area of the actual
hanger's roof.

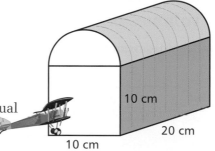

10 cm

20 cm

10 cm

4 A model of the Great Pyramid of Giza has a square base of
side 12 cm and a vertical height of 8 cm. The actual
pyramid is 160 metres high.
a Calculate the total area of the four sides of the model.
b Calculate the actual surface area of the sides of the Great
Pyramid.

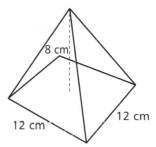

8 cm

12 cm

12 cm

5 An ice-cream cone has a diameter of 3 cm and a slant
height of 5 cm.
Outside the ice-cream shop, there is a large model of a cone
similar to the ones sold. It has a slant height of 75 cm.
The formula for the curved surface of a cone, A, is given by
$A = \pi r s$, where r is the radius and s is the slant height of
the cone.
Calculate:
a the radius
b the curved surface area of the display model.

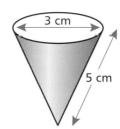

3 cm

5 cm

6 The surface area of a large beach ball is 7850 cm².
A smaller ball has a surface area $\frac{1}{4}$ of the large ball.
Note that every sphere is similar to every other sphere!
a What is the reduction scale factor for:
  i the areas   **ii** the radii of the balls?
b The surface area of a sphere, A, is given by $A = 4\pi r^2$.
  Calculate the radius of:
  i the larger ball using this formula
  ii the smaller ball by considering the reduction factor.

7 A physics teacher has two sizes of equiangular triangular prisms.
The length of a side of the base of the first is 6 cm and its height is 4 cm.
The length of a side of the base of the other is 4 cm.
a Calculate the area of the base of the first prism.
b Calculate the total surface area of the first prism.
c Calculate the total surface area of the second prism.

10 Volumes of similar solids

Two cuboids are similar (with an enlargement factor of k).

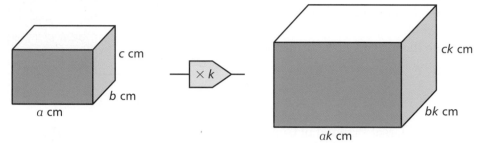

For the smaller cuboid: $V = abc \, \text{cm}^3$
For the larger cuboid: $V = akbkck \, \text{cm}^3 = abck^3 \, \text{cm}^3$

The larger cuboid has a volme which is k^3 times as big as the smaller.

> The scale factor for volume is the cube of the scale factor for length.

This is true of all similar solids.

Example
A bottle of juice of height 18 cm holds 440 ml of liquid.
A free sample is given away in a similar shaped bottle
of height 3 cm.
What volume of liquid is in the free sample?

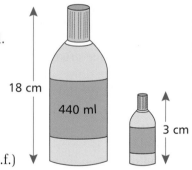

The reduction factor (length) $= \frac{3}{18} = \frac{1}{6}$
The reduction factor (volume) $= (\frac{1}{6})^3 = \frac{1}{216}$

The volume of the sample $= \frac{1}{216} \times 440 = 2.04$ ml (to 3 s.f.)

Exercise 10.1

1 A cuboid measures 2 cm high by 3 cm broad by 4 cm long.
   Another similar cuboid has a height of 3 cm.
   Calculate the volume of this second cuboid.

2 A hotel has two swimming pools which are similar in shape.
   The children's pool is 0·5 m deep and holds 6 m³ of water.
   The adults' pool is 1·5 m deep.
   a Calculate the volume of the adults' pool.
   b The surface of the adult pool is 108 m².
      What is the area of the surface of the children's pool?

3 A shop selling chest freezers claims that one freezer is similar to a smaller one.
   The larger one is 2 m wide, while the smaller is 1·5 m wide.
   The volume of the smaller freezer is 180 000 cm³ and the larger is 240 000 cm³.
   Is the shop's claim justified?
   Give a reason for your answer.

4 Two bins outside a sports hall are similar.
The larger has a volume of 125 000 cm³ and the
smaller 64 000 cm³.
The height of the larger bin is 50 cm.
What is the height of the smaller bin?

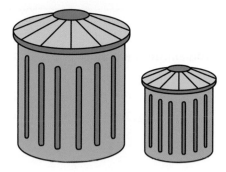

5 A builder required 2·512 m³ of concrete to make a cylindrical column of height
5 metres.
  a How much concrete will be needed to construct a similar column which is only
    3 metres high?
  b What is the radius of this second column?

6 A paper cup can hold 18 cm³ and is 4 cm high.
Another similar cup can hold 27 cm³.
What will be the height of this cup?

4 cm

7 Two medicine bottles are similar in shape.
The first holds 160 ml and is 6 cm high.
The second is 4 cm high.
  a What is the volume of the second bottle?
  b The first bottle costs £2·80.
    The chemist decides to sell the smaller bottle for 75p.
    Is the second bottle a good buy?
    Give a reason for your answer.

Brainstormer

The volume and surface area of a sphere are given by $\frac{4}{3}\pi r^3$ and $4\pi r^2$ respectively.
  a Find the radius of the sphere when the volume numerically equals the surface
    area.
  b Find the ratio of volume to surface area when the radius is:
    i doubled   **ii** trebled   **iii** halved
    iv divided by 3   **v** multiplied by n **vi** divided by n.

◀◀ RECAP

You should be able to use scales expressed as **ratios** or **representative fractions**.

Example 1   On a map with scale 1 : 150 000, the distance between two towns is 3 cm. What is the actual distance between the towns?

     Answer   1 cm : 150 000 cm

$\Rightarrow$ 3 cm : 450 000 cm = 3 cm : 4·5 km

The actual distance between the towns is 4·5 km.

You should be able to use scale factors or ratios to solve problems in similar triangles. Corresponding sides are in proportion.
Remember you can find corresponding sides by finding the equal angles.
Corresponding sides are opposite the equal angles.

Example 2   The diagram shows two similar triangles.
A pair of parallel lines has been marked.
Find the value of *x*.

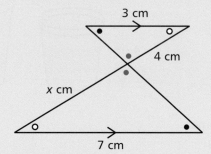

Equal angles are marked using the properties of parallel lines.

Enlargement factor $= \frac{7}{3}$

$\Rightarrow x = 4 \times \frac{7}{3} = 9\frac{1}{3}$ cm

You should be able to use scale factors to solve problems on:

- areas of similar shapes
- surface areas of similar solids.

> The scale factor for area is the square of the scale factor for length.

You should also be able to use scale factors to solve problems on volumes of similar solids.

> The scale factor for volume is the cube of the scale factor for length.

1 Work out these actual distances using the representative fractions given.
   a 6 cm on a map with a scale of 1 : 25 000
   b 8 cm on a map with a scale of 1 : 100 000
   c 3·4 cm on a map with a scale of 1 : 50 000
   d 2·6 cm on a map with a scale of 1 : 400 000

2 Each diagram shows a pair of similar triangles.
   Calculate the labelled lengths. Parallel lines have been marked.

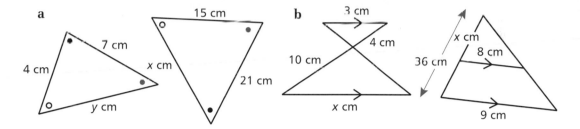

3 In each case two similar triangles are shown.
   Parallel lines have been indicated.
   Calculate the lengths marked.

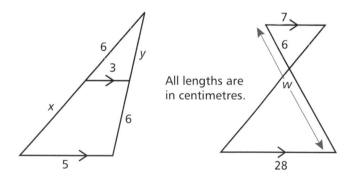

All lengths are
in centimetres.

4 Two similar barrels are sold for holding water.
   The height of the larger is 40 cm while that of the smaller is 30 cm.
   The larger barrel has a volume of 22 400 cm³.
   Calculate the volume of the smaller barrel.

REVISE

5 The design of a box for an Easter egg is a complicated affair.
The two boxes shown are mathematically similar.

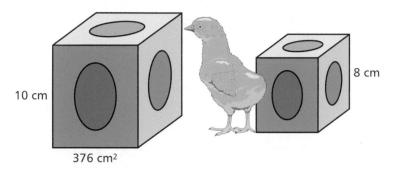

The larger has a height of 10 cm and uses 376 cm² of card for its construction.
The smaller has a height of 8 cm.
What area of card is needed for its construction?

6 A cylindrical chimney with a diameter of 1 metre has a curved surface of 6·28 m².
A smaller, similar chimney has a height of 50 cm.
a Calculate:
   i the height of the larger chimney
   ii the scale factor (reduction)
   iii the surface area of the smaller chimney.
b What is the area of the top of
   i the larger
   ii the smaller chimney?

4 Formulae

Srinivasa Ramanujan was born in 1887 into a poor family in South India.

At the age of 16 he found a maths book which contained hundreds of theorems. He became so absorbed in the book that he failed his school exams.

Ramanujan discovered some astonishing formulae. One of the most famous is a formula that counts the number of partitions of a number.

Srinivasa Ramanujan

There are exactly seven 'partitions' of the number 5. These are just diffferent ways of writing 5 as the sum of positive integers:

$$5$$
$$4 + 1$$
$$3 + 2$$
$$3 + 1 + 1$$
$$2 + 2 + 1$$
$$2 + 1 + 1 + 1$$
$$1 + 1 + 1 + 1 + 1$$

1 Review

Reminders:

1. By convention, calculations are done in a fixed order. The mnemonic BODMAS helps us to remember this order: **B**rackets; **o**f; **d**ivision/**m**ultiplication; **a**ddition/**s**ubtraction. Other operations will be fitted into this arrangement as we meet them.

2. Squares and square roots are performed before division and multiplication.

   Example 1   If $x = 3$ then $x^2 = 3^2 = 3 \times 3 = 9$

   Example 2   If $x = 9$ then $\sqrt{x} = \sqrt{9} = 3$ (since $3^2 = 9$)

   Example 3   If $x = 16$ then $x^2 + \sqrt{x} = 16^2 + \sqrt{16} = 256 + 4 = 260$

Exercise 1.1

1 Calculate:

a $3 + 2 \times 4$ **b** $14 - 2 \times 6$ **c** $10 + \frac{1}{2}$ of 8 **d** $\frac{1}{3}$ of $6 + 9$

e $4 \times 5 - 7$ **f** $6 + 4 - 2$ **g** $6 - 4 + 2$ **h** $6 - (4 + 2)$

i $\dfrac{2 + 6}{4}$ **j** $\dfrac{9}{2 + 1}$ **k** $\frac{1}{2}$ of $\frac{4}{2}$ **l** $\frac{15}{5} - 2$

m 5^2 **n** $30 - 4^2$ **o** $\frac{1}{2}$ of 6^2 **p** $9 + 2^2$

q $\sqrt{49}$ **r** $\sqrt{81}$ **s** $\frac{1}{10}$ of $\sqrt{100}$ **t** $\sqrt{3 + 22}$

u $\sqrt{6^2 + 8^2}$ **v** $9^2 + \sqrt{9}$ **w** $\dfrac{5 - \sqrt{4}}{6}$ **x** $\dfrac{5 + 4}{5 - 2}$

2 If $a = 9$ and $b = -3$ find the value of:

a $2a$ **b** $2b$ **c** $a + b$ **d** $a - b$

e ab **f** $\dfrac{a}{b}$ **g** a^2 **h** b^2

i $-b$ **j** $a + 2b$ **k** $2a - 3b$ **l** $a - b^2$

m $(a + b)^2$ **n** $2a^2$ **o** $4b^2$ **p** $(2b)^2$

q $\sqrt{a}$ **r** $\sqrt{1 - b}$ **s** $\dfrac{a - b}{a + b}$ **t** $\dfrac{b^2}{a}$

u $\dfrac{b}{\sqrt{a}}$ **v** $\dfrac{b}{a + b}$ **w** $a(a + b)$ **x** $\dfrac{b}{a}(b - a)$

3 Remove the brackets:

a $5(x - 2)$ **b** $12(a + b)$ **c** $k(m + 2n)$ **d** $x(x - y)$

e $-(w - z)$ **f** $-3(2c - 3d)$ **g** $5m(3 - 2m)$ **h** $-4n(m - 3n)$

4 Multiply out then simplify:

a $8 - 3(x - 1)$ **b** $k - 2(k + 3)$ **c** $2 - 5(2 - n)$

d $3k - (2k - 5)$ **e** $-4(2 - x) + 3(x - 3)$ **f** $-(a + b) - 3(a - b)$

5 Solve:

a $5x = 15$ **b** $-3y = 33$ **c** $2a - 5 = 7$

d $\dfrac{w}{3} = 5$ **e** $-3 = \dfrac{k}{2}$ **f** $7(x + 2) = 21$

g $2y - 3 = y + 4$ **h** $6 - 3x = 2x - 4$ **i** $a(a + 2) = a^2 - 4$

6 For each situation, write down the formula you would use for calculating the stated quantity.

a The area, A cm², of a circle of radius r cm

b The circumference, C cm, of a circle of diameter D cm

c The circumference, C cm, of a circle of radius r cm

d The perimeter, P cm, of a square of side x cm

e The area, A cm², of a square of side x cm

f The perimeter, P cm, of a rectangle with sides of length m cm and n cm

g The area, A cm², of a rectangle with sides of length m cm and n cm

7 A cube has an edge of k cm.
Write down a formula for:
 a its volume, V cm³
 b its surface area, A cm²
 c the total length of its edges, E cm.

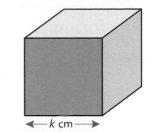

8 A cuboid has dimensions l cm $\times$ b cm $\times$ h cm.
Write down a formula for:
 a its volume, V cm³
 b its surface area, A cm²
 c the total length of its edges, E cm.

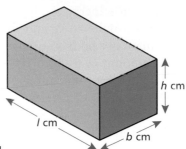

9 a The nth term of a sequence is given by the formula:

$$n\text{th term} = 5n - 2$$

Find:
 i the 4th term
 ii the difference between the 9th and 10th terms
 iii which term has a value of 68 (hint: write an equation and solve it).

b For the sequence: 3, 10, 17, 24, 31, …
 i find the nth term formula
 ii use the formula to find the 50th term of the sequence
 iii by forming an equation, find which term has a value of 157.

Challenge

Below are ten **formulae for A**, ten pairs of **values for m and n** and ten **values for A**.
Your task is to make ten correct triples.

If the triple is correct then substituting the **values for m and n** into the **formula for A** will produce the **value for A**.

There is only one way to make ten correct triples.

$A = m - n^2$	$m = 7$, $n = 2$	$A = 7$	$A = m + n$	$m = 2$, $n = 1$	$A = 3$
$A = m^2 - n^2$	$m = 3$, $n = 3$	$A = 5$	$A = m^2 + n$	$m = 6$, $n = 4$	$A = 2$
$A = mn$	$m = 2$, $n = 4$	$A = 8$	$A = m^2 - n$	$m = 5$, $n = 2$	$A = 1$
$A = m - n$	$m = 6$, $n = 5$	$A = 9$	$A = \dfrac{8}{m + n}$	$m = 1$, $n = 1$	$A = 6$
$A = m + n^2$	$m = 6$, $n = 2$	$A = 10$	$A = \dfrac{m + n}{2}$	$m = 3$, $n = 2$	$A = 4$

2 Using formulae

Example

a You can estimate the surface area of your skin, A m², using the formula:

$$A = \sqrt{\frac{hw}{3600}}$$

where h is your height in centimetres and w is your
weight in kilograms.
Mhairi weighs 70 kg and she is 1·8 metres tall.
Estimate her body's surface area.

b The amount of water, L litres, you require each day is
related to your body area by the formula $L = 1·5A$
where A m² is your body's surface area.
How much water does Mhairi need per day?

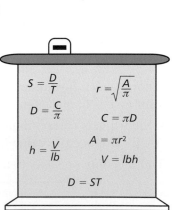

a $A = \sqrt{\dfrac{180 \times 70}{3600}} = \sqrt{3·5} = 1·87$ m² (to 2 d.p.)   **b** $L = 1·5 \times 1·87 = 2·8$ litres (to 1 d.p.)

Exercise 2.1

1 Use the formulae $A = \sqrt{\dfrac{hw}{3600}}$ and $L = 1·5A$ to calculate:

  i the body surface area, A m²
  ii the minimum daily fluid requirement, L litres, for these people.
    a Alison who is 176 cm in height and weighs 74 kg
    b Ricky who has a height of 185 cm and weighs 55 kg
    c Gabrielle who weighs 40 kg and is 1·25 m tall

2 To answer each question:
  i choose the correct formula
  ii calculate the answer, to the required degree of
  accuracy.
    a Find the area, to 1 decimal place, of a circle with
    radius 8·2 cm.
    b Find the distance covered, to the nearest
    kilometre, if you travel at 35 km/h for $2\frac{1}{2}$ hours.
    c Find the radius of a circle, to the nearest
    millimetre, that has an area of 200 cm².
    d Find the volume of a cuboid, to the nearest cubic
    centimetre, with dimensions 8·2 cm × 10·3 cm × 1·8 cm.
    e What is your average speed, to the nearest mile per hour, if you travel 255 miles
    in 4 hours?
    f What is the circumference of a circular plate, to the nearest centimetre, that has
    a diameter of 23·7 cm?
    g Find the height, to the nearest centimetre, of a cuboid with length 8 cm, breadth
    6·5 cm and volume 240 cm³.
    h What is the diameter, to the nearest millimetre, of a coin with circumference
    5·8 cm?

The formulae shown:

$$S = \frac{D}{T} \qquad r = \sqrt{\frac{A}{\pi}}$$

$$D = \frac{C}{\pi} \qquad C = \pi D$$

$$A = \pi r^2$$

$$h = \frac{V}{lb} \qquad V = lbh$$

$$D = ST$$

3 Use one of the given formulae for finding the area of each shape.

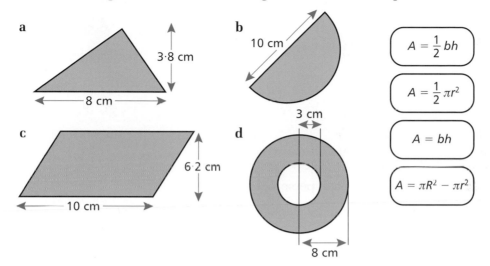

$$A = \frac{1}{2} bh$$

$$A = \frac{1}{2} \pi r^2$$

$$A = bh$$

$$A = \pi R^2 - \pi r^2$$

4 Use the given formulae to calculate the volume to 3 s.f.

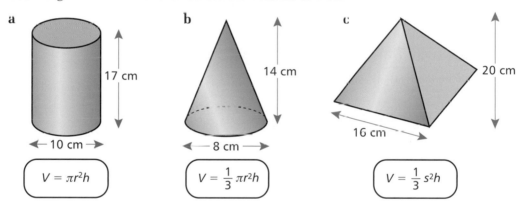

Exercise 2.2

1 At a keep-fit class they work out your body mass index, B, to decide if you are the correct weight for your height.

$$B = \frac{w}{h^2} \quad \text{where } w \text{ is weight in kilograms and } h \text{ is height in metres.}$$

The following table is then consulted.

The body mass index		
Condition	Women	Men
Underweight	<19·1	<20·7
Ideal weight	19·1–25·8	20·7–26·4
Slightly overweight	25·8–27·3	26·4–27·8
Overweight	27·3–32·3	27·8–31·1
Obese	>32·3	>31·1

a Calculate the body mass index, *B*, for each of these people.

Mr Henderson
82 kg
1·78 m

Mr Lewis
102 kg
1·80 m

Ms Lawson
65 kg
1·85 m

Mrs Smith
77 kg
1·72 m

b Use this index to describe each person's condition.

2 The power, *P* units, of a lens can be calculated using the formula:

$P = \dfrac{100}{f}$ where *f* is the focal length in centimetres.

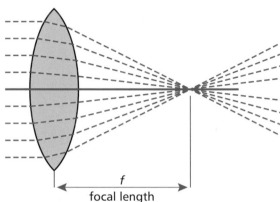

f
focal length

a i Which lens is more powerful –
lens A, focal length 25 cm, or lens B, focal length 10 cm?

ii By how many units?

b The magnification, *M*, of a telescope is given by:

$M = \dfrac{f_o}{f_e}$ Eyepiece → 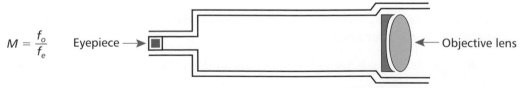 ← Objective lens

where f_o is the focal length of the objective lens and f_e is the focal length of the eyepiece.

i If a telescope is made using the two lenses in part **a** above, which lens would you use for the eyepiece and what magnification would it have?

ii Which of the two telescopes described in the table has the greater magnification?

	Objective lens	Eyepiece
Telescope A	500 mm	8 mm
Telescope B	381 mm	6 mm

(The figures give the focal lengths of the lenses.)

3 The formula for changing degrees Celsius to degrees Fahrenheit is $F = \dfrac{9}{5}C + 32$.

  a Which is warmer, London at 20 °C or New York at 70 °F?

  b Water freezes at 0 °C and boils at 100 °C, giving a difference between these two points of 100 °C.

    Find this difference in degrees Fahrenheit.

  c Another temperature scale, *Kelvin*, is often used by scientists.

    The formula for changing °C to °K is $K = C + 273 \cdot 15$.

    The lowest possible theoretical temperature is 0 °K.

    What is this temperature measured in degrees Fahrenheit?

4 Logs are cut up into boards.

The useable volume, V, of a log is measured in 'board feet'.

Look at these two formulae.

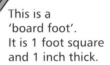

Formula 1 (used from 1825 to 1950): $V = L\left(\dfrac{D-4}{4}\right)^2$

Formula 2 (used from 1950): $V = \dfrac{L\,(0 \cdot 79D^2 - 2D - 4)}{16}$

In each case D is the end diameter in inches and
L is the log's length in feet.

This is a
'board foot'.
It is 1 foot square
and 1 inch thick.

Compare the estimation of 'board feet' given by the two formulae for:

  a a 20 foot log with diameter 15 inches

  b a 12 foot log with diameter 24 inches.

5 When a tuning fork is hit, it vibrates
with a frequency, f, measured in
cycles per second (c/s).
This frequency depends on the width,
w cm, and the length, L cm, of the fork.

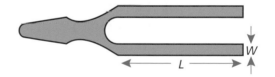

The frequency is calculated using the formula $f = \dfrac{85\,000w}{L^2}$

Match each tuning fork with its correct note.

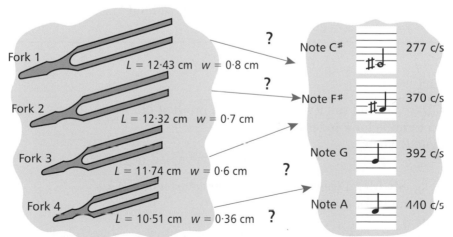

Fork 1 $L = 12 \cdot 43$ cm $w = 0 \cdot 8$ cm   ?   Note C♯   277 c/s

Fork 2 $L = 12 \cdot 32$ cm $w = 0 \cdot 7$ cm   ?   Note F♯   370 c/s

Fork 3 $L = 11 \cdot 74$ cm $w = 0 \cdot 6$ cm   ?   Note G   392 c/s

Fork 4 $L = 10 \cdot 51$ cm $w = 0 \cdot 36$ cm   ?   Note A   110 c/s

Challenge

If there is a wind blowing on a cold day, it can feel a lot colder than the actual air temperature.
This effect is called *wind chill*, *W*, and can be calculated using the formula:

$$W = 0.045(5.27\sqrt{V} + 10.45 - 0.28V)(T - 33) + 33$$

W is measured in degrees Celsius. *V* is the wind speed in km/h and *T* is the air temperature in degrees Celsius.
a How cold does it feel (*W* °C) if the air temperature is 5 °C and the wind is blowing at 25 km/h?
b Which situation below feels colder and by how many degrees?
Situation 1: wind speed 10 km/h; air temperature −30 °C
Situation 2: wind speed 60 km/h; air temperature −15 °C

3 Making formulae

Example 1

GARAGE CHARGES
Call-out charge: £80
Labour charge per hour   £20
Parts charged at cost

Find a formula for the total cost, £*T*, of a breakdown that needs *h* hours to fix with parts costing £*P*.

Each hour costs £20.   (2 hours: 2 × £20; 3 hours: 3 × £20; etc.)
⇒   *h* hours costs *h* × £20 = £20*h* with the other charges added on
⇒   total cost = labour + parts + call-out
⇒   *T* = 20*h* + *P* + 80

Exercise 3.1

1 I buy a CD for £*m* and sell it for £*n*, making a profit of £*P*.
  a Make a formula for *P* in terms of *m* and *n*.
  b Calculate *P* when:
    i *m* = 5, *n* = 7   **ii** *m* = 3·75, *n* = 8·25

2 I weigh *x* kg. My friend weighs *y* kg. Our average weight is *A* kg.
  a Write a formula for *A* in terms of *x* and *y*.
  b Calculate *A* if:
    i *x* = 48 and *y* = 53   **ii** *x* = 65·2 and *y* = 53·4

3 My salary is £23 000 and I have been promised *annual* increases of £1500.

  a Find a formula for my salary, £*S*, after *y* years.

  b Calculate *S* if:

    i $y = 2$

    ii $y = 6$

4 I started my shopping trip with £250. On average, each hour I spent £30.

  a Write a formula for the money I have left, £*M*, after *h* hours.

  b Calculate *M* if:

    i $h = 3$

    ii $h = 6$

5 I had my kitchen re-floored and was charged £25 per hour by the tile fitter. The tiles cost me £22 per m².

  a Write down a formula for the total cost, £*C*, if the fitter took *h* hours and my kitchen floor area is *A* m².

  b Calculate *C* if:

    i $h = 5$ and $A = 15$

    ii $h = 8$ and $A = 32$

6 To finance my new car I pay a deposit of £*D* and make *m* monthly payments of £*a*.

  a Write a formula for calculating the total price, £*P*, of the car.

  b Calculate *P* if:

    i $D = 4500$, $m = 30$ and $a = 250$ **ii** $D = 2000$, $m = 48$ and $a = 215$

7 a Find a formula for the number of matches, *M*, needed to make design *n*.

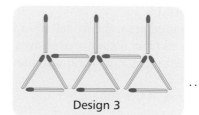

Design 1   Design 2   Design 3   ...

The pattern of designs continues...

  b How many matches are needed for:

    i design 17

    ii design 56?

8 For each sequence find a formula for T_n, the *n*th term, and use it to find the required term:

  a 11, 19, 24, 33, 42, ...   find T_{17} (the 17th term: use $n = 17$)

  b 6, 15, 24, 33, 42, ...   find T_{23}

  c 2, 14, 26, 38, 50, ...   find T_{47}

  d 17, 40, 63, 86, 109, ...   find T_{100}

Challenge

Here is a pentagon.
Pick a vertex and draw in the diagonals from it.

The pentagon is divided into three triangles.

The sum of the angles of the triangles is the sum of the angles of the pentagon.

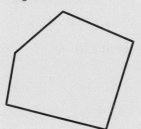
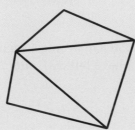
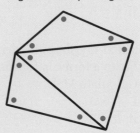

The sum of the angles of the pentagon is therefore $3 \times 180° = 540°$.

a Use the same method to find the sum of the angles in:
  i a quadrilateral   **ii** a hexagon   **iii** a heptagon.

b Find a formula for the sum of the angles, $S°$, of a polygon with n sides.

c If a polygon has all sides equal and all angles equal it is called **regular**.
What is the size of each angle in:
  i a regular quadrilateral   **ii** a regular hexagon   **iii** a regular heptagon?

d Find a formula for the size of each angle, $A°$, of a regular polygon with n sides.

e Use the formula in **d** to find the size of each angle in:
  i a regular octagon   **ii** a regular dodecagon ($n = 12$)

Example 2

Find a formula for:

a the shaded area, $A\,\text{cm}^2$, left after four quarter circles are removed from a square as shown

b the perimeter, $P\,\text{cm}$, of the shaded area.

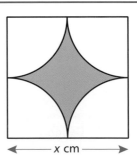

a The area of the square $= x^2\,\text{cm}^2$

The radius of each quarter circle $= \dfrac{x}{2}$

So the area of a *quarter* circle $= \dfrac{1}{4} \times \pi\left(\dfrac{x}{2}\right)^2$

The shaded area $=$ area of the square $-$ 4 quarter circles

$$\Rightarrow \quad A = x^2 - 4 \times \frac{1}{4} \times \pi\left(\frac{x}{2}\right)^2 = x^2 - \pi\left(\frac{x}{2}\right)^2 = x^2 - \frac{1}{4}\pi x^2 = x^2\left(1 - \frac{1}{4}\pi\right)$$

b Perimeter $= 4 \times$ quarter circumference $=$ whole circumference

$$= \pi \times x$$

So $P = \pi x\,\text{cm}$

Exercise 3.2

1 For each shaded shape find a formula for:
  i the perimeter, P cm   **ii** the area, A cm²

a ← 2x cm →

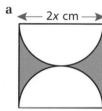

A square with
two touching
semicircles

b ← x cm →

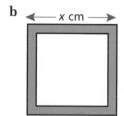

A square with a
1 cm wide border
on the inside

c ← 2x cm →

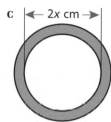

A circle with a
1 cm wide border
on the outside

d

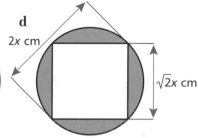

2x cm

√2x cm

A circle with a square
whose vertices are on
the circumference

2 For each shaded shape show all the necessary steps to produce the given formula.
P cm is the perimeter of the shaded shape.
A cm² is the area of the shaded shape.
(All dimensions are in centimetres.)

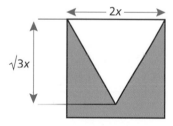

← 2x →

√3x

a Show that:
  i $P = 10x$
  ii $A = (4 - \sqrt{3})x^2$

A square with an equilateral
triangle removed as shown

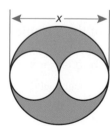

← x →

b Show that:
  i $P = 2\pi x$
  ii $A = \dfrac{\pi x^2}{8}$

A circle with two touching
circles removed from inside

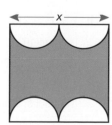

← x →

c Show that:
  i $P = (2 + \pi)x$
  ii $A = \left(\dfrac{8 - \pi}{8}\right)x^2$

A square with four identical
touching semicircles
removed from inside

4 Changing the subject of a formula

$A = \pi r^2$ is a formula for finding A when r is known.

A is called the **subject** of the formula.

If we know A and want to find r, we can change the formula to make r the subject.

$$A = \pi r^2$$

$$\Rightarrow \quad r^2 = \frac{A}{\pi} \qquad \text{(dividing both sides by } \pi)$$

$$\Rightarrow \quad r = \sqrt{\frac{A}{\pi}} \qquad \text{(taking the square root of both sides)}$$

Example

The perimeter, P cm, of a rectangle can be calculated using the formula $P = 2(l + b)$ where l cm and b cm are the length and breadth of the rectangle respectively. Make the length, l cm, the subject of the formula.

$$P = 2(l + b)$$

$$\Rightarrow \quad (l + b) = \frac{P}{2} \qquad \text{(dividing both sides by 2)}$$

$$\Rightarrow \quad l = \frac{P}{2} - b \qquad \text{(subtracting } b \text{ from both sides)}$$

Exercise 4.1

1 In each case change the subject of the formula to x by following the hint.

a $m = \dfrac{x}{k}$

(multiply by k)

b $q = r + x$

(subtract r)

c $a = b - x$

(add x; subtract a)

d $a = x^2$

(take the square root)

e $A = bx$

(divide by b)

f $k - x = m$

(add x; subtract m)

g $c = \dfrac{d}{x}$

(multiply by x; divide by c)

h $x + n = 2m$

(subtract n)

i $M = 2x^2$

(divide by 2; square root)

j $\dfrac{R}{x} = T$

(multiply by x; divide by T)

k $x^2 + 3 = n$

(subtract 3; square root)

l $3x + 4 = k$

(subtract 4; divide by 3)

m $\sqrt{x} = k$

(square both sides)

n $w = \sqrt{x + 2}$

(square both sides; subtract 2)

o $3\sqrt{x} = m$

(divide by 3; square)

2 Change the subject to P in each case:

a $m = n - P$ **b** $k = \dfrac{P}{5}$ **c** $a = b + P$ **d** $w = P^2$

e $w = \dfrac{P^2}{2}$ **f** $c = \dfrac{2}{P}$ **g** $Q = 3P^2$ **h** $a = 1 + \dfrac{P}{2}$

i $e = \dfrac{x}{P}$ **j** $y = 2P - 3$ **k** $r = QP$ **l** $x = 5 + yP$

3 For each formula:

i change the subject to the indicated letter

ii use this new formula to do the required calculation.

a $A - lb$ **i** change to b
      ii find b if $A = 8{\cdot}6$ and $l = 4$

b $V = lbh$ **i** change to h
      ii find h if $V = 13{\cdot}8$, $l = 2{\cdot}3$ and $b = 1{\cdot}5$

c $F = mc$ **i** change to m
      ii find m if $F = 4{\cdot}2$ and $c = 3{\cdot}5$

d $K = \dfrac{P}{Q}$ **i** change to Q
      ii find Q if $P = 2{\cdot}25$ and $k = 1{\cdot}5$

e $P = 2(x + y)$ **i** change to x
      ii find x if $P = 10$ and $y = 3{\cdot}4$

f $A = \dfrac{bh}{2}$ **i** change to b
      ii find b if $A = 13{\cdot}6$ and $h = 8{\cdot}5$

4 Electric current, I, is measured in amperes.
Electrical resistance, R, is measured in ohms.
The 'force', E, of a current is measured in volts.
Electrical power, W, is measured in watts.
Here are four formulae connecting these quantities:

$$W = EI \qquad E = IR \qquad R = \dfrac{E^2}{W} \qquad I = \sqrt{\dfrac{W}{R}}$$

Find a formula for:

a R in terms of E and I

b I in terms of E and W

c E in terms of R and W

d W in terms of I and R

e R in terms of I and W.

5 A little trick! Some formulae are of the form $A = BC$ or $C = \dfrac{A}{B}$.

These can be entered in a triangle to help you change the subject.

Hide what you want
to reveal a formula for it.

$A = BC$

$B = \dfrac{A}{C}$

$C = \dfrac{A}{B}$

Make a triangle for each of these formulae and write down the two other versions.

a The area, A cm², of a rectangle is given by $A = lb$ where l cm is the length and b cm is the breadth.

b The current, I amperes, is given by $I = \dfrac{E}{R}$ where E is the 'force' in volts and R is the resistance in ohms.

c The magnification, M, of a telescope is given by $M = \dfrac{F}{f}$ where F cm is the focal length of the objective lens and f cm is the focal length of the eyepiece.

d The volume, V cm³, of a prism is given by $V = Al$ where A cm² is the area of the base and l cm is the length of the prism.

e Using suitable units, the energy, E, contained in a mass, m, is given by $E = mc^2$ where c is the speed of light in a vacuum.
(Treat c^2 as a single object.)

Challenge

Draw a number of vertices (dots), V of them, and join them up with some edges (lines), E of them.

All the vertices should be connected by at least one edge, so that no dot is isolated. Edges cannot cross each other.

What you have drawn is a **connected graph** and it divides the page into several enclosed regions or areas, R of them.

Note that $V = 5$, $E = 6$, $R = 2$, and that $V + R - E = 1$.

a Draw several more connected graphs.
   Does the formula in the example still hold?

b Change the subject of the formula to:
   i V **ii** R **iii** E.

Example

c i How many regions would you expect if there were 7 vertices and 7 edges?
   ii How many edges would you expect if there were 7 vertices and 3 regions?
   iii How many vertices are needed for a connected graph of 12 edges and 5 regions?

d Can you draw an example of each graph in part **c** above?

Harder examples

The formula for changing degrees Celsius to degrees Fahrenheit is $F = \dfrac{9}{5}C + 32$.
Here are the steps to change the subject:

$$F - 32 = \frac{9}{5}C \qquad \text{(subtract 32 from both sides)}$$

$$9C = 5(F - 32) \qquad \text{(multiply both sides by 5)}$$

$$C = \frac{5}{9}(F - 32) \qquad \text{(divide both sides 9)}$$

Exercise 4.2

1 a The surface area, A cm², of a sphere is given by the formula:

$$A = 4\pi r^2 \text{ where } r \text{ is the radius in centimetres.}$$

   i Change the subject of the formula to r.
   ii Calculate the diameter of a sphere with a surface area of 30 cm².

b The volume, V cm³, of a sphere is given by:

$$V = \tfrac{4}{3}\pi r^3$$

   i Change the subject of the formula to r.
   ii Calculate the radius of a sphere with a volume of 1 litre (1000 cm³).

2 Change the subject of each formula to x:

a $p = q - 3x$ **b** $y = \frac{1}{2}x + 3$ **c** $\dfrac{3x + 5}{2} = y$ **d** $A = 2(x - 3)$

e $\frac{1}{3}(x + 5) = B$ **f** $\dfrac{5 - x}{6} = k$ **g** $\dfrac{2}{x + 1} = w$ **h** $6 = \dfrac{a}{1 - x}$

i $a(b - x) = c$ **j** $\dfrac{a}{b + x} = c$ **k** $a = \dfrac{b}{c - x}$ **l** $\dfrac{a}{b + cx} = d$

3 The volume, V cm^3, of a cone is given by the formula:

$$V = \frac{1}{3}\pi r^2 h$$

where r cm is the base radius and h cm is the vertical height.
  a Change the subject of this formula to r.
  b Calculate the radius of a 15 cm tall cone that has a volume of $1\frac{1}{2}$ litres (1500 cm^3).

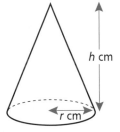

4 The formula $A = \sqrt{\dfrac{hw}{3600}}$ calculates the body surface area, A m^2, of a person with height h cm and weight w kg.
  a Change the subject of this formula to h.
  b Estimate, using this formula, the height in centimetres of a 60 kg man with a surface body area of 1·74 m^2.

5 Rocket science!
The area, A inches^2, of each fin is given by:

$$A = 0.17h(d + 0.5) \text{ for a 3-fin rocket}$$
$$A = 0.13h(d + 0.5) \text{ for a 4-fin rocket.}$$

  a Change the subject of each formula to h.
  b You are building a rocket with a fin area of 30 inches^2 and with a diameter of 6 inches. Which would be higher and by how much: a 3-fin or a 4-fin rocket?

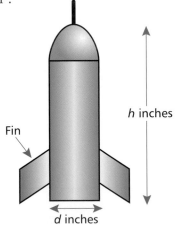

Fin

h inches

d inches

6 The diagrams show two solids called a **paraboloid** and a **neiloid**.
The volume, V cm^3, is given by the indicated formula.
  a Write each formula with r as the subject.
  b A paraboloid and a neiloid have the same volume of 500 cm^3 and the same height of 10 cm.
  Which has the greater base radius and by how much?

A paraboloid

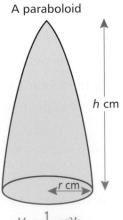

h cm

r cm

$$V = \frac{1}{2}\pi r^2 h$$

A neiloid

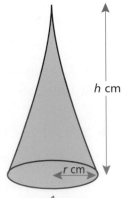

h cm

r cm

$$V = \frac{1}{4}\pi r^2 h$$

7 An indication of the use of each formula is given.
In each case a different use is required.
Change the subject to fit that new use.

a Used to calculate the gravitational force, F, between two
masses, M and m, that are a distance d apart.
Now to be used to find the distance.

$$F = \frac{GMm}{d^2}$$

b Used to calculate the time, T, for one swing of a pendulum
with length l.
Now to be used to find the length.

$$T = 2\pi\sqrt{\frac{l}{g}}$$

c Used to calculate the frequency, f, of the note produced by
a tuning fork of width w and length l.
Now to be used to find the length.

$$f = \frac{85\,000w}{l^2}$$

d Used to calculate the useable volume, V, of a log with
diameter D and length L.
Now to be used to find the diameter.

$$V = L\left(\frac{D-4}{4}\right)^2$$

e Used to calculate the length, L, of a golf drive where U is
the speed of the golf ball.
Now to be used to find the speed.

$$L = \frac{5}{4}U - 27$$

f Used to calculate the surface area, A, of a cylinder
with radius r and height h.
Now to be used to find the height.

$$A = 2\pi r(r + h)$$

Challenge

Astronaut Andy weighs w kg at sea level.
At a height of h km, Andy weighs w_h kg, given by the
formula:

The Earth

$$w_h = \left(\frac{R}{R+h}\right)^2 w \quad \text{where } R \text{ is the radius of the Earth.}$$

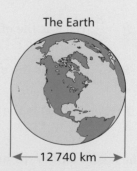

12 740 km

a Suppose Andy travels to a height where his weight is
one quarter of his weight at sea level.

Write down the value of: $\dfrac{w_h}{w}$ (Hint: divide both sides of
the formula by w.)

b Change the subject of the formula to h using the value you obtained in **a**.

c Write down the value of the height in **a**.

d Do some calculations to find the height at which Andy's weight is halved.

5 Understanding formulae

Example 1
The circumference of a circle is given by $C = \pi D$. What happens if we double D?

Original circle: $C = \pi D$. New circle: $C = \pi(2D) = 2\pi D$ = twice the original circumference.

Example 2
The length of a rectangle is given by $l = \dfrac{A}{b}$. What happens if we double b?

Original rectangle: $l = \dfrac{A}{b}$. New rectangle: $l = \dfrac{A}{2b} = \dfrac{1}{2} \times \dfrac{A}{b}$ = half the original length.

Example 3
The area of a circle is given by $A = \pi r^2$. What happens if we double r?

Original circle: $A = \pi r^2$.

New circle: $A = \pi(2r)^2 = 4\pi r^2$ = four times the original area.

Exercise 5.1

1 In each case, carry out the action indicated to find the required effect.

   a $D = ST$. What is the effect on D of doubling T?

   b $S = \dfrac{D}{T}$. What is the effect on S of doubling T?

   c $W = EI$. What is the effect on W of halving E?

   d $R = \dfrac{E}{I}$. What is the effect on R of halving I? $\left(\text{Hint: } \dfrac{E \times 2}{\frac{1}{2}I \times 2}\right)$

   e $R = \dfrac{E^2}{W}$. What is the effect on R of doubling E?

   f $I = \sqrt{\dfrac{W}{R}}$. What is the effect on I of multiplying W by 9?

2 The coefficient of restitution e of a golf ball can be found by dropping it onto a hard surface. The formula used is $e = \dfrac{\sqrt{h_2}}{\sqrt{h_1}}$ where h_1 is the height from which the ball is dropped and h_2 is the height it reaches after its first bounce.

   Two golf balls are dropped from the same height.
   After the first bounce one ball reaches a height four times that of the other.
   How do their coefficients of restitution compare?

3 The frequency, f, of the note sounded when a tuning fork is hit is given by:

$$f = \frac{85\,000w}{L^2}$$

where w is the width of each prong and L is the length of the fork.

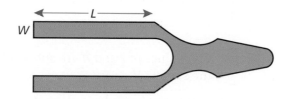

a Jasmine owns an unfamiliar tuning fork. She compares it with an 'A' tuning fork.
Its prongs are six times as wide and it is three times as long.
How does its frequency compare with that of the note A?

b She knows the frequency of three notes:

A	444
F	352
D	296

Identify the note of the unknown fork.

4 Astronomers photographing distant objects such as stars and planets calculate the
exposure time, T minutes, using the formula $T = \dfrac{f^2}{SB}$ where f is the 'f-number' of the
lens, S is the film's 'ISO speed' and B is the brightness factor of the object.

a If the 'f-number' is increased from 1·4 to 5·6 (i.e. multiplied by 4), what effect will
this have on a 2 minute exposure time?

b If the 'f-number' is decreased from 5·6 to 2·8 and at the same time the film's 'speed'
is doubled, what effect will this have on an hour long exposure?

◀◀ RECAP

A formula shows how various quantities are related to each other. One of the quantities appears on the left of the formula and is called the **subject**.

The right-hand side will then show the calculation required using the other quantities to find the subject.

Example
The subject is V. $V = \pi r^2 h$ The values of r and h are substituted to calculate the subject V.

You will have become familiar with many formulae during your maths course. Here are some examples:

Formulae about circles:
$$C = \pi D$$
$$A = \pi r^2$$

Formulae about cuboids:
$$A = 2lb + 2lh + 2bh$$
$$V = lbh$$

Formulae about triangles:
$$A = \tfrac{1}{2} bh$$
$$a^2 = b^2 + c^2$$

Formulae about distance, speed and time:
$$D = ST$$
$$S = \frac{D}{T}$$
$$T = \frac{D}{S}$$

You can change the subject of a formula from one quantity to another.
This is done by treating the formula like an equation and carrying out the same actions to both sides. The aim is to isolate the required quantity.

Example
Make b the subject of this formula: $a = \sqrt{b^2 + c^2}$

$$a^2 = b^2 + c^2 \qquad \text{(squaring both sides)}$$
$$\Rightarrow \quad a^2 - c^2 = b^2 \qquad \text{(subtracting } c^2 \text{ from both sides)}$$
$$\Rightarrow \quad \sqrt{a^2 - c^2} = b \qquad \text{(taking the square root of both sides)}$$
$$\Rightarrow \quad b = \sqrt{a^2 - c^2}$$

Altering some quantities in a formula by, for instance, multiplying or dividing them by some factor will have an effect on the subject. You should be able to calculate that effect.

1 Calculate, to 3 significant figures where necessary:
   a the circumference of a circle of radius 10 cm ($C = 2\pi r$)
   b the area of a rectangle, 5 cm by 8 cm ($A = lb$)
   c the area of a triangle of base 15 cm and altitude 9 cm ($A = \frac{1}{2}ab$)
   d the perimeter of a rectangle 5 cm by 8 cm ($P = 2(l + b)$)
   e the area of a circle of radius 10 cm ($A = \pi r^2$)

2 The electrical resistance, R ohms, of L metres of wire with a circular cross-section of radius r mm is given by:

$$R = \frac{L}{2r^2}$$

   a Calculate R when $L = 200$ and $r = 0.5$.
   b Which has more resistance,
      a 100 metre length of wire with 0·5 mm radius or
      a 400 metre length of wire with 1 mm radius?

3 Three hundred identical wooden cylinders are to be painted. The surface area, A cm², of a cylinder with radius r cm and height h cm is given by:

$$A = 2\pi r(h + r)$$

Calculate the total surface area to be painted.
Give your answer, in m², to the nearest m².

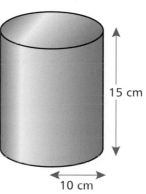

15 cm

10 cm

REVISE

4 I earn £27 000 per year and have been promised *annual* increases of £1200.
   a Find a formula for my annual salary, £S, after n years.
   b Calculate S when $n = 5$.

5 **a** Find a formula for the number of matches, M, needed to make design n.

Design 1   Design 2   Design 3

   b How many matches are needed to make design 37?

6 Six identical cheeses fit together into a cylindrical container with diameter D cm and height h cm.
   a Make a formula for the volume, V cm³, of the six pieces.
   b Give the formula for the volume, v cm³, of one piece of cheese.

7 In each case change the subject of the formula to y:

   a $k = \dfrac{v}{b}$

   b $r = \dfrac{c}{y}$

   c $2(m + y) = 3$

   d $y^2 - 3 = B$

   e $5\sqrt{y} = w$

   f $Q = 2 + \dfrac{y}{3}$

5 Equations and inequations

The ancient Egyptians and fractions
These are examples of Eygptian hieroglyphs for fractions:

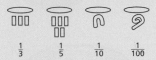

$\frac{1}{3}$ $\frac{1}{5}$ $\frac{1}{10}$ $\frac{1}{100}$

⬭ was the hieroglyph for mouth. When placed above a number it meant 'part'.

Liquid volumes were measured in fractions of the *heqat*, which was about $4\frac{3}{4}$ litres.

These 'fractions' were depicted using parts of the eye of the falcon god Horus:

Eye of Horus

1 Review

◀◀ Exercise 1.1

1 Solve these equations:

a $2x = 6$	**b** $3y = 30$	**c** $5a = -15$	**d** $8w = -32$
e $-3x = 9$	**f** $-2k = 14$	**g** $-m = 14$	**h** $-n = 16$
i $-2y = -10$	**j** $-7w = -28$	**k** $-c = -11$	**l** $-5f = -25$

2 Now solve these. You should expect fractions as solutions in most cases.

a $3x = -2$	**b** $2x = -3$	**c** $-4x = 5$	**d** $-7a = -3$
e $-6 = -4a$	**f** $-8 = -6k$	**g** $-10 = 4m$	**h** $14 = -4n$
i $5k + 2 = 0$	**j** $6n - 2 = 0$	**k** $7m + 3 = 0$	**l** $8 + 12x = 0$
m $6 - 8y = 0$	**n** $6x - 2 = 1$	**o** $4w + 1 = -1$	**p** $7x + 2 = x$

3 Multiply out and simplify where possible:

a $4(x + 3)$	**b** $-3(y - 2)$	**c** $-(2x + 3)$	**d** $-7(2 - y)$
e $2 + 3(w + 2)$	**f** $5 - 3(x + 1)$	**g** $9 + 2(k - 3)$	**h** $6 - 3(m + 1)$
i $2x + 2(3x - 1)$	**j** $5n - 3(n - 1)$	**k** $4m - 2(m + 1)$	**l** $8w - 3(2w - 3)$

4 Write down the lowest common multiple (LCM) of:

a 2 and 3	**b** 4 and 6	**c** 3 and 9	**d** 4 and 10
e 14 and 21	**f** 10 and 15	**g** 16 and 24	**h** 18 and 28
i 2 and x	**j** x^2 and $2x$	**k** $3y$ and $2x$	**l** $4x^2$ and $2y$

5 Multiply out the brackets and simplify:

a $(x - 2)(x + 3)$	**b** $(y - 2)(2y + 1)$	**c** $(2w - 1)(w - 3)$
d $(m - 5)^2$	**e** $3(m - 1)^2$	**f** $5 - (x - 1)^2$
g $2x - (5 - x)^2$	**h** $(n + 1)^2 - (n - 1)^2$	**i** $(3m + 4)^2 - (2m - 3)^2$

2 Brushing up those equation solving skills

Reminder of the equation solving steps

Step 1   Remove any brackets.

Step 2   Simplify each side of the equation if possible.

Step 3   Carry out balancing actions to each side of
the equation to find the solution.

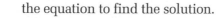
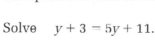

Example 1   Solve $y + 3 = 5y + 11$.

$$y + 3 = 5y + 11$$

$$\Rightarrow \quad y + 3 - y = 5y + 11 - y \qquad \text{(subtracting } y \text{ from both sides)}$$

$$\Rightarrow \quad\quad\quad 3 = 4y + 11$$

$$\Rightarrow \quad 3 - 11 = 4y + 11 - 11 \qquad \text{(subtracting 11 from both sides)}$$

$$\Rightarrow \quad\quad -8 = 4y$$

$$\Rightarrow \quad\quad -\frac{8}{4} = \frac{4y}{4} \qquad \text{(dividing both sides by 4)}$$

$$\Rightarrow \quad\quad\quad y = -2$$

Example 2   Solve $4x - 5(2x - 1) = 2(1 - x)$.

$$4x - 5(2x - 1) = 2(1 - x)$$

$$\Rightarrow \quad 4x - 10x + 5 = 2 - 2x \qquad \text{(removing the brackets)}$$

$$\Rightarrow \quad\quad -6x + 5 = 2 - 2x \qquad \text{(simplifying the left-hand side)}$$

$$\Rightarrow \quad\quad\quad -4x = -3 \qquad \text{(add } 2x \text{ and subtract 5 from both sides)}$$

$$\Rightarrow \quad\quad\quad x = \frac{-3}{-4} = \frac{3}{4} \qquad \text{(dividing both sides by } -4)$$

Exercise 2.1

1 Solve these equations:

    a $4(x + 2) = 12$ **b** $5x + 2 = 3x - 4$ **c** $x - 3 = 5x + 9$

    d $5 - 3x = x + 1$ **e** $5x = 8 - 3x$ **f** $3x = 5x - 8$

    g $-2x = x - 6$ **h** $-x = -3x + 8$ **i** $4 - 12x = x + 17$

    j $5x - 7 = -13 - x$ **k** $-3x - 2 = -5x - 6$ **l** $-x + 14 = -7x + 2$

2 Solve these equations by first removing the brackets:

    a $4(x + 2) = 12$ **b** $-6 = 2(y - 3)$ **c** $8 = 4(n - 2)$

    d $3(m + 2) = -6$ **e** $5(k - 8) = -10$ **f** $9(w + 3) = -27$

    g $5(y + 2) + 10 = 0$ **h** $14 + 7(w - 3) = 0$ **i** $0 = 9 - 3(2 - x)$

    j $4(3x - 2) - 16$ **k** $10 = 5(8 - 2w)$ **l** $3(2k + 5) - 45 = 0$

    m $2(4 + 5m) + 42 = 0$ **n** $7 + 2(n + 3) = 23$ **o** $5 - (r + 2) = -1$

    p $15 = 7(5 - c) + 8$ **q** $4 = 8 - 2(q - 3)$ **r** $20 - 3(x + 2) = 8$

3 Find the solution to each of these equations:

a $f - 3(f + 2) = 4(f - 3)$ **b** $2x + 3(2 - x) = x$

c $2(y + 3) - 3y = 6 + y$ **d** $3x - 2 = 12 - 2(x + 2)$

e $16 + m = 5m + 3(5 - 2m)$ **f** $2x - 2(2x + 3) = 2(5 - 2x)$

g $3(k + 2) - 5(k + 1) = k + 4$ **h** $3(w + 4) - 2(5 - 2w) = 3(2w + 1)$

i $8(3 - y) - (2y + 4) = 2(12 - y)$ **j** $12n - 7(n + 2) = 9(1 - 3n) - 7$

k $3(6w + 1) - 5(9 - 8w) = 30w$ **l** $12a - (5 + 7a) = 3(4a - 1) - 4a$

4 For each rectangle:

   i make an equation   **ii** solve the equation   **iii** calculate the length and breadth.

All measurements are in centimetres.

a

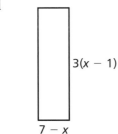

$x - 2$

$x + 6$

The perimeter is 48 cm.

b

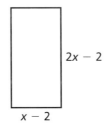

$2x - 2$

$x - 2$

The perimeter is 34 cm.

c

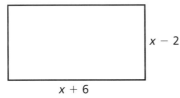

$2(x + 1)$

$3x$

The perimeter is 34 cm.

d

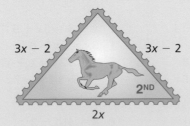

$3(x - 1)$

$7 - x$

The perimeter is $(5x + 3)$ cm.

e

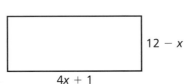

$12 - x$

$4x + 1$

The perimeter is $2(6x + 1)$ cm.

f

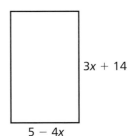

$3x + 14$

$5 - 4x$

The perimeter is $10(5 + x)$ cm.

Challenges

A Stamping ground

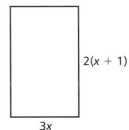

$3x - 2$ $3x - 2$ $5 - 2x$

2ND

$2x$ $x + 1$

All lengths are in centimetres.

The rectangular stamp has a perimeter 1 cm longer than the triangular stamp.

a Form an equation and solve it to find the dimensions of each stamp.

b Which stamp has the greater area and by how much?

B Suffering from panes

Size I x

$2x + 1$

Size II $2x$

$x + 1$

Sizes are in metres.

Three panes of glass of size I have exactly the same area as two panes of glass of size II.
Find the dimensions of each size.

Example 3 Solve $x^2 - 3 = (x + 2)(x - 1)$.

$$x^2 - 3 = (x + 2)(x - 1)$$

$\Rightarrow \quad x^2 - 3 = x^2 + x - 2$ (multiplying out the brackets)

$\Rightarrow \quad -3 = x - 2$ (subtracting x^2 from both sides)

$\Rightarrow \quad -1 = x$ (adding 2 to both sides)

$\Rightarrow \quad x = -1$

Example 4 Find the lengths of the sides of this triangle.

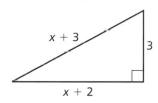

$x + 3$

3

$x + 2$

Lengths are in centimetres.

Using Pythagoras' theorem:

$$(x + 3)^2 = (x + 2)^2 + 3^2$$

$\Rightarrow \quad x^2 + 6x + 9 = x^2 + 4x + 4 + 9$ (multiplying out the brackets)

$\Rightarrow \quad 6x + 9 = 4x + 13$ (subtracting x^2 from both sides)

$\Rightarrow \quad 2x = 4$ (subtracting $4x$ and 9 from both sides)

$\Rightarrow \quad x = 2$

Now substitute $x = 2$ in the side length expressions: $x + 3 = 5$; $x + 2 = 4$.

The sides are 3 cm, 4 cm and 5 cm.

Exercise 2.2

1 Solve these equations:

a $x^2 - 7 = (x - 1)(x - 2)$

b $x(x - 1) = (x - 2)(x + 2)$

c $(x - 5)(x + 3) = (x - 2)(x - 3)$

d $(x + 2)(x - 5) = (x + 1)(x - 1)$

e $(x + 2)(x + 7) = (x + 3)(x + 10)$

f $(x + 3)(x - 7) = (x - 1)(x + 5)$

g $(2x + 1)(x \quad 3) = 2x(x + 2)$

h $(x + 2)^2 + (x + 4)^2 = 2x^2 - 4$

i $x^2 + (x - 3)^2 = (2x + 3)(x - 2)$

j $x^2 - (x - 1)^2 = x + 4$

k $(2x + 3)^2 - 2(x - 1)^2 = x(2x + 9)$

l $3(x - 1)(x + 2) - (x + 3)^2 = 2(x + 3)^2 - 3$

2 Find the lengths of the sides of these triangles. (All measurements are in centimetres.)

a

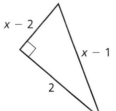

b

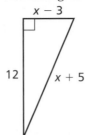

$x - 3$

12

$x + 5$

c

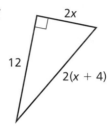

12

$x - 1$

$x + 5$

d

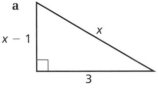

$x - 2$

$x - 1$

2

e

4

$2x + 1$

$2x - 1$

f

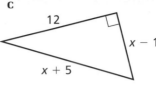

$2x$

12

$2(x + 4)$

---

3 Equations with fractions

Equation solving – taking that extra step

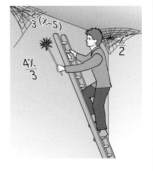

Step 1   Remove any fractions first.

Step 2   Remove any brackets.

Step 3   Simplify each side of the equation if possible.

Step 4   Carry out balancing actions to each side of the equation to find the solution.

Example 1   Solve $\dfrac{y}{5} = \dfrac{1}{3}$.

$$\dfrac{y}{5} = \dfrac{1}{3}$$

$$\Rightarrow \quad 5 \times \dfrac{y}{5} = 5 \times \dfrac{1}{3} \qquad \text{(multiplying both sides by 5)}$$

$$\Rightarrow \qquad y = \dfrac{5}{3}$$

Example 2   Solve $m = \dfrac{7}{3}(m + 2)$.

$$m = \dfrac{7}{3}(m + 2)$$

$$\Rightarrow \quad 3 \times m = 3 \times \dfrac{7}{3}(m + 2) \qquad \text{(multiplying both sides by 3)}$$

$$\Rightarrow \qquad 3m = 7(m + 2)$$

$$\Rightarrow \qquad 3m = 7m + 14 \qquad \text{(multiplying out the brackets)}$$

$$\Rightarrow \quad -4m = 14 \qquad \text{(subtracting } 7m \text{ from both sides)}$$

$$\Rightarrow \qquad m = \dfrac{14}{-4} = -\dfrac{7}{2} \qquad \text{(dividing both sides by } -4)$$

Example 3   Solve $x - 2 = \dfrac{3}{4}x$.

$$x - 2 = \frac{3}{4}x$$

$\Rightarrow \quad 4(x - 2) = 4 \times \dfrac{3}{4}x$ (multiplying both sides by 4)

$\Rightarrow \quad 4x - 8 = 3x$ (multiplying out the brackets)

$\Rightarrow \quad 4x = 3x + 8$ (adding 8 to each side)

$\Rightarrow \quad x = 8$ (subtracting 3x from both sides)

Exercise 3.1

1 Do these calculations. Express your answers in their simplest form.

a $\frac{1}{2} \times 4$ **b** $\frac{1}{3} \times 6$ **c** $\frac{1}{4} \times 6$ **d** $\frac{1}{8} \times 6$

e $9 \times \frac{1}{3}$ **f** $10 \times \frac{1}{6}$ **g** $12 \times \frac{1}{9}$ **h** $3 \times \frac{1}{4}$

i $2 \times \frac{2}{5}$ **j** $5 \times \frac{3}{7}$ **k** $8 \times \frac{7}{10}$ **l** $7 \times \frac{3}{14}$

2 Write these in simplest form:

a $\dfrac{x}{5} \times 10$ **b** $12 \times \dfrac{y}{8}$ **c** $\dfrac{1}{2} \times \dfrac{x}{2}$ **d** $3 \times \dfrac{a}{12}$

e $w \times \dfrac{3}{w}$ **f** $x \times \dfrac{2}{x}$ **g** $\dfrac{x}{4} \times \dfrac{8}{x}$ **h** $\dfrac{1}{a} \times \dfrac{2a}{3}$

i $\dfrac{1}{3} \times \dfrac{6y}{5}$ **j** $12 \times \dfrac{3x}{4}$ **k** $16 \times \dfrac{5w}{4}$ **l** $2w \times \dfrac{17}{2w}$

3 Solve these equations:

a $\dfrac{y}{3} = 2$ **b** $\dfrac{x}{4} = 1$ **c** $\dfrac{k}{6} = \dfrac{1}{3}$ **d** $\dfrac{m}{4} = \dfrac{1}{2}$

e $\dfrac{x}{8} = \dfrac{3}{4}$ **f** $\dfrac{n}{12} = \dfrac{2}{3}$ **g** $\dfrac{w}{10} = \dfrac{3}{5}$ **h** $\dfrac{x}{6} = \dfrac{3}{2}$

i $\dfrac{2}{3} = \dfrac{y}{9}$ **j** $4 = \dfrac{2r}{3}$ **k** $\dfrac{3n}{4} = 1$ **l** $\dfrac{1}{2} = \dfrac{3e}{4}$

m $\dfrac{5k}{2} = 10$ **n** $8 = \dfrac{2r}{7}$ **o** $\dfrac{1}{2} = \dfrac{5a}{6}$ **p** $\dfrac{4n}{5} = 8$

4 Solve:

a $2 = \frac{1}{3}(y + 2)$ **b** $7 = \dfrac{x + 4}{3}$ **c** $\frac{1}{5}(m - 2) = 1$ **d** $\dfrac{n + 1}{4} = 3$

e $\frac{1}{6}(y - 3) = 2$ **f** $\dfrac{k - 1}{2} = \dfrac{1}{2}$ **g** $w = \frac{1}{2}(w + 3)$ **h** $n = \dfrac{n + 6}{3}$

i $\frac{1}{5}(x + 9) = 2x$ **j** $\frac{1}{4}(y + 22) = 3y$ **k** $\dfrac{a + 20}{7} = 3a$ **l** $\frac{2}{3}(k + 1) = k$

m $y = \frac{3}{4}(y + 3)$ **n** $m = \frac{6}{7}(m + 5)$ **o** $\frac{5}{6}(w + 3) = w$

5 a i I have £x and spend £300. How much do I now have?

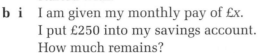

   ii I now spend half of this remaining money, leaving me with £200.

     Set up an equation for this situation.

   iii Solve this equation to find the amount of money I started with.

b i I am given my monthly pay of £x.

   I put £250 into my savings account.

   How much remains?

   ii $\frac{2}{3}$ of this remaining money is for spending, after which I will have only £180 left.

     Set up an equation for this situation.

   iii Solve the equation and find how much I am paid each month.

6 Solve these equations:

a $x - 5 = \frac{2}{3}x$ **b** $k - 3 = \frac{3}{4}k$ **c** $\frac{1}{2}m = 2m - 3$

d $3n - 8 = \frac{1}{3}n$ **e** $\frac{3}{5}y = 2y - 7$ **f** $\frac{2}{5}w = 3w - 26$

g $m - 10 = \frac{2}{7}m$ **h** $\frac{5}{6}x = 4x - 19$ **i** $\frac{2}{9}a = a - 7$

7 A variety of wheat is known to grow at the rate of 6 cm per week.

Three different crops of this variety are grown.

The results are described below.

For each crop:

i set up an equation that describes the situation

   (let the starting height be x cm)

ii solve the equation to find the starting height.

Crop a	**Crop b**	**Crop c**
After 4 weeks this crop is $1\frac{2}{3}$ times its starting height.	After 3 weeks the wheat is only $1\frac{1}{2}$ times its starting height.	After 5 weeks these plants are $2\frac{1}{3}$ times their starting height.

8 A company designs storage tanks.

For each tank described below:

i set up an equation to represent the situation

   (let the tank start with x litres)

ii solve the equation to find the original volume of liquid in the tank.

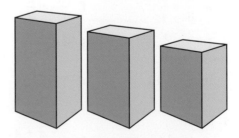

ºIn each case the tank loses 12 litres each hour through evaporation.

Tank a	**Tank b**	**Crop c**
After 3 hours this tank contains $\frac{2}{3}$ of its original volume.	After 5 hours this tank has lost half of its original volume.	After 8 hours only $\frac{2}{5}$ of the original contents of the tank remain.

Challenges

A Will and Harry

Uncle Harry died leaving a considerable sum of money. He left two-thirds of it to charity.

A further £20 000 went to his nephew.

The remainder, which was actually one quarter of the total amount, was split equally between his nephew's three daughters.

How much did Uncle Harry leave and how much did each of his nephew's daughters receive?

(Hint: suppose he left £x. Now set up an equation.)

B The multiplying microbes

A microbiologist was studying the extraordinary speed at which a bacteria culture was growing.

She checked the numbers in the culture after each hour.

At the first check about 1200 new bacteria had appeared.

At the second check the numbers had increased by $\frac{3}{5}$ since the first check.

She now estimated that there were four times the starting number.

How many are there now?

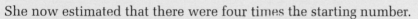

4 More equations with fractions: using the LCM

Example 1   Solve $\dfrac{x}{4} - \dfrac{x}{6} = 3$.

(Note: the LCM of the denominators 4 and 6 is 12.)

$$\frac{x}{4} - \frac{x}{6} = 3$$

$$\Rightarrow \quad 12\left(\frac{x}{4} - \frac{x}{6}\right) = 12 \times 3 \qquad \text{(multiplying each side by 12)}$$

$$\Rightarrow \quad 12 \times \frac{x}{4} - 12 \times \frac{x}{6} = 36$$

$$\Rightarrow \quad 3x - 2x = 36$$

$$\Rightarrow \quad x = 36$$

Example 2   Solve $\dfrac{y-1}{2} - \dfrac{y+1}{3} = 2$.

(Note: the LCM of the denominators 2 and 3 is 6.)

$$\frac{y-1}{2} - \frac{y+1}{3} = 2$$

$$\Rightarrow \quad 6\left(\frac{y-1}{2} - \frac{y+1}{3}\right) = 6 \times 2 \qquad \text{(multiplying each side by 6)}$$

$$\Rightarrow \quad 6 \times \frac{y-1}{2} - 6 \times \frac{y+1}{3} = 6 \times 2$$

$$\Rightarrow \qquad 3(y-1) - 2(y+1) = 12$$

$$\Rightarrow \qquad 3y - 3 - 2y - 2 = 12$$

$$\Rightarrow \qquad y - 5 = 12$$

$$\Rightarrow \qquad y = 17$$

Exercise 4.1

1 Solve these equations:

a $\dfrac{y}{2} - \dfrac{y}{3} = 7$ **b** $\dfrac{m}{3} - \dfrac{m}{4} = 1$ **c** $\dfrac{w}{3} + \dfrac{w}{2} = 4$

d $\dfrac{k}{2} - \dfrac{k}{5} = 6$ **e** $\dfrac{a}{3} = 1 + \dfrac{a}{4}$ **f** $\dfrac{k+1}{3} = \dfrac{k-1}{4}$

g $\dfrac{x-4}{5} = \dfrac{x+3}{2}$ **h** $\dfrac{f}{2} = \dfrac{2f-1}{3}$ **i** $\dfrac{3y+1}{5} = \dfrac{2y}{3}$

j $\dfrac{8-c}{2} = \dfrac{2c+2}{5}$ **k** $\dfrac{3h+6}{5} = \dfrac{h+2}{7}$ **l** $\dfrac{3-x}{3} = \dfrac{1-x}{2}$

2 Solve:

a $\frac{1}{2}(x-2) - \frac{1}{4}x = 16$ **b** $\frac{1}{2}(a-4) + \frac{1}{5}a = 5$ **c** $\frac{1}{3}(x+2) - \frac{1}{4}x = 1$

d $\frac{1}{4}(y+1) - \frac{1}{5}(y-5) = 2$ **e** $\frac{1}{6}(a+4) - \frac{1}{8}(a-4) = 2$ **f** $\frac{1}{3}(x-2) = 5 + \frac{1}{4}(x-6)$

g $\frac{1}{3}(n-3) = 15 - \frac{1}{2}(n-8)$ **h** $10 - \frac{1}{3}(x+2) = \frac{1}{4}(x+21)$ **i** $\frac{1}{3}(x+9) - \frac{1}{5}(x-6) = 7$

3 a In each case write an expression, in terms of x, for the amount of liquid left in the container.

     i A flask holds x litres. 3 litres are added, then half of the contents is poured away.

     ii A tumbler holds the same amount as the flask (x litres). 4 litres are poured out, then $\frac{2}{3}$ of what is left is also poured out.

   b There is more left in the flask above than in the tumbler.

     Also the difference between the amounts left in each container is 8 litres.

     Use this information to set up an equation.

     Solve this equation to find out how much was originally in each container.

4 The owner of a hotel decides to reduce the number of bedrooms by ten to create a new leisure and fitness area. After the change, on average only $\frac{2}{3}$ of the bedrooms were occupied each night.

This was actually $\frac{3}{5}$ of the original number of bedrooms.

a How many bedrooms did the hotel originally have? (Hint: suppose it originally had x bedrooms.)

b How many does it now have?

5 a Iain runs a d km race at an average speed of 15 km/h. Write an expression, in terms of d, for the time he took in hours.

b Martin ran the same race at an average speed of 20 km/h. How long did he take?

c Martin was 12 minutes ahead of Iain at the end.

Explain why $\frac{d}{15} - \frac{d}{20} = \frac{1}{5}$.

d Solve this equation to find the length, in kilometres, of the race.

Memory jogger

6 A motorist travelled to Glasgow, taking 3 hours to get there. On the return journey she took 5 hours. Her average speed was 15 mph slower returning compared to going. How far did she travel in total? (Hint: let the distance she travelled be d miles. Now find expressions for the two speeds there and back.)

5 Inequalities

Reminders:

The inequality signs are:
$>$ 'is greater than' $\geqslant$ 'is greater than or equal to'
$<$ 'is less than' $\leqslant$ 'is less than or equal to'.

A statement that uses an inequality sign is called an **inequality**.

You can use inequalities such as $A \geqslant 2$ and $10 < t < 15$ to describe situations:

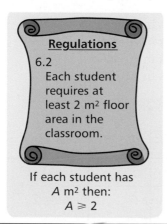

Regulations

6.2
Each student requires at least 2 m² floor area in the classroom.

If each student has A m² then:
$A \geqslant 2$

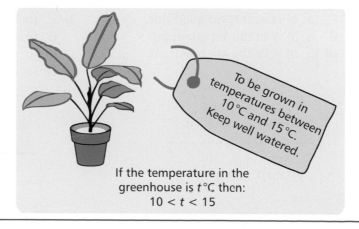

To be grown in temperatures between 10 °C and 15 °C. Keep well watered.

If the temperature in the greenhouse is t °C then:
$10 < t < 15$

Exercise 5.1

1 Write an inequality to describe each of these situations.
  a Traffic speed, v mph, in Zone A must not exceed 20 mph.
  b The total weight, W tonnes, on a trailer must be below 5 tonnes.
  c The minimum wage is £6·50 per hour. Iain earns £M per hour.
  d Daily intake of water, L litres, must be at least 4 litres.
  e I have to grow my plant in a temperature, t °C, that is at least 16 °C but not more than 25 °C.
  f The cost, £C, of a seat at the theatre is from £3·50 to £15.

2 There are three zones in a neighbourhood, defined by the maximum speed allowed.

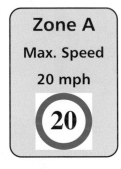
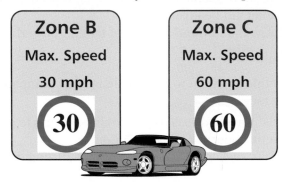

Write an inequality for each the following situations where the speed of each car is V mph. The car is travelling:
  a in Zone A
  b in Zone C
  c in Zone B at a speed not allowed in Zone A
  d in Zone C at a speed not allowed in Zone B
  e in Zone B and breaking the speed limit
  f in Zone A, breaking the speed limit at a speed that would be allowed in Zone B.

3 For each point (a, b) write two inequalities, one for a and one for b.
For example: if (a, b) is in the 2nd quadrant then $a < 0$ and $b > 0$.

  a i (a, b) is in the 3rd quadrant.
    ii (a, b) is in the 1st quadrant.
    iii (a, b) is in the 4th quadrant.

  b i (a, b) is in Area E.
    ii (a, b) is in Area F.
    iii (a, b) is in Area G.
    iv (a, b) is in Area H.

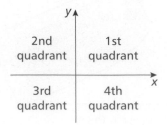

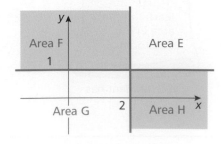

4 For each pair of points, use their relative positions in the diagram to write a pair of inequalities, one using the two *x* coordinates and one using the two *y* coordinates.

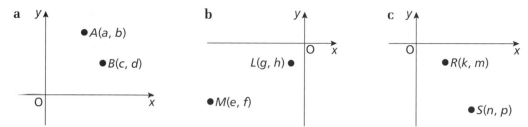

a *y* ↑

● *A*(*a*, *b*)

● *B*(*c*, *d*)

O ————— *x*

b *y* ↑

O *x*

L(*g*, *h*) ●

● *M*(*e*, *f*)

c *y* ↑

O ————— *x*

● *R*(*k*, *m*)

● *S*(*n*, *p*)

5 Describe each situation using an inequality.

a *x* kg / 8 kg

b 3 kg / *y* kg

c 2 kg *m* kg / 6 kg

Challenges

A The two balance puzzle

From the two pictures is

$$c > b \text{ or } b > c?$$

Show your reasoning!

B The three balance puzzle

From the information in the three pictures can you deduce whether $d < c$ or $d > c$? (Remember to show all your reasoning.)

109

6 Solving inequalities (inequations)

The balancing actions that you used to solve equations can be used to solve inequalities. But you have to be careful when:

- multiplying throughout by a negative amount

4 is greater than 2 but multiplying both quantities by -2 gives -8 is less than -4

$$a > b \Rightarrow -a < -b$$

- dividing throughout by a negative amount

4 is greater than 2 but dividing both quantities by -2 gives -2 is less than -1

> Multiplying or dividing both sides by a **negative** number requires a switch to the inequality sign.

Example 1   Solve $5x + 2 \leqslant -3$ choosing solutions from $\{-3, -2, -1, 0, 1, 2, 3\}$.

$$5x + 2 \leqslant -3$$

$\Rightarrow \qquad 5x \leqslant -5$ (subtracting 2 from both sides)

$\Rightarrow \qquad x \leqslant -1$ (dividing both sides by 5)

$\Rightarrow$ solutions are $-3, -2$ and -1 (picking from the given list)

Example 2   Solve $8 - 3m < 2$.

$$8 - 3m < 2$$

$\Rightarrow \qquad -3m < -6$ (subtracting 8 from both sides)

$\Rightarrow \qquad \dfrac{-3m}{-3} > \dfrac{-6}{-3}$ (dividing both sides by -3; note the change of sign)

$\Rightarrow \qquad m > 2$

Notes: **a** $-a < b \Rightarrow a > -b$ (multiplying both sides by -1)

                b $a < x \Rightarrow x > a$ (and vice versa)

Exercise 6.1

1 Solve these inequalities. Write your solutions in the form $x \leqslant 2$ or $x > 3$, etc.

   a $x + 3 \geqslant 2$ **b** $y - 4 < 3$ **c** $k + 1 \leqslant -3$ **d** $4 > x + 1$

   e $-2 \leqslant y - 5$ **f** $3 \geqslant w - 2$ **g** $n - 2 > 0$ **h** $c + 5 \geqslant 7$

   i $-2 > k + 1$ **j** $-8 \geqslant 4 + t$ **k** $3 + x \leqslant 5$ **l** $m + 7 > -3$

   m $a + 2 \leqslant 2$ **n** $3 \geqslant z + 9$ **o** $k + 3 < 10$ **p** $x - 7 \leqslant 1$

2 Solve these inequations.
Remember to switch the inequality sign when you divide both sides by a negative number.

a $2x \leqslant 8$　　**b** $5x < 15$　　**c** $4k > 16$　　**d** $7m \geqslant -14$

e $-7m \geqslant 14$　　**f** $-2x < 8$　　**g** $-3y \leqslant 9$　　**h** $-4k > -8$

i $-8n \geqslant -8$　　**j** $4t > 2$　　**k** $5w \leqslant 1$　　**l** $6a \geqslant 3$

m $-4c < 2$　　**n** $-2e > 1$　　**o** $-6x \geqslant -3$　　**p** $-12w \leqslant 18$

3 Write each of these inequations with the letter first, for example $2 \leqslant x$ becomes $x \geqslant 2$, etc.

a $2 < y$　　**b** $-3 > k$　　**c** $-4 \leqslant w$　　**d** $7 \geqslant m$

e $-11 \leqslant y$　　**f** $12 \geqslant n$　　**g** $\frac{1}{2} < c$　　**h** $-\frac{2}{3} > y$

4 Solve:

a $-8 > 2x$　　**b** $6 < 3y$　　**c** $9 \geqslant 3k$　　**d** $21 \leqslant 7n$

e $12 \geqslant 4x$　　**f** $-10 \leqslant 5n$　　**g** $10 \geqslant -5w$　　**h** $16 > -4k$

i $-12 \leqslant -12m$　　**j** $-8 < -2x$　　**k** $15 \leqslant -3k$　　**l** $-25 \geqslant -5n$

5 In each case:
i solve the inequation
ii list the possible solutions choosing from $\{-3, -2, -1, 0, 1, 2, 3\}$.

a $2x - 1 \geqslant -3$　　**b** $3x + 4 < 1$　　**c** $5x - 2 > 8$　　**d** $2x - 6 \leqslant -6$

e $8x - 3 > 5$　　**f** $4x + 2 < 10$　　**g** $3x + 2 \leqslant -4$　　**h** $10x - 15 \geqslant 15$

i $3x + 6 \geqslant 6$　　**j** $7x + 10 \leqslant -4$　　**k** $5x - 3 > -13$　　**l** $11x - 7 \leqslant -40$

6 Solve these (remember if $-x > 2$ then $x < -2$, etc.):

a $8 - x < 10$　　**b** $4 - k \geqslant 3$　　**c** $8 - m \leqslant 6$　　**d** $4 - 2k \leqslant 6$

e $2 - 5x < 12$　　**f** $4 - 3y \geqslant -2$　　**g** $26 - 13x \leqslant 0$　　**h** $9 - 4w > 11$

i $3 - 2a \geqslant -5$　　**j** $13 - 6x < 1$　　**k** $15 - p \leqslant 5$　　**l** $12 - 2x > 11$

Example 3　　Solve　$5(x - 1) - 8x \geqslant -11$.

$$5(x - 1) - 8x \geqslant -11$$
$$\Rightarrow \quad 5x - 5 - 8x \geqslant -11 \quad \text{(removing the brackets)}$$
$$\Rightarrow \quad -3x - 5 \geqslant -11 \quad \text{(simplifying the left-hand side)}$$
$$\Rightarrow \quad -3x \geqslant -6 \quad \text{(adding 5 to both sides)}$$
$$\Rightarrow \quad x \leqslant \frac{-6}{-3} \quad \text{(dividing both sides by } -3 \dots \text{ switch sign)}$$
$$\Rightarrow \quad x \leqslant 2$$

Exercise 6.2

1 Solve these inequations. Write your solutions in the form $x \leqslant 2$ or $x > 3$, etc.

a $4x + 3 > 2x + 9$　　**b** $4w + 5 < 2w + 17$　　**c** $8y - 2 > 3y - 17$

d $10m - 3 \geqslant 14 - 7m$　　**e** $8 - 3k \geqslant 4 - k$　　**f** $13 - 6n \leqslant 2n + 9$

g $3(a + 2) + 1 > 13$　　**h** $5(k - 2) - 1 < -16$　　**i** $3 - 2(x + 4) \geqslant -3x$

j $6 - 5(x - 2) \leqslant 3x$　　**k** $6x \geqslant 12 - 3(x - 5)$　　**l** $2 - x \geqslant 4(2 - x)$

m $3(2 - x) \leqslant 5(1 - x)$　　**n** $2 - (3 - x) \geqslant -2(x + 3)$　　**o** $1 - 3x \leqslant 2(x + 5) - 2x$

p $3(5 - 2x) + x \geqslant 3 - 2x$　　**q** $12 + 3(2 - 3x) \leqslant 4 - 5x$　　**r** $8 - 2(x + 4) \geqslant 11 - (2 - x)$

Challenge

The integer hunt

x, y and z are positive integers. Can you find them using these clues?

Clue 1: $5 - 4(z + 2) > 5(1 - z)$

Clue 2: $2(8 - y) + 3y < 3(2 + y)$

Clue 3: $2(x - 1) - 3x < 2(10 - 2x) + x$

Clue 4: $y + 4(1 - 2y) > 4(y - 13) - 3y$

Clue 5: $2(6 - z) - (z - 3) > 2 - (z + 7)$

Clue 6: $3(9 - x) - (x - 3) < 3 - x$

◀◀ RECAP

When multiplying two fractions, multiply numerators and multiply denominators,

e.g. $\dfrac{2}{3} \times \dfrac{5}{7} = \dfrac{2 \times 5}{3 \times 7} = \dfrac{10}{21}$

Fractions are 'simplified' by dividing numerators and denominators by any common factor,

e.g. $\dfrac{16}{28} = \dfrac{16 \div 4}{28 \div 4} = \dfrac{4}{7}$

When solving equations, in general, you should follow this advice:

Step 1 Remove any fractions (multiply both sides by the LCM of the denominators).

Step 2 Remove any brackets.

Step 3 Simplify each side if possible.

Step 4 Carry out the same simplifying actions to both sides while maintaining the balance.

The inequality signs are:

$<$ 'is less than'

$\leqslant$ 'is less than or equal to'

$>$ 'is greater than'

$\geqslant$ 'is greater than or equal to'.

Inequalities, or inequations, can be solved using the guidance above for equations. However, be careful when:

- multiplying throughout by a negative amount,

  e.g. $-x < 3 \Rightarrow x > -3$ (multiplying both sides by -1)

- dividing throughout by a negative amount,

  e.g. $-3x > -6 \Rightarrow x < 2$ (dividing both sides by -3)

$a < x \Rightarrow x > a$ (and vice versa)

1 Solve these equations:

   a $4x = 10 - x$ **b** $6(y - 3) = 18$ **c** $6 - (k + 2) = 5$

   d $11 - 2(m - 3) = 15$ **e** $y - 2 = 2y + 3(y - 2)$

2 The perimeter of this rectangle is 34 cm.

   a Make an equation to represent this situation.

   b Solve the equation.

   c Find the dimensions of the rectangle.

All lengths in centimetres

$3x - 2$

$2(x + 2)$

3 For the right-angled triangle shown:

   a use Pythagoras' theorem to make an equation

   b solve the equation

   c calculate the perimeter and area of the triangle.

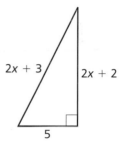

$2x + 3$ $2x + 2$ 5

4 Solve these equations:

   a $x^2 - 3 = (x - 3)(x + 6)$ **b** $(x + 3)^2 - (x - 1)(x - 2) = 10x + 2$

5 Write in simplest form:

   a $\dfrac{x}{6} \times 12$ **b** $12 \times \dfrac{4x}{3}$ **c** $8\left(\dfrac{3x}{4} - \dfrac{1}{2}\right)$

6 Solve these equations:

   a $\dfrac{x}{12} = \dfrac{5}{6}$ **b** $\frac{2}{3}(y + 3) = y$ **c** $\dfrac{x}{2} + \dfrac{x}{3} = 5$

   d $\dfrac{2k - 4}{4} = \dfrac{3k}{8}$ **e** $\frac{1}{3}(y - 1) - \frac{1}{4}y = 2$

7 **a** There were x people at the concert.
      250 walked out after 5 minutes.
      How many remained?

   b Another $\frac{1}{6}$ of those remaining left at the
      interval, leaving an audience of 350.
      Set up an equation that describes this
      situation.

   c Solve the equation to find the size of the
      audience at the start of the concert.

8 Describe each of the given price ranges using an inequality.
   Let £P be the price of an item.

Range A
Iesss than £10

Range B
From £10 up to a
maximum of £25

Range C
Greater than £25

REVISE

9 Solve these inequations:

a $m - 3 \geqslant 2$ **b** $-2 < k + 3$ **c** $3n \leqslant -9$ **d** $-5x > -10$

e $12x - 2 > 22$ **f** $12 - 2y < 2$ **g** $8 - 3(x - 2) \geqslant -x$

10 If n is a positive integer find the possible values for n given that:
$3(n - 2) > n - 4$ and $7 - 4n \geqslant 3(1 - n)$

REVISE

6 Algebraic fractions

The ancient Egyptians worked with fractions.

However, with a few exceptions, each fraction had a numerator of 1.

Fractions like these can be expressed as the sum of two other such 'Egyptian fractions'.

Here is one method they used:

$\dfrac{1}{a}$ ... square the denominator ... a^2.

Factorise it, say, $a^2 = a_1 \times a_2$.

The required sum is: $\dfrac{1}{a} = \dfrac{1}{a + a_1} + \dfrac{1}{a + a_2}$

You might be able to prove this later.

1 Review

Reminders:

- A fraction in the form $\dfrac{a}{b}$ where a and b are whole numbers can be thought of as $a \div b$.

  a is called the numerator and b the denominator. ($b \neq 0$)

- Fractions which have the same value are called equivalent fractions, e.g. $\dfrac{2}{5} = \dfrac{4}{10} = \dfrac{8}{20} = 0.4$

- We can find equivalent fractions by multiplying the numerator and denominator by a common factor, e.g. $\dfrac{2}{5} = \dfrac{2 \times 3}{5 \times 3} = \dfrac{6}{15}$

- Before adding or subtracting two fractions we must find equivalent fractions with a common denominator,

  e.g. $\dfrac{1}{3} + \dfrac{2}{5} = \dfrac{5}{15} + \dfrac{6}{15} = \dfrac{11}{15}$

- Multiply fractions by multiplying numerators and multiplying denominators, e.g. $\dfrac{2}{5} \times \dfrac{3}{4} = \dfrac{2 \times 3}{5 \times 4} = \dfrac{6}{20}$

◄◄ Exercise 1.1

1 List all the factors of:

    a 24   **b** $12x$ **c** a^2b **d** $4xy$

2 Factorise:

    a $2x + 12y$ **b** $12x - 4xy$ **c** $6xy - 3x^2$

    d $x^2 - 4$ **e** $4x^2 - 1$ **f** $9x^2 - 49y^2$

    g $x^2 + 3x + 2$ **h** $x^2 - x - 6$ **i** $6x^2 + 7x + 2$

3 **a** Find a fraction equivalent to $\frac{3}{4}$ with a denominator of 24.

    b Find a fraction equivalent to $\frac{4}{5}$ with a numerator of 36.

    c Express both $\frac{4}{5}$ and $\frac{2}{7}$ as fractions with the lowest possible common denominator.

    d Find a fraction that lies between $\frac{2}{3}$ and $\frac{3}{4}$ by first expressing these as fractions with a common denominator (not the lowest).

4 Calculate:

    a i $2x + 3x$ **ii** $\frac{2}{5} + \frac{3}{5}$ **iii** $\frac{2}{7} + \frac{3}{7}$ **iv** $\frac{2}{9} + \frac{3}{9}$

    b i $6x - 4x$ **ii** $\frac{6}{5} - \frac{4}{5}$ **iii** $\frac{6}{7} - \frac{4}{7}$ **iv** $\frac{6}{9} - \frac{4}{9}$

    c i $2x \times 3x$ **ii** $\frac{2}{5} \times \frac{3}{5}$ **iii** $\frac{2}{3} \times \frac{3}{3}$ **iv** $\frac{2}{7} \times \frac{3}{7}$

    d i $6x \div 4x$ **ii** $\frac{6}{5} \div \frac{4}{5}$ **iii** $\frac{6}{7} \div \frac{4}{7}$ **iv** $\frac{6}{9} \div \frac{4}{9}$

5 Calculate each of the following by first expressing the fractions with a common denominator:

    a $\frac{1}{5} + \frac{1}{4}$ **b** $\frac{2}{5} + \frac{3}{4}$ **c** $\frac{2}{3} + \frac{2}{7}$ **d** $\frac{3}{5} + \frac{5}{8}$

    e $\frac{1}{4} - \frac{1}{8}$ **f** $\frac{7}{8} - \frac{1}{3}$ **g** $\frac{2}{3} - \frac{2}{7}$ **h** $\frac{5}{9} - \frac{1}{2}$

6 Calculate:

    a $\frac{1}{5} \times \frac{1}{4}$ **b** $\frac{2}{5} \times \frac{3}{4}$ **c** $\frac{2}{3} \times \frac{2}{7}$ **d** $\frac{3}{5} \times \frac{5}{8}$

7 $\frac{1}{5} \div \frac{3}{4} = \frac{1}{5} \times \frac{4}{3} = \frac{4}{15}$. In a similar fashion calculate:

    a $\frac{1}{5} \div \frac{1}{4}$ **b** $\frac{2}{5} \div \frac{3}{4}$ **c** $\frac{2}{3} \div \frac{2}{7}$ **d** $\frac{3}{5} \div \frac{5}{8}$

8 One half of the pupils at Springfield High bring a packed lunch. One fifth of them don't take lunch. The rest take school lunches. What fraction of the pupils take school lunches?

9 $\frac{2}{3}$ of the cars in a car park are less than 3 years old. Of these $\frac{1}{2}$ are black.

    What fraction of cars in the car park are black and less than 3 years old?

10 A vehicle travelled $\frac{3}{5}$ km in $\frac{2}{3}$ of a minute. Using the formula $S = D \div T$, calculate the vehicle's speed in kilometres per minute.

2 Equivalent fractions

We know that any fraction of the form $\dfrac{x}{x}$ equals 1. $(x \div x = 1)$

We know that multiplying any quantity by 1 produces a quantity of the same value.

Thus multiplying any fraction by a fraction of the form $\dfrac{x}{x}$ will produce a fraction of the same value ... an **equivalent fraction**.

Example 1   Find three fractions equivalent to $\dfrac{3}{5}$.

$$\frac{3}{5} = \frac{3x}{5x} = \frac{3(x+1)}{5(x+1)} = \frac{3(x^2 - 2x + 1)}{5(x^2 - 2x + 1)} \qquad \frac{3}{5} \text{ is called the simplest form.}$$

Example 2   Find two fractions equivalent to $\dfrac{3x}{y+2}$.

$$\frac{3x}{y+2} = \frac{3x^2}{x(y+2)} = \frac{3x(x-2)}{(y+2)(x-2)} \qquad \frac{3x}{y+2} \text{ is called the simplest form.}$$

Example 3   Express $\dfrac{3}{x}$ and $\dfrac{x}{2y}$ with the simplest common denominator.

It is easy to turn both denominators, x and $2y$, into $2xy$ by multiplication

$$\frac{3}{x} = \frac{3 \times 2y}{x \times 2y} = \frac{6y}{2xy} \text{ and } \frac{x}{2y} = \frac{x \times x}{2y \times x} = \frac{x^2}{2xy}$$

... and we have the required result.

Example 4   Express $\dfrac{4x + 12}{7x + 21}$ in its simplest form.

Factorise both numerator and denominator and look for factors

common to both $\dfrac{4x + 12}{7x + 21} = \dfrac{4(x+3)}{7(x+3)} = \dfrac{4}{7} \cdot \dfrac{(x+3)}{(x+3)} = \dfrac{4}{7}.$

Exercise 2.1

1 Write down equivalent fractions to $\dfrac{1}{x}$ by multiplying numerator and denominator by

   a 2   **b** $x + 1$ **c** x

2 Which of the following are equivalent to $\dfrac{2a}{b}$?

   a $\dfrac{2a^2}{ab}$ **b** $\dfrac{6a}{3b}$ **c** $\dfrac{2a + 4}{b + 2}$ **d** $\dfrac{2a + 4}{b + 4}$

   e $\dfrac{2a(a + 1)}{b(a + 1)}$ **f** $\dfrac{2a^2 + 2a}{ab + b}$ **g** $\dfrac{2ab}{b^2}$

3 Express each pair of fractions with the simplest common denominator.

a $\dfrac{1}{x}$ and $\dfrac{1}{2}$ **b** $\dfrac{2}{x}$ and $\dfrac{1}{3x}$ **c** $\dfrac{1}{x}$ and $\dfrac{1}{y}$ **d** $\dfrac{2}{3x}$ and $\dfrac{1}{3y}$

e $\dfrac{1}{x^2}$ and $\dfrac{1}{x}$ **f** $\dfrac{2}{x+1}$ and $\dfrac{1}{x}$ **g** $\dfrac{1}{x+1}$ and $\dfrac{2}{x}$ **h** $\dfrac{2}{x+1}$ and $\dfrac{3}{4}$

i $\dfrac{1}{x+2}$ and $\dfrac{1}{x+1}$ **j** $\dfrac{2}{1-x}$ and $\dfrac{3}{x}$ **k** $\dfrac{x}{7}$ and $\dfrac{1}{x}$ **l** $\dfrac{2x}{3}$ and $\dfrac{4x+1}{5}$

4 Simplify each of the fractions.

a $\dfrac{12}{48}$ **b** $\dfrac{34}{51}$ **c** $\dfrac{x}{2x}$ **d** $\dfrac{5}{25x}$

e $\dfrac{10x}{30x}$ **f** $\dfrac{ax}{bx}$ **g** $\dfrac{x^2}{4x}$ **h** $\dfrac{3x^2}{6x}$

i $\dfrac{4pq}{12q}$ **j** $\dfrac{8ab}{ab^2}$ **k** $\dfrac{5x}{3x^2y}$ **l** $\dfrac{6abc}{12a^2c}$

m $\dfrac{5x^2y}{25xy^2}$ **n** $\dfrac{18pq^2r^2}{27p^2q^2}$ **o** $\dfrac{12e^2f^2}{2ef^3}$

5 Now simplify each of these.

a $\dfrac{a(b+1)}{3(b+1)}$ **b** $\dfrac{4(x+y)}{5(x+y)}$ **c** $\dfrac{x(3-y)}{4(3-y)}$ **d** $\dfrac{(x-1)(x+1)}{3(x-1)}$

e $\dfrac{(x-1)(x+2)}{(x+2)(x-3)}$ **f** $\dfrac{a^2}{a^2(a+1)}$ **g** $\dfrac{p-q}{(p+q)(p-q)}$ **h** $\dfrac{(x-4)(3-x)}{4(x-4)(3-x)}$

6 Factorise the numerator and/or the denominator and then simplify:

a $\dfrac{4x+12}{x+3}$ **b** $\dfrac{5x-5}{x-1}$ **c** $\dfrac{2-3x}{14-21x}$ **d** $\dfrac{a-6}{4a-24}$

e $\dfrac{6x+3}{10x+5}$ **f** $\dfrac{6x+4}{15x+10}$ **g** $\dfrac{24-18x}{16-12x}$ **h** $\dfrac{x^2+3x}{2x+6}$

Further factorising

Remember when factorising that we should consider:

- common factors   e.g. $3x+6 = 3(x+2)$
- the difference of two squares   e.g. $4x^2 - 49 = [2x]^2 - [7]^2 = (2x-7)(2x+7)$
- trinomials   e.g. $x^2 - 3x - 4 = (x-4)(x+1)$

Example 5   Express in its simplest form.
Factorise both numerator and denominator and look for common factors.

$$\frac{4x^2 - 8x}{x^2 - x - 2} = \frac{4x(x-2)}{(x+1)(x-2)} = \frac{4x}{(x+1)} \cdot \frac{(x-2)}{(x-2)} = \frac{4x}{(x+1)}$$

Exercise 2.2

1 Factorise, then simplify (look for common factors and the difference of two squares):

a $\dfrac{4x + 4}{x^2 - 1}$ **b** $\dfrac{1 - x}{1 - x^2}$ **c** $\dfrac{x^2 + 2x}{x^2 - 4}$ **d** $\dfrac{4a + 1}{16a^2 - 1}$

e $\dfrac{8x + 12}{4x^2 - 9}$ **f** $\dfrac{4x^2 - 36}{x - 3}$ **g** $\dfrac{3 - 12x^2}{4 - 8x}$ **h** $\dfrac{x^4 - 1}{x^2 - 1}$

2 Factorise, then simplify (look for common factors and trinomials):

a $\dfrac{3x + 6}{x^2 + 3x + 2}$ **b** $\dfrac{x - 3}{x^2 \; 2x - 3}$ **c** $\dfrac{3x - 6}{x^2 + 2x - 8}$ **d** $\dfrac{2x + 1}{2x^2 - x - 1}$

e $\dfrac{4 - x}{4 + 3x - x^2}$ **f** $\dfrac{6 - 3x}{6 - x - x^2}$ **g** $\dfrac{7 - 14x}{1 - 3x + 2x^2}$ **h** $\dfrac{x^2 + 2x}{x^2 + 7x + 10}$

i $\dfrac{x - x^2}{5 - 6x + x^2}$ **j** $\dfrac{2x^2 - 5x + 2}{2x^2 - x}$ **k** $\dfrac{x^2 - 5x + 6}{xy - 3y}$ **l** $\dfrac{3x^2 - 4x + 1}{3x^2 - x}$

3 Simplify this 'mixed bag':

a $\dfrac{x^2 - x - 20}{x^2 - 4x - 5}$ **b** $\dfrac{4x^2 - 1}{2x^2 - 3x + 1}$ **c** $\dfrac{12 + 13x + 3x^2}{3 - 2x - x^2}$ **d** $\dfrac{3x^2 - x - 2}{x^2 - 1}$

e $\dfrac{2x^2 - x - 1}{6x^2 + 5x + 1}$ **f** $\dfrac{4x^2 - x - 3}{4x^2 + 7x + 3}$ **g** $\dfrac{3 - 4x - 4x^2}{3x + 2x^2}$ **h** $\dfrac{15 - 17x - 4x^2}{3x - 4x^2}$

i $\dfrac{1 - x^4}{x^2 + x - 2}$ **j** $\dfrac{x^4 - 16}{x^2 + 6x - 16}$ **k** $\dfrac{x - x^3}{x^2 - x}$ **l** $\dfrac{x^3 - 9x}{x^3 + x^2 - 2x}$

---

3 Multiplying fractions

With numbers, we multiply fractions by multiplying numerators and multiplying denominators, e.g.

$$\frac{2}{5} \times \frac{3}{4} = \frac{2 \times 3}{5 \times 4} = \frac{6}{20}$$

In algebra we follow the same strategy.

Example 1   Calculate $\dfrac{x}{y} \times \dfrac{2}{y}$.

$$\frac{x}{y} \times \frac{2}{y} = \frac{x \times 2}{y \times y} = \frac{2x}{y^2}$$

We can sometimes simplify.

Example 2   Calculate and simplify $\dfrac{2(x + 1)}{x^2} \times \dfrac{x}{3(x + 1)}$.

$$\frac{2(x + 1)}{x^2} \times \frac{x}{3(x + 1)} = \frac{2x(x + 1)}{3x^2(x + 1)} = \frac{2}{3x}$$

Where we have to, we can create a denominator of 1.

Example 3   Calculate $2x \times \dfrac{4}{3y}$.

$$2x \times \frac{4}{3y} = \frac{2x}{1} \times \frac{4}{3y} = \frac{2x \times 4}{1 \times 3y} = \frac{8x}{3y}$$

Exercise 3.1

1 Multiply and simplify where possible.

a $\dfrac{a}{2} \times \dfrac{b}{2}$ **b** $\dfrac{x}{y} \times \dfrac{2}{3}$ **c** $\dfrac{3}{t} \times \dfrac{k}{4}$ **d** $\dfrac{1}{x} \times \dfrac{1}{y}$

e $\dfrac{2}{k} \times \dfrac{3}{1}$ **f** $\dfrac{m}{4} \times \dfrac{n}{4}$ **g** $\dfrac{2m}{3} \times \dfrac{3n}{5}$ **h** $\dfrac{2}{3m} \times \dfrac{4}{n}$

i $\dfrac{2}{k} \times \dfrac{k}{4}$ **j** $\dfrac{3t}{2} \times \dfrac{6}{t}$ **k** $\dfrac{4}{5t} \times \dfrac{15}{2t}$ **l** $\dfrac{3x}{7} \times \dfrac{35}{x^2}$

m $\dfrac{y}{3x} \times \dfrac{6x}{2y}$ **n** $\dfrac{b^2}{ax} \times \dfrac{a^2}{by}$ **o** $\dfrac{y}{x^2} \times \dfrac{4}{xy}$ **p** $\dfrac{4y}{3x} \times \dfrac{3x}{4y}$

q $3x \times \dfrac{1}{x}$ **r** $4y \times \dfrac{3}{y}$ **s** $2t \times \dfrac{5}{t}$ **t** $2z \times \dfrac{z}{4}$

u $\dfrac{3}{k} \times 4k^2$ **v** $\dfrac{1}{2g} \times 6g$ **w** $\dfrac{3}{mn} \times 5n^2$ **x** $k \times \dfrac{1}{k}$

2 Find an expression for the area of each shape.

a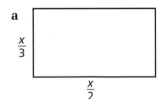

b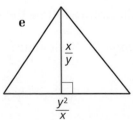

c 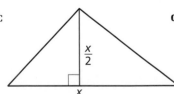   **d** 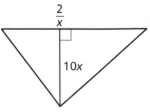   **e**

3 Multiply:

a $\dfrac{x+1}{4} \times \dfrac{16x}{x+1}$ **b** $\dfrac{3y-1}{2} \times \dfrac{y+1}{3y-1}$ **c** $\dfrac{z(z+1)}{z+2} \times \dfrac{z+2}{z+1}$

d $\dfrac{(y+2)^2}{4-x} \times \dfrac{x}{y+2}$ **e** $\dfrac{x+4}{x} \times \dfrac{x}{(x+4)(y+4)}$ **f** $\dfrac{y}{(2y+5)^2} \times \dfrac{(2y+5)y}{2-x}$

4 Show that $\dfrac{2x + 2}{4x + 6} \times \dfrac{2x + 3}{x + 1} = 1$.

5 The larger triangle (base x cm and altitude y cm) has been reduced, by scale factor $\dfrac{1}{k}$, to the smaller triangle (base p cm and altitude q cm).

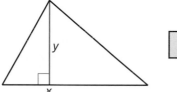 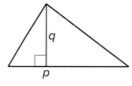

  a **i** Express p in terms of x and k.
     ii Express q in terms of y and k.
  b **i** Express the area of the large triangle in terms of x and y.
     ii Express the area of the small triangle in terms of x and y.
  c By what factor has the area been reduced?

6 a Copy and complete the multiplication table.
  b $a > b$. Which of the following is true?

$\times$	$\dfrac{1}{a}$	$\dfrac{1}{b}$	$\dfrac{1}{c}$	c	b	a
$\dfrac{1}{a}$						
$\dfrac{1}{b}$						
$\dfrac{1}{c}$						
c						
b						
a						

    i $\dfrac{1}{a} > \dfrac{1}{b}$

    ii $\dfrac{1}{a} < \dfrac{1}{b}$

  c In the table, $a > b > c$.

    What can you say about $\dfrac{1}{a}, \dfrac{1}{b}, \dfrac{1}{c}$?

  d Comment on the entries:
    i in the diagonal which comes down from left to right
    ii in the diagonal which goes up from left to right
    iii either side of the diagonal in (i)
    iv either side of the diagonal in (ii).

7 Calculate and simplify:

  a $\dfrac{3}{a} \times \dfrac{a}{3}$ **b** $\dfrac{3}{4} \times \dfrac{4}{3}$ **c** $\dfrac{p}{q} \times \dfrac{q}{p}$ **d** $\dfrac{x + 1}{x - 3} \times \dfrac{x - 3}{x + 1}$

  e $a \times \dfrac{1}{a}$ **f** $4f \times \dfrac{1}{4f}$ **g** $\dfrac{1}{2g} \times 2g$ **h** $\dfrac{1}{x + 3} \times (x + 3)$

8 $\dfrac{b}{a}$ is called the **reciprocal** of $\dfrac{a}{b}$.

  Comment on the product of a number and its reciprocal.

4 Dividing fractions

We see that $a \div a = 1 = a \times \dfrac{1}{a}$.

Similarly $K \div a = \dfrac{K}{a} = K \times \dfrac{1}{a}$.

> Dividing by a number is the same as multiplying by its reciprocal.

Example 1 Calculate: **a** $\dfrac{2}{3} \div 3$ **b** $\dfrac{2}{x} \div 3$ **c** $\dfrac{2}{3} \div \dfrac{3}{x}$

a $\dfrac{2}{3} \div 3 = \dfrac{2}{3} \div \dfrac{3}{1} = \dfrac{2}{3} \times \dfrac{1}{3} = \dfrac{2}{9}$

b $\dfrac{2}{x} \div 3 = \dfrac{2}{x} \div \dfrac{3}{1} = \dfrac{2}{x} \times \dfrac{1}{3} = \dfrac{2}{3x}$

c $\dfrac{2}{3} \div \dfrac{3}{x} = \dfrac{2}{3} \times \dfrac{x}{3} = \dfrac{2x}{9}$

Example 2 Calculate: $\dfrac{x}{x+1} \div \dfrac{4x}{x+1}$

$$\dfrac{x}{x+1} \div \dfrac{4x}{x+1} = \dfrac{x}{x+1} \times \dfrac{x+1}{4x} = \dfrac{x(x+1)}{4x(x+1)} = \dfrac{1}{4}$$

Exercise 4.1

1 Calculate:

a $\dfrac{3}{4} \div 6$ **b** $\dfrac{3}{x} \div 9$ **c** $\dfrac{3x}{2} \div 12$ **d** $\dfrac{3}{4} \div x$

e $\dfrac{5}{8} \div 20$ **f** $\dfrac{4}{y} \div 20$ **g** $\dfrac{2y}{5} \div 6$ **h** $\dfrac{1}{7} \div t$

2 Simplify:

a $\dfrac{2}{3} \div \dfrac{4}{9}$ **b** $\dfrac{3}{x} \div \dfrac{5}{2x}$ **c** $\dfrac{1}{y} \div \dfrac{y}{2}$ **d** $\dfrac{3}{2d} \div \dfrac{5}{3d}$

e $\dfrac{2x}{5y} \div \dfrac{4x}{15y}$ **f** $\dfrac{6y^2}{x} \div \dfrac{6y}{3x^2}$ **g** $\dfrac{a}{b^2} \div \dfrac{3}{2b}$ **h** $\dfrac{1}{x^3} \div \dfrac{1}{3x^2}$

i $\dfrac{ab}{c} \div \dfrac{ab^2}{c^2}$ **j** $\dfrac{3a}{xy} \div \dfrac{4ab}{x^2y}$ **k** $\dfrac{1}{abc} \div \dfrac{a}{bc}$ **l** $\dfrac{s}{td} \div \dfrac{s}{td}$

3 Simplify, factorising if necessary:

a $\dfrac{2x + 4}{5} \div \dfrac{3x + 9}{10}$

b $\dfrac{6}{5a + 15} \div \dfrac{7}{6a + 18}$

c $\dfrac{z - 1}{k} \div \dfrac{2(z - 1)}{5}$

d $\dfrac{x^2 - 1}{2x + 4} \div \dfrac{x + 1}{5x + 10}$

e $\dfrac{y^2 - 49}{x + 2} \div \dfrac{y - 7}{x^2 - 4}$

f $\dfrac{1 - a}{2t - 6} \div \dfrac{1 - a^2}{(t - 3)^2}$

g $\dfrac{x^2 + x - 2}{x^2 + 4x + 3} \div \dfrac{x^2 + 6x + 8}{x^2 + 7x + 12}$

h $\dfrac{x^2 - 1}{x^2 + 4x - 5} \div \dfrac{2x^2 + 10x + 8}{x^2 + 9x + 20}$

4 Last week a payout of £14 million pounds was shared amongst x people.
This week the same amount was shared amongst one more person.
  a Write an expression in terms of x for what one share was worth last week.
  b Write an expression in terms of x for what one share is worth this week.
  c Express this week's share as a fraction of last week's share.

5 The diagram shows two rectangles.

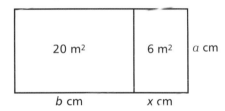

  a $A = lb \Rightarrow l = A \div b$. Express a in terms of x.
  b Express b in terms of
    i a
    ii x.

6 a Copy and complete the division table.

÷	$\dfrac{1}{a}$	$\dfrac{1}{b}$	$\dfrac{1}{c}$	c	b	a
$\dfrac{1}{a}$						
$\dfrac{1}{b}$						
$\dfrac{1}{c}$						
c						
b						
a						

  b Describe the diagonals of the table.
  c Compare this table with the multiplication table.

5 Adding fractions

In the same way that we can only add like terms in algebra, we can only add like fractions, i.e. fractions with the same denominator. When the denominators differ, we can change each fraction for a more suitable equivalent fraction.

Example 1   Evaluate $\dfrac{3}{4} + \dfrac{5}{7}$.

$$\dfrac{3}{4} + \dfrac{5}{7} = \dfrac{3 \times 7}{4 \times 7} + \dfrac{5 \times 4}{7 \times 4} = \dfrac{21}{28} + \dfrac{20}{28} = \dfrac{41}{28}$$

Compare this with $21x + 20x = 41x$ where x stands for 'twenty eighths'.

Example 2   Simplify $\dfrac{3}{x} + \dfrac{5}{y}$

$$\dfrac{3}{x} + \dfrac{5}{y} = \dfrac{3 \times y}{x \times y} + \dfrac{5 \times x}{y \times x} = \dfrac{3y}{xy} + \dfrac{5x}{xy} = \dfrac{3y + 5x}{xy}$$

Example 3   Simplify $\dfrac{3}{x} + \dfrac{x}{4}$.

$$\dfrac{3}{x} + \dfrac{x}{4} = \dfrac{3 \times 4}{x \times 4} + \dfrac{x \times x}{4 \times x} = \dfrac{12}{4x} + \dfrac{x^2}{4x} = \dfrac{12 + x^2}{4x}$$

Exercise 5.1

1 Simplify:

a $\dfrac{x}{3} + \dfrac{x}{4}$ **b** $\dfrac{a}{2} + \dfrac{a}{5}$ **c** $\dfrac{t}{4} + \dfrac{t}{7}$ **d** $\dfrac{2k}{3} + \dfrac{3k}{2}$ **e** $\dfrac{x}{5} + \dfrac{5x}{6}$

f $\dfrac{x}{3} + \dfrac{1}{4}$ **g** $\dfrac{p}{5} + \dfrac{3}{4}$ **h** $\dfrac{q}{3} + \dfrac{5}{7}$ **i** $\dfrac{2g}{7} + \dfrac{5}{8}$ **j** $\dfrac{n}{3} + \dfrac{4}{9}$

k $\dfrac{x}{5} + \dfrac{3}{y}$ **l** $\dfrac{w}{4} + \dfrac{1}{v}$ **m** $\dfrac{2r}{5} + \dfrac{2}{k}$ **n** $\dfrac{3h}{4} + \dfrac{3}{2g}$ **o** $\dfrac{3m}{7} + \dfrac{8}{9n}$

p $\dfrac{x}{2} + \dfrac{5}{x}$ **q** $\dfrac{t}{5} + \dfrac{1}{t}$ **r** $\dfrac{2p}{9} + \dfrac{7}{p}$ **s** $\dfrac{5g}{8} + \dfrac{1}{3g}$ **t** $\dfrac{6h}{11} + \dfrac{2}{5h}$

2 Simplify:

a $\dfrac{1}{x} + \dfrac{2}{y}$ **b** $\dfrac{3}{a} + \dfrac{2}{b}$ **c** $\dfrac{5}{t} + \dfrac{1}{w}$ **d** $\dfrac{4}{y} + \dfrac{2}{z}$ **e** $\dfrac{1}{m} + \dfrac{1}{n}$

f $\dfrac{3}{2x} + \dfrac{1}{y}$ **g** $\dfrac{1}{3e} + \dfrac{1}{2f}$ **h** $\dfrac{1}{4r} + \dfrac{2}{5t}$ **i** $\dfrac{7}{y} + \dfrac{8}{3x}$ **j** $\dfrac{9}{t} + \dfrac{2}{3v}$

k $\dfrac{2}{x} + \dfrac{5}{3x}$ **l** $\dfrac{1}{2g} + \dfrac{3}{5g}$ **m** $\dfrac{5}{4k} + \dfrac{1}{k}$ **n** $\dfrac{9}{2b} + \dfrac{2}{7b}$ **o** $\dfrac{2}{3x} + \dfrac{3}{5x}$

p $\dfrac{3x}{3y} + \dfrac{5}{2y}$ **q** $\dfrac{x}{y} + \dfrac{y}{x}$ **r** $\dfrac{3a}{2b} + \dfrac{b}{3a}$ **s** $\dfrac{6k}{h} + \dfrac{3}{5k}$ **t** $\dfrac{4v}{t} + \dfrac{3v}{5t}$

3 When two resistors are connected in parallel
their total resistance, R ohms, can be worked
out using the formula

$$\frac{1}{R} = \frac{1}{R_1} + \frac{1}{R_2}$$

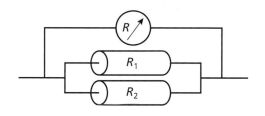

where R_1 and R_2 are the resistances of the
individual resistors.

a Add $\dfrac{1}{R_1} + \dfrac{1}{R_2}$.

b By taking the reciprocal of your answer, find a formula for R.

c What is the total resistance when a 5 ohm and a 7 ohm resistor are connected in
parallel?

4 A concave mirror is used in optics.

The distance to the object, u cm, the distance
to the image, v cm, and the focal distance, f cm,
are related by the equation:

$$\frac{1}{f} = \frac{1}{u} + \frac{1}{v}$$

a Add $\dfrac{1}{u} + \dfrac{1}{v}$.

b Find a formula with f as the subject.

c Calculate f when $u = 5$ and $v = 6$.

Example 4 Simplify $\dfrac{3}{x-1} + \dfrac{x}{x+1}$.

$$\frac{3}{x-1} + \frac{x}{x+1} = \frac{3(x+1)}{(x-1)(x+1)} + \frac{x(x-1)}{(x+1)(x-1)} = \frac{3x+3}{x^2-1} + \frac{x^2-x}{x^2-1}$$

$$= \frac{3x+3+x^2-x}{x^2-1} = \frac{x^2+2x+3}{x^2-1}$$

Example 5 Simplify $3 + \dfrac{a}{a+2}$.

$$3 + \frac{a}{a+2} = \frac{3}{1} + \frac{a}{a+2} = \frac{3(a+2)}{(a+2)} + \frac{a}{(a+2)}$$

$$= \frac{3a+6+a}{a+2} = \frac{4a+6}{a+2}$$

Exercise 5.2

1 Simplify:

a $\dfrac{x+1}{2} + \dfrac{x+3}{5}$ **b** $\dfrac{2w-3}{3} + \dfrac{w}{4}$ **c** $\dfrac{g+4}{4} + \dfrac{g-2}{2}$

d $\dfrac{k+5}{3} + \dfrac{2k}{5}$ **e** $\dfrac{2r}{7} + \dfrac{r-6}{3}$ **f** $\dfrac{3h-1}{2} + \dfrac{h-1}{3}$

g $\dfrac{p}{5} + \dfrac{1}{p+1}$ **h** $\dfrac{t}{2} + \dfrac{3}{t-1}$ **i** $\dfrac{3f}{5} + \dfrac{1}{2f+3}$

j $\dfrac{3}{x} + \dfrac{1}{x+1}$ **k** $\dfrac{1}{x-1} + \dfrac{1}{x+1}$ **l** $\dfrac{1}{x+3} + \dfrac{1}{x-2}$

m $\dfrac{3}{1-x} + \dfrac{2}{2x-1}$ **n** $\dfrac{7}{2x+3} + \dfrac{3}{x-5}$ **o** $\dfrac{1}{2x+3} + \dfrac{4}{3x+1}$

p $\dfrac{x}{1-2x} + \dfrac{x}{2x-1}$ **q** $\dfrac{x}{x-1} + \dfrac{1-x}{x-5}$ **r** $\dfrac{x+1}{x-1} + \dfrac{x}{x+1}$

2 Simplify:

a $4 + \dfrac{x+2}{7}$ **b** $5 + \dfrac{k}{k+1}$ **c** $\dfrac{t+1}{t-1} + 1$

d $\dfrac{2m-1}{m} + 2$ **e** $7 + \dfrac{3}{2d-1}$ **f** $\dfrac{1}{x+1} + 1$

3 In the introduction a method for expressing one Egyptian fraction as the sum of two others was described.

It showed that $\dfrac{1}{a} = \dfrac{1}{a+a_1} + \dfrac{1}{a+a_2}$ where $a_1 a_2 = a^2$.

a Add $\dfrac{1}{a+a_1} + \dfrac{1}{a+a_2}$.

b If $a_1 a_2 = a^2$ simplify the denominator by substitution and taking out a common factor.

c Simplify the fraction.

4 A boat with an outboard motor can travel at v mph in still water.
It travels 5 miles up river and then 5 miles back.
Over this stretch the river has a current of 3 mph.

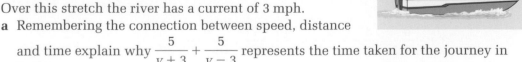

a Remembering the connection between speed, distance and time explain why $\dfrac{5}{v+3} + \dfrac{5}{v-3}$ represents the time taken for the journey in hours.

b Simplify this expression.

c Calculate the time taken if the boat can manage 15 mph in still water.

5 **a** Copy and complete the addition table.
b Describe the diagonals of the table.
c Describe the symmetry in the table.

+	$\dfrac{1}{a}$	$\dfrac{1}{b}$	$\dfrac{1}{c}$	c	b	a
$\dfrac{1}{a}$						
$\dfrac{1}{b}$						
$\dfrac{1}{c}$						
c						
b						
a						

6 Subtracting fractions

In a similar manner to addition, we can only subtract like fractions. When the denominators differ, we can change each fraction for a more suitable equivalent fraction.

Example 1 Evaluate $\dfrac{7}{8} - \dfrac{3}{7}$.

$$\frac{7}{8} - \frac{3}{7} = \frac{7 \times 7}{8 \times 7} - \frac{3 \times 8}{7 \times 8} = \frac{49}{56} - \frac{24}{56} = \frac{25}{56}$$

Example 2 Simplify $\dfrac{3}{x} - \dfrac{1}{x+5}$.

$$\frac{3}{x} - \frac{1}{x+5} = \frac{3(x+5)}{x(x+5)} - \frac{1 \times x}{(x+5)x} = \frac{3x + 15 - x}{x(x+5)} = \frac{2x - 15}{x(x+5)}$$

Exercise 6.1

1 Simplify:

a $\dfrac{a}{2} - \dfrac{a}{3}$ **b** $\dfrac{b}{3} - \dfrac{b}{7}$ **c** $\dfrac{c}{6} - \dfrac{c}{8}$ **d** $\dfrac{4a}{5} - \dfrac{2a}{7}$ **e** $\dfrac{g}{8} - \dfrac{3g}{10}$

f $\dfrac{t}{2} - \dfrac{1}{7}$ **g** $\dfrac{u}{3} - \dfrac{5}{6}$ **h** $\dfrac{v}{4} - \dfrac{1}{2}$ **i** $\dfrac{4h}{3} - \dfrac{3}{4}$ **j** $\dfrac{k}{5} - \dfrac{2}{7}$

k $\dfrac{a}{3} - \dfrac{2}{b}$ **l** $\dfrac{c}{5} - \dfrac{3}{d}$ **m** $\dfrac{3e}{4} - \dfrac{5}{f}$ **n** $\dfrac{7y}{3} - \dfrac{1}{5x}$ **o** $\dfrac{b}{2} - \dfrac{1}{4c}$

p $\dfrac{a}{3} - \dfrac{7}{a}$ **q** $\dfrac{b}{4} - \dfrac{2}{b}$ **r** $\dfrac{3c}{2} - \dfrac{1}{2c}$ **s** $\dfrac{9k}{4} - \dfrac{2}{5k}$ **t** $\dfrac{h}{5} - \dfrac{15}{h}$

2 Simplify:

a $\dfrac{1}{a} - \dfrac{1}{b}$ **b** $\dfrac{2}{x} - \dfrac{5}{y}$ **c** $\dfrac{3}{k} - \dfrac{7}{t}$ **d** $\dfrac{5}{p} - \dfrac{6}{k}$ **e** $\dfrac{3a}{b} - \dfrac{2}{c}$

f $\dfrac{1}{2a} - \dfrac{1}{3a}$ **g** $\dfrac{6}{a} - \dfrac{7}{4a}$ **h** $\dfrac{2b}{a} - \dfrac{a}{b}$ **i** $\dfrac{5}{g} - \dfrac{3}{g^2}$ **j** $\dfrac{4}{2k} - \dfrac{1}{k^2}$

k $5 - \dfrac{1}{a}$ **l** $1 - \dfrac{2}{3k}$ **m** $\dfrac{3}{2g} - 4$ **n** $\dfrac{3}{5r} - 7$ **o** $k - \dfrac{1}{k}$

p $3t - \dfrac{3}{t}$ **q** $\dfrac{x}{y} - x$ **r** $4p - \dfrac{p}{4}$ **s** $3f - \dfrac{3f}{2g}$ **t** $\dfrac{4a}{b^2} - \dfrac{b}{a^2}$

3 The magnification of a lens can be worked out using the formula $\dfrac{u + v}{u} - 1$ where v cm is the distance to the image and u cm is the distance to the object.

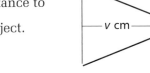

a Simplify the expression $\dfrac{u + v}{u} - 1$.

b What is the magnification when $u = 4$ and $v = 12$?

c The focal distance, f cm, is related to u and v by the relation $\dfrac{1}{v} - \dfrac{1}{u} = \dfrac{1}{f}$.

   i Simplify the left-hand side of this equation.

   ii By considering reciprocals, find a formula for calculating f.

4 Adam and Bill ran a race. Bill was given a 2 metre start. Adam finished the race in 10 seconds. Bill took 12 seconds. Let the distance that Adam ran be x metres.

a Write an expression for the distance Bill ran.

b Write an expression for the speed in metres per second of

   i Adam

   ii Bill.

c Give a simplified expression for how much faster Adam was.

d If Adam was 2 metres per second faster than Bill, how long was the race?

5 Terri needed to buy some calculators. She had £20. Before she could buy them the price went up by a pound. Let the price before the increase be £x.

a Write an expression for the price after the increase.

b Write an expression for the number of calculators that could be bought for £20

   i before the increase

   ii after the increase.

c Give a simplified expression for how many more could be bought before the increase.

d If, in fact, the difference is only one calculator, what was the cost of the calculator before the increase?

Exercise 6.2

1 Simplify:

a $\dfrac{x-1}{3} - \dfrac{x+2}{4}$ 　　 b $\dfrac{k+2}{2} - \dfrac{k}{3}$ 　　 c $\dfrac{h+1}{2} - \dfrac{h-1}{4}$ (Be careful!)

d $\dfrac{f}{3} - \dfrac{f+1}{5}$ 　　 e $\dfrac{2b}{6} - \dfrac{b-1}{3}$ 　　 f $\dfrac{4m}{5} - \dfrac{3-m}{2}$ 　　 g $\dfrac{d}{3} - \dfrac{1}{d+1}$

h $\dfrac{a}{9} - \dfrac{2}{a-1}$ 　　 i $\dfrac{4v+1}{2} - \dfrac{2}{3v}$ 　　 j $\dfrac{1}{a} - \dfrac{1}{a+1}$ 　　 k $\dfrac{1}{p} - \dfrac{1}{p+1}$

l $\dfrac{1}{h-3} - \dfrac{1}{h+1}$ 　　 m $\dfrac{x}{x-2} - \dfrac{x}{x-1}$ 　　 n $\dfrac{x+1}{x+2} - \dfrac{x}{x+1}$ 　　 o $\dfrac{g-1}{2g+1} - \dfrac{g-1}{3g+1}$

2 Simplify:

a $5 - \dfrac{v+2}{4}$ 　　 b $1 - \dfrac{a}{a-1}$ 　　 c $\dfrac{b+2}{b-1} - 1$

d $\dfrac{k}{k+1} - 2$ 　　 e $3 - \dfrac{x-1}{2+x}$ 　　 f $\dfrac{d}{3d-4} - d$

3 Simplify:

a $\dfrac{5}{x} + \dfrac{x}{2} - \dfrac{x+1}{3}$ 　　 b $\dfrac{1}{x} - \dfrac{1}{x+1} + \dfrac{1}{2}$ 　　 c $\dfrac{1}{x} - \dfrac{1}{2} + \dfrac{1}{3}$

d $\dfrac{x}{y} + \dfrac{y}{x} - 2$ 　　 e $\dfrac{3}{x} + 2 - \dfrac{1}{2+x}$ 　　 f $\dfrac{1}{3} + \dfrac{k}{2k+1} - k$

4 A container is 120 mm high. It is designed to hold thinner boxes.

Initially the thinner boxes were to be x mm high.

We would then expect $\dfrac{120}{x}$ boxes to fit in the container.

a Write an expression for the number of boxes we could fit in the container if they were:

　i 2 mm thinner 　　 ii 2 mm thicker.

b Write an expression for the difference between these two cases.

c If you can actually get five more boxes to fit in with the thinnest option, form an equation and solve it to find x.

d How many boxes fit in the container with the thinnest option?

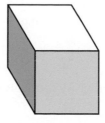

5 The cuboids shown have a curious property. The volume of one is numerically equal to the area of the base of the other and vice versa.

a Write down an expression for the height of each cuboid.

b Write an expression for the difference in heights.

c If this difference is 1·5 cm, calculate the value of x.

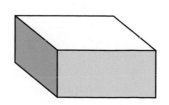

Volume = 2 cm³
Area of base = x cm²

Volume = x cm³
Area of base = 2 cm²

6 a Copy and complete the subtraction table.

$-$	$\dfrac{1}{a}$	$\dfrac{1}{b}$	$\dfrac{1}{c}$	c	b	a
$\dfrac{1}{a}$						
$\dfrac{1}{b}$						
$\dfrac{1}{c}$						
c						
b						
a						

b Describe the diagonals of the table.

c Describe the symmetry in the table.

◀◀ RECAP

Equivalent fractions

- Any fraction of the form $\dfrac{x}{x}$ equals 1. $(x \div x = 1)$

- Multiplying any fraction by a fraction of the form $\dfrac{x}{x}$ will produce a fraction of the same value, i.e. an equivalent fraction.

Factorising

We can simplify a fraction by dividing the numerator and denominator by a common factor.

We should factorise the numerator and denominator to find common factors.

Remember when factorising that we should consider:

- common factors   e.g. $3x + 6 = 3(x + 2)$

- the difference of two squares   e.g. $4x^2 - 49 = [2x]^2 - [7]^2 = (2x - 7)(2x + 7)$

- trinomials   e.g. $x^2 - 3x - 4 = (x - 4)(x + 1)$

Multiplying fractions

We multiply fractions together by multiplying numerators and multiplying denominators, e.g.

$$\frac{x}{y} \times \frac{2}{y} = \frac{x \times 2}{y \times y} = \frac{2x}{y^2}$$

Dividing fractions

Dividing by a number is the same as multiplying by its reciprocal, e.g.

$$\frac{2}{3} \div \frac{3}{x} = \frac{2}{3} \times \frac{x}{3} = \frac{2x}{9}$$

Adding fractions

We can only add *like* fractions, i.e. fractions with the same denominator. When the denominators differ, we can change each fraction for a more suitable equivalent fraction, e.g.

$$\frac{3}{x} + \frac{5}{y} = \frac{3 \times y}{x \times y} + \frac{5 \times x}{y \times x} = \frac{3y}{xy} + \frac{5x}{xy} = \frac{3y + 5x}{xy}$$

Subtracting fractions

We can only subtract *like* fractions, i.e. fractions with the same denominator. When the denominators differ, we can change each fraction for a more suitable equivalent fraction, e.g.

$$\frac{3}{x} - \frac{5}{y} = \frac{3 \times y}{x \times y} - \frac{5 \times x}{y \times x} = \frac{3y}{xy} - \frac{5x}{xy} = \frac{3y - 5x}{xy}$$

1 Write down equivalent fractions to $\dfrac{3}{y}$ by multiplying numerator and denominator by:

a 5   **b** y **c** x

2 Which of the following are equivalent to $\dfrac{x}{3y}$?

a $\dfrac{x^2}{3xy}$ **b** $\dfrac{3x}{6y}$ **c** $\dfrac{x+4}{3y+4}$

d $\dfrac{x^2+x}{3xy+3y}$ **e** $\dfrac{ax}{3by}$ **f** $\dfrac{x-xy}{3y-3y^2}$

3 Factorise and simplify:

a $\dfrac{x^2+4x-5}{x^2+x-2}$ **b** $\dfrac{9x^2-1}{3x^2+11x-4}$ **c** $\dfrac{2x^2-x-1}{2x^2+x}$

4 Multiply:

a $\dfrac{x}{3}\times\dfrac{y}{5}$ **b** $\dfrac{x}{2y}\times\dfrac{3y}{2x}$ **c** $2k\times\dfrac{5}{3k}$ **d** $\dfrac{x+2}{x}\times\dfrac{3}{x+2}$

5 Divide:

a $\dfrac{3x}{7y}\div\dfrac{2x}{5y}$ **b** $\dfrac{4a}{b^2c}\div\dfrac{2a^2}{bc^2}$ **c** $k\div\dfrac{1}{k}$ **d** $\dfrac{x-1}{x+1}\div\dfrac{x^2-1}{x}$

6 Add:

a $\dfrac{x}{5}+\dfrac{x}{7}$ **b** $\dfrac{g}{3}+\dfrac{5}{g}$ **c** $\dfrac{3}{m}+\dfrac{4}{n}$

d $\dfrac{2}{x+1}+\dfrac{3}{x-1}$ **e** $x+\dfrac{1}{x}$

7 Subtract:

a $\dfrac{a}{3}-\dfrac{a}{7}$ **b** $\dfrac{1}{x}-\dfrac{1}{y}$ **c** $\dfrac{3}{x}-4$

d $\dfrac{1}{x-1}-\dfrac{1}{x+1}$ **e** $g-\dfrac{2}{g}$

8 On the flat Sophie manages 10 km per hour.
Going uphill for 9 km she only averaged $(10-x)$ km/h.
However coming back down again she managed $(10+x)$ km/h.

a Form an expression in x for:
   i   the time taken to go up the hill
   ii the time taken to go down the hill
   iii the total time for the round trip.

b If this total time is 5 hours calculate Sophie's average speed going uphill.

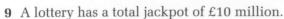

9 A lottery has a total jackpot of £10 million.

a If there are x winners, write an expression for the prize that each person receives.
b Write an expression for the size of the prize if there had been four fewer winners.
c Write an expression for the difference in the sizes of the prizes.
d If the difference is £8 million, what is the value of x?

REVISE

7 The straight line

A grid system was used for surveying in the ancient Egyptian and Roman worlds.

However, it wasn't until Descartes published *La Géométrie* in 1637 that its use in bridging the gap between Geometry and Algebra became widespread. It is said that Descartes got his idea as he lay in bed and watched a fly buzzing around one corner of the ceiling.

1 Review

The **cartesian** plane (named after Descartes) is divided into four **quadrants** by the *x* and *y* **axes**.

The axes intersect at the **origin**.

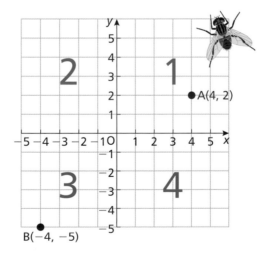

Every point in the plane can be described in terms of its distance from the origin
i in the *x*-direction and   **ii** in the *y*-direction.

These two distances are called the **coordinates** of the point,
e.g. A is the point (4, 2) ... 4 is the *x* coordinate of A.

B(−4, −5) is in the third quadrant. Its *x* coordinate is negative which indicates it is to the left of the origin; its *y* coordinate is negative which indicates it is below the origin.

◀◀ **Exercise 1.1**

1 The diagram shows some points plotted on a cartesian coordinate grid.

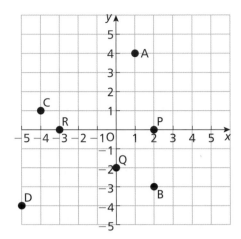

  a State the coordinates of A.

  b In which quadrant lies   **i** B   **ii** C   **iii** D?

  c Between which quadrants do we find   **i** P   **ii** Q   **iii** R?

  d How far apart are   **i** P and B   **ii** P and R?

2 The diagram shows the points A(−2, −5) and B(4, 1).

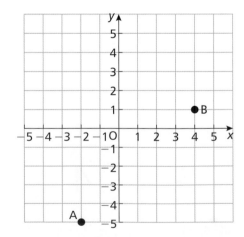

  a Make a copy of the diagram and draw the line which passes through A and B.

  b C(1, 2) lies above this line.

    Where are   **i** P(0, −3)   **ii** Q(2, −2)   **iii** R (2, 0) in relation to the line?

  c Which of the following describes both A and B?

    i Its y coordinate is a quarter of its x coordinate.

    ii Its y coordinate is 3 more that its x coordinate.

    iii Its y coordinate is 3 less than its x coordinate.

  d Which of the other named points fits the same description?

3 a Shade in the triangle that is bounded by the lines described as passing through:
  i $(-3, -5)$ and $(4, 2)$
  ii $(-1, 5)$ and $(-1, -1)$
  iii $(-3, 4)$ and $(5, 0)$.
b State the coordinates of the vertices of the triangle.
c Inside or what? Where do each of the following points lie in relation to the triangle?
  i $A(-2, 0)$ **ii** $B(1, 1)$ **iii** $C(1, 2)$

4 a i Draw a line parallel to the y axis passing through $(-3, -4)$.
  ii Draw a line parallel to the x axis passing through $(1, -3)$.
  iii Draw a line perpendicular to the x axis passing through $(3, 1)$.
  iv Draw a line perpendicular to the y axis passing through $(-2, 3)$.
b The four lines define a square. What are the coordinates of the vertices of the square?

5 A study compared the latitude a person lives at with their width as measured at their hips in centimetres. The table shows the averages for various locations.

Latitude (°N)	0	35	30	65	70	10	40	25	10	60	20	55
Width at hips (cm)	24	27	27·5	29	30	25	28	26	26	28	27	29·5

a Use a grid like the one shown to draw a scatter graph of the data.

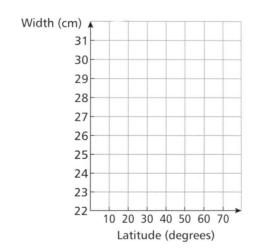

b Draw a best-fitting straight line to represent the 'relation' between *latitude* and *width*.
c Describe any correlation.
d Calculate the mean
  i latitude
  ii width, and plot this mean point on your grid.
e Where does this mean point lie in relation to your line?

2 Naming grid lines

Four points have been marked on the grid opposite.
They are $(-5, 2)$, $(-3, 2)$, $(1, 2)$ and $(4, 2)$.
Note that the y coordinate of each is 2.
Each point lies on the line which is 2 units up from the origin.

For obvious reasons, this grid line is referred to as the line $y = 2$.
$y = 2$ is referred to as the **equation** of the line.
Each of the other gridlines can be similarly identified.

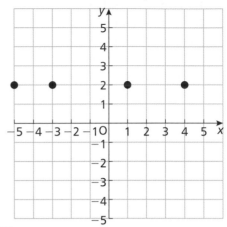

Examples

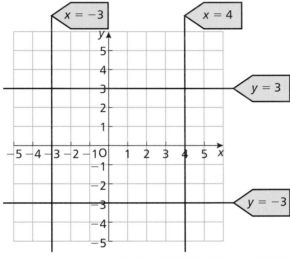

Exercise 2.1

1 Name each grid line highlighted.

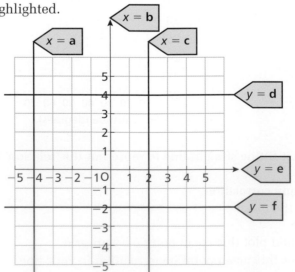

2 a What is special about the line $y = 0$?
 b What is special about the line $x = 0$?
 c Copy and complete these statements.
   i A line with equation of the form $x = a$ is perpendicular to the … axis.
   ii A line with equation of the form $x = a$ is … to the y axis.
   iii A line with equation of the form $y = a$ is perpendicular to the … axis.
   iv A line with equation of the form $y = a$ is … to the x axis.

3 a i The lines $x = 4$ and $x = 6$ are the margins of a strip on the grid.
     How wide is the strip?
   ii The lines $x = 1$ and $x = 6$ are the margins of another strip on the grid.
     How wide is the strip?
   iii The lines $x = a$ and $x = b$ are the margins of a strip on the grid.
     Write down a formula for how wide the strip is in terms of a and b.
   iv Test your formula for a strip with margins $x = -4$ and $x = 4$.
 b i The lines $y = 3$ and $y = 5$ are the margins of a strip on the grid.
     How wide is the strip?
   ii The lines $y = a$ and $y = b$ are the margins of a strip on the grid.
     Write down a formula for how wide the strip is in terms of a and b.
   iii Test your formula for a strip with margins $y = -1$ and $y = 5$.

4 a i Where does the grid line $y = 6$ intersect the line $x = 4$?
   ii Where does the grid line $y = -2$ intersect the line $x = 3$?
   iii Where does the grid line $y = 0$ intersect the line $x = 0$?
 b At which point do the lines $x = a$ and $y = b$ intersect?

5 a How wide is a strip with margins $y = -2$ and $y = 3$?
 b How wide is a strip with margins $x = -4$ and $x = 5$?
 c These four lines will define a rectangle.
   What is the area of the rectangle?
 d What is the area of the rectangle trapped between:
   i $x = 1$, $x = 5$, $y = -2$ and $y = 6$
   ii $x = -1$, $x = -2$, $y = 3$ and $y = 0$
   iii $x = -1$, $x = -2$, $y = -5$ and $y = -1$?
 e Calculate the area trapped between the four lines in the diagram.

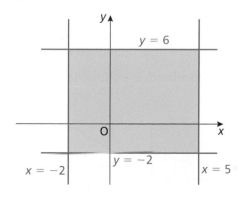

6 Name the line:

   a parallel to the *x* axis passing through (4, 3)

   b parallel to the *y* axis passing through (5, 1)

   c perpendicular to the *x* axis passing through (6, −1)

   d perpendicular to the *y* axis passing through (−4, 0)

   e passing through (3, 7) and (5, 7)

   f passing through (−2, 5) and (−2, 8)

   g parallel to $y = 6$ and passing through (−2, −4)

   h parallel to $x = 1$ and passing through (2, −9)

   i perpendicular to $y = -2$ and passing through (1, 4)

   j perpendicular to $x = 5$ and passing through (−6, −7).

7 **a** Define the quadrilateral below by mentioning the grid lines on which its edges lie.

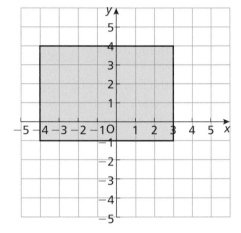

   b Describe its diagonals as best you can. (So that someone else could draw them without using this picture.)

Brainstormer

Find a formula for the area trapped between $x = a$, $x = b$, $y = c$ and $y = d$.

---

3 Gradient

There are many ways of describing the slope of a hill.
Most depend on giving the ratio of height climbed to distance travelled.

In mathematics the slope, or **gradient**, is defined in the following manner:

$$\text{Gradient} = \frac{\text{vertical height}}{\text{horizontal distance}}$$

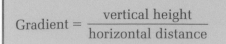

Vertical height

1 in 12

Horizontal distance

Example   A hill has a gradient of $\frac{1}{12}$.

    a How much does it rise over a horizontal distance of 72 m?
    b Over what horizontal distance will it rise by 10 m?

    a $\dfrac{\text{rise}}{72} = \dfrac{1}{12} \Rightarrow \text{rise} = \dfrac{1 \times 72}{12} = 6$ metres

    b $\dfrac{10}{\text{horizontal}} = \dfrac{1}{12} \Rightarrow \text{horizontal} = \dfrac{10 \times 12}{1} = 120$ metres

Notes
- When a surface is horizontal, the vertical rise is zero. Hence the gradient is zero.
- When a surface is vertical, the horizontal distance is zero. We can't divide by zero, so for a vertical surface the gradient is **undefined**.

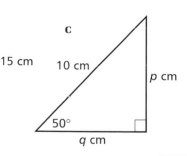

Exercise 3.1

1 State the gradient of each slope as a common fraction expressed in its simplest form.

a
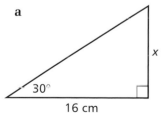
6 m
30 m

b

6 m
4 m

c
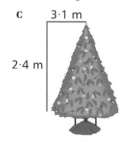
3·1 m
2·4 m

d
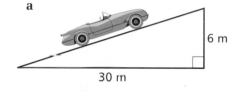
150 m
200 m

e
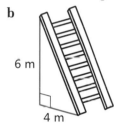

f
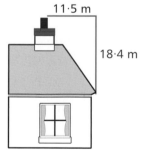
11·5 m
18·4 m

2 i Use trigonometry to establish the size of the marked length.
  ii Work out the gradient of the slope (to 1 d.p.)

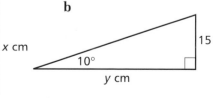

a
x cm
30°
16 cm

b
15 cm
10°
y cm

c
10 cm
p cm
50°
q cm

3 Conditions on the ski runs one day were such that skiers were advised to keep off slopes where the gradient was greater than 0·45.
Which of these runs are safe?

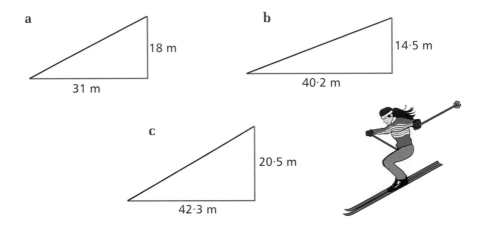

a   18 m   31 m

b   14·5 m   40·2 m

c   20·5 m   42·3 m

4 A diver descends on a path which has a gradient of $\frac{1}{10}$.

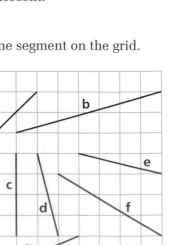

a At what depth is the diver after he has travelled a horizontal distance of 60 m?

b How far has he travelled horizontally when he has reached a depth of 12·5 m?

c Another diver descends with a gradient of 10.
At what depth is the diver after he has travelled a horizontal distance of 60 m?

d A third diver goes down without any horizontal motion.
Describe the gradient of his descent.

5 Describe the gradient of each line segment on the grid.

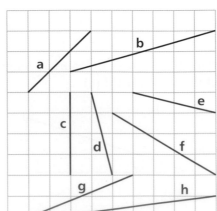

a   b   c   d   e   f   g   h

4 Gradients on a coordinate grid

When lines are drawn on a grid, the amount that they slope up to the right is referred to as the gradient of the line.

In mathematics the gradient is defined as $\dfrac{\text{change in the } y\text{-direction}}{\text{corresponding change in the } x\text{-direction}}$

Example 1   Calculate the gradient of:
   a the line AB   **b** the line CD.

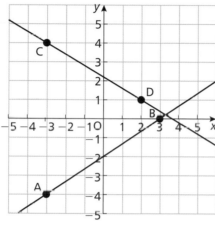

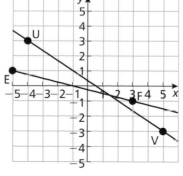

a The gradient of AB $= \dfrac{\text{change in } y}{\text{change in } x} = \dfrac{4}{6} = \dfrac{2}{3}$

b The gradient of CD $= \dfrac{\text{change in } y}{\text{change in } x} = \dfrac{-3}{5}$

Note that CD is sloping down from left to right. It has a **negative** gradient.

Exercise 4.1

1 Calculate the gradient of each line.

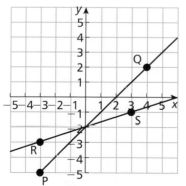

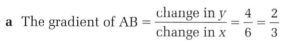

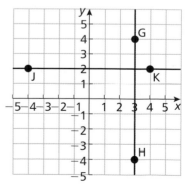

2 Two pairs of lines are drawn on the grids below.

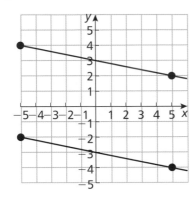 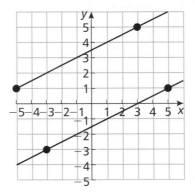

  a What can be said about each pair of lines?
  b Calculate the gradient of each line.
  c Comment on this statement:
     Lines with the same gradient are parallel.

3 In this diagram only one pair is parallel.
  Calculate the gradient of each line to
  identify the parallel pair.

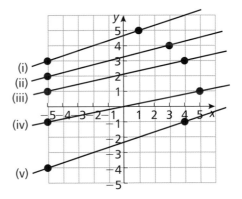

4 The gradient of AB is usually written as m_{AB}.
  A pentagon has vertices A$(-2, 3)$, B$(3, 4)$, C$(4, -2)$, D $(-2, -5)$ and E$(-4, -1)$.
  a Plot the points on a grid.
  b Calculate the value of:
     i m_{AB} **ii** m_{BC} **iii** m_{CD} **iv** m_{DE} **v** m_{EA}.
  c Comment on the gradient of BE, m_{BE}.

5 The diagram shows two sections of road.
  Both these sections should be straight.
  a Calculate the value of:
     i m_{EF}
     ii m_{FG}
     iii m_{UV}
     iv m_{VW}.
  b Only one of the sections EFG and UVW
     is a straight line.
     Which one?

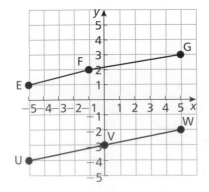

6 A quadrilateral has vertices P(9, 1), Q(10, 4), R(6, 5) and S(5, 2).
  a Plot the points P, Q, R and S.
  b Calculate:
    i m_{PQ} **ii** m_{RS} **iii** m_{QR} **iv** m_{PS}.
  c What kind of quadrilateral is PQRS?

The gradient formula

The diagram shows the points $A(x_1, y_1)$ and $B(x_2, y_2)$.
The change in x from A to B $= x_2 - x_1$
The change in y from A to B $= y_2 - y_1$

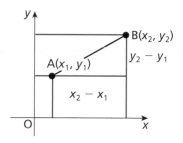

$$\Rightarrow \quad m_{AB} = \frac{y\text{-change}}{x\text{-change}}$$

$$\Rightarrow \quad m_{AB} = \frac{y_2 - y_1}{x_2 - x_1}$$

Example   Find the gradient of the line passing through A(−4, 5) and B(−2, 9).

$$m_{AB} = \frac{y_2 - y_1}{x_2 - x_1} = \frac{9 - 5}{-2 - (-4)} = \frac{4}{2} = 2$$

Exercise 4.2

1 Use the gradient formula to calculate the gradient of the line joining:
  a A(2, 3) to B(5, 12)   **b** C(3, 1) to D(9, 9)   **c** E(2, 3) to F(5, 12)
  d G(−2, 1) to H(−6, 7)   **e** I(2, −5) to J(−5, −1)   **f** K(−6, 0) to L(3, −9)
  g M(−3, −2) to N(−3, 4)   **h** O(7, 6) to P(3, 6)   **i** Q(−4, −3) to R(−3, −4)

2 A quadrilateral has vertices M(2, 2), N(4, −3), P(−2, −2) and Q(−4, 3).
  a Calculate the gradient of each side.
  b What kind of quadrilateral do you think it is?
  c Plot the points. Do you wish to refine your answer?

3 a i Plot the point A(1, 2).
    ii Plot B, the point which is 2 to the right and 1 up from A.
    iii Draw the line passing through A with a gradient of $\frac{1}{2}$.
  b i Plot the point P(3, 4).
    ii Plot Q, the point which is 1 to the right and 2 up from P.
    iii Draw the line passing through P with a gradient of 2.
  c Draw the line passing through R(0, 4) with a gradient of $\frac{1}{3}$.
  d Draw the line passing through S(0, −2) with a gradient of $\frac{2}{3}$.

4 A(−4, 2), B(2, 3), C(3, −3) and D(−3, −4) are the vertices of a square.

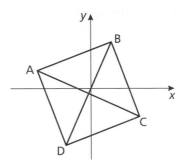

    a i Calculate the gradient of AB and BC.
      ii At what angle do AB and AC meet?
      iii Calculate the gradient of AC and BD.
      iv At what angle do AC and BD meet?
    b Repeat the above for a square with vertices A(3, 3), B(5, −3), C(−1, −5) and D(−3, 1).
    c Comment on your answers.

---

5 The equation of a line

Consider the set of points where the y coordinate is double the x coordinate: $y = 2x$. We can make a table of values.

Use some convenient values of x and work out the corresponding values of y.

x	0	1	2	3	4
y = 2x	0	2	4	6	8

Plot the points on a coordinate grid.

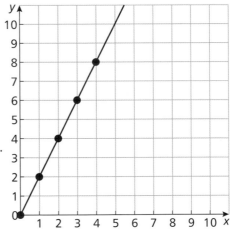

You should notice that the points all lie on a straight line.
We give this line the name $y = 2x$.
$y = 2x$ is referred to as the equation of the line.

Picking any two points, say (1, 2) and (4, 8), gives us the gradient of the line

$$m = \frac{8 - 2}{4 - 1} = \frac{6}{3} = 2.$$

Exercise 5.1

1 For each equation:
    i copy and complete the table
    ii draw a graph
    iii calculate the gradient of the line.

a

x	0	1	2	3	4
y = 3x	0				

b

x	0	1	2	3	4
y = x	0				

c

x	0	2	4	6	8
y = ½x	0				

d

x	0	3	6	9	12
y = ⅓x	0				

2 **a** Draw the graph of $y = 4x$.
  b Calculate its gradient.
  c Draw the graph of $y = \frac{1}{4}x$.
  d Calculate its gradient.
  e How can you find the gradient of a line when its equation is of the form $y = mx$?
  f What point do all these lines pass through?

3 Write down the equation of the line which passes through the origin with a gradient of:
  a 3   **b** 5   **c** -4 **d** $\frac{1}{5}$ **e** $-\frac{2}{3}$

4 **a** Copy and complete this table.
  b Draw the line with equation $y = x + 1$.
  c What is its gradient?
  d Where does it cut the y axis?

x	0	1	2	3	4
$y = x + 1$	1				

5 For each equation:
  i draw the line   **ii** give its gradient   **iii** state where it cuts the y axis.
    a $y = 2x + 1$
    b $y = 3x + 2$
    c $y = 2x - 3$
    d $y = \frac{1}{2}x + 4$
    e $y = \frac{1}{4}x - 1$

6 Consider the straight line with equation $y = mx + c$.
  a When $x = 0$, what is the value of y?
  b The point where the line cuts the y axis is called the y-intercept.
    State the coordinates of the y-intercept.
  c State the y-intercept of the following lines.
    i $y = 3x + 2$ **ii** $y = 2x + 1$ **iii** $y = 3x - 2$ **iv** $y = x - 4$
    v $y = 2x$ **vi** $y = -x - 1$ **vii** $y = \frac{1}{2}x + 5$ **viii** $y = -\frac{1}{2}x + 4$
  d By considering gradients, state two pairs of parallel lines.

---

6 The line $y = mx + c$

An equation of the form $y = mx + c$ represents a line with a gradient of m and a y-intercept of $(0, c)$.

Example 1   **a** Sketch the line with equation $y = 3x + 2$.
          b Indicate where it cuts the x axis.

          a By inspection we see the gradient is 3
             and the line cuts the y axis at $(0, 2)$.
          b It cuts the x axis when $y = 0$
             $\Rightarrow \quad 3x + 2 = 0$
             $\Rightarrow \quad 3x = -2$
             $\Rightarrow \quad x = -\frac{2}{3}$

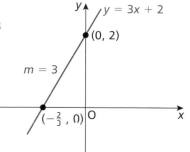

Exercise 6.1

1 Write down the gradient and y-intercept of each line.
   a $y = 2x + 3$ **b** $y = -3x - 6$ **c** $y = \frac{2}{5}x + 1$ **d** $y = -x$

2 Make a sketch of each of the following lines, showing the y-intercept and marking the gradient.
   a $y = x + 4$ **b** $y = 2x - 3$ **c** $y = -x + 5$
   d $y = -3x - 1$ **e** $y = \frac{1}{3}x + 1$ **f** $y = -\frac{2}{3}x - 2$

3 State the equation of the line in each sketch.

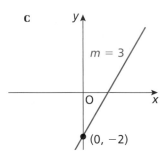

4 Calculate the gradient then give the equation of each line.

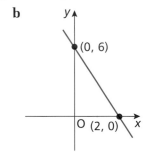 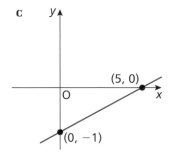

5 Write down the equation of the line with:
   a a gradient of 5, cutting the y axis at -4 **b** a gradient of -2, passing through $(0, 6)$
   c a gradient of $\frac{1}{3}$, cutting the y axis at -1 **d** a gradient of $-\frac{4}{5}$, passing through $(0, 3)$.

6 **a** A line passes through $(0, 1)$ and $(4, 3)$.
      i Calculate its gradient.   **ii** State its y-intercept.   **iii** State its equation.
   b In a similar way, find the equation of the line passing through:
      i $(0, 2)$ and $(6, 8)$ **ii** $(0, -1)$ and $(2, 7)$ **iii** $(0, -3)$ and $(1, -7)$ **iv** $(0, 5)$ and $(4, 3)$

7 A line passes through $(0, 1)$ and $(5, 11)$.
   a Find its equation.
   b Use the equation to see which of these points lie on it.
      i $(80, 161)$ **ii** $(-30, -59)$ **iii** $(-12, -25)$
   c Is the point $(20, 42)$ above or below the line?

8 Two lines are parallel. One has the equation $y = 3x + 5$.
   a The second line passes through $(0, -2)$. What is its equation?
   b Describe the position of each of these points as above both lines, below both lines or between the lines.
      i $(-1, 3)$ **ii** $(2, 6)$ **iii** $(3, 6)$ **iv** $(-4, -8)$

Finding the equation given two points

If we know two points then there is only one line that can be drawn through those two points.

So we should be able to get the equation of the line from the two points.

Example Find the equation of the line which passes through $(2, 5)$ and $(-1, -4)$.

The line has an equation of the form $y = mx + c$.

$$m = \frac{5 - (-4)}{2 - (-1)} = \frac{9}{3} = 3$$

So the line has an equation of the form $y = 3x + c$.

Since $(2, 5)$ is on the line, we have $x = 2$ when $y = 5$.
Substituting into the equation gives: $5 = 3 \times 2 + c$
$\Rightarrow \quad 5 = 6 + c$
$\Rightarrow \quad c = -1$

Thus the equation of the line is $y = 3x - 1$.

Exercise 6.2

1 Find the equation of the line passing through:
 a $(1, 5)$ and $(-2, -1)$　　　　**b** $(-2, 5)$ and $(1, -1)$
 c $(4, -2)$ and $(-2, -5)$　　　**d** $(8, 4)$ and $(4, 3)$
 e $(6, -1)$ and $(12, -3)$　　　**f** $(5, -5)$ and $(10, -7)$

2 A line passes through the points $(4, 2)$ and $(3, 3)$.
 a Find the equation of the line.
 b Is the point $(12, 6)$ above, on or below the line?
 c Where are the following points in relation to the line?
  i $(14, -7)$
  ii $(-13, 19)$
  iii $(-10, 15)$

3 A line has a gradient of 6 and passes through $(1, 1)$.
 a Given that its equation is of the form $y = 6x + c$, calculate the value of c.
 b State the equation of the line.

4 Find the equation of each of the following lines.
 a It has a gradient of 5 and passes through $(2, 7)$.
 b It has a gradient of -1 and passes through $(5, -1)$.
 c It is parallel to $y = 3x$ and passes through $(3, 3)$.
 d It is parallel to $y = \frac{1}{2}x$ and passes through $(4, 5)$.

5 Find the equation of each labelled line.

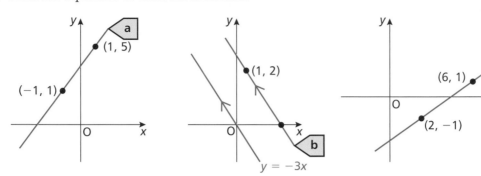

Challenge

Spreadsheets can be used to convert equations into tables of values then into graphs.

Follow these steps to create a graph of $y = 3x + 4$.

In A1 type: -4
In A2 type: $=A1+1$
Fill down to A8.
In B1 type: $=3*A1+4$
Fill down till B8.

Your spreadsheet should look like the one shown here.

Highlight the filled cells and select the 'Make chart' option. From the gallery of possible charts, select an XY line.

	A	B
1	−4	−8
2	−3	−5
3	−2	−2
4	−1	1
5	0	4
6	1	7
7	2	10
8	3	13

Experiment with other equations.

7 Modelling situations

Many situations in real life produce straight line graphs.

We can therefore find the equation of the line and use it to model the situation.

Example   The graph shows an electrician's charging system.
He has a call-out charge plus an hourly rate.

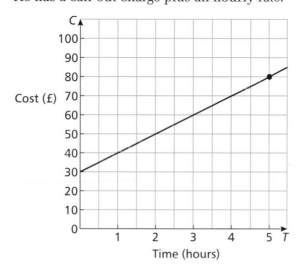

Cost (£)

Time (hours)

a Work out the equation of the line.
b Calculate the cost of a 10 hour job.

a Pick two convenient points on the line, say (0, 30) and (5, 80).
Thinking of the form $y = mx + c$:

- the y-intercept gives $c = 30$
- $m = \dfrac{80 - 30}{5 - 0} = \dfrac{50}{5} = 10$
- the label on the x axis is T; the label on the y axis is C.

So the equation $y = mx + c$ becomes: $C = 10T + 30$
b When $T = 10$, $C = 10 \times 10 + 30 = 130$
The cost of a 10 hour job is £130.

Exercise 7.1

1 The graph shows the charges, £C, for
bike hire for T hours.
  a Identify two convenient points on
     the line.
  b Calculate the gradient of the line.
  c Write down the equation of the line,
     expressing C in terms of T.
  d Use the equation to calculate the
     cost of hiring a bike for 12 hours.

Cost (£)

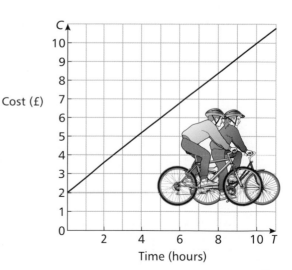

Time (hours)

2 A heating engineer drains a water tank.
The graph shows how the volume
drops over time.

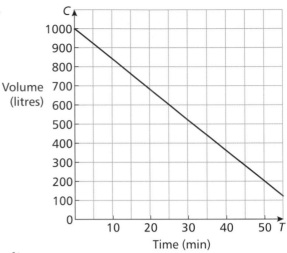

a Identify two convenient points on the line.
b Calculate the gradient of the line.
c Write down the equation of the line, expressing *V* in terms of *T*.
d Use the equation to calculate when the tank will be empty.

3 The Premium Chocolates Shop charges £1 for the box then 20p per
chocolate.

a Draw a graph to show the charges for 0 up to 10 chocolates.
b Work out the equation of the line, expressing the cost, £*C*, in
terms of the number, *N*, of chocolates bought.

4 Simon took out a loan. He had to pay back £10 000.
Each year he paid back £1000.
a Draw a graph to show how the loan decreases over time.
b Work out the equation of the graph, expressing the amount due, £*D*, in terms of the
time, *T* years.
c When will he pay back the loan?

---

8 The best-fitting straight line revisited

In Chapter 6 of Book S3[3] we examined scatter diagrams and their use in
highlighting possible connections. This table gives information on the shoe size and
height of 20 students.

Pupil	1	2	3	4	5	6	7	8	9	10
Shoe size	7·5	8·5	7	8	10	6·5	6	8·5	9·5	8
Height (cm)	135	143	131	145	160	135	120	147	152	140

Pupil	11	12	13	14	15	16	17	18	19	20
Shoe size	9	7	9	7·5	6	8	8·5	9·5	10	6·5
Height (cm)	150	135	152	143	129	137	150	155	155	126

A scatter diagram can be constructed, and a best-fitting straight line drawn.

By inspection, two convenient (well separated) points on the line can be identified, say (6, 125) and (9, 150).

From this we get a gradient of:

$$m = \frac{150 - 125}{9 - 6} = \frac{25}{3} = 8\frac{1}{3}$$

Using $y = 8\frac{1}{3}x + c$ and the point (6, 125) gives:

$$125 = 8\frac{1}{3} \times 6 + c$$
$$\Rightarrow \quad 125 = 50 + c$$
$$\Rightarrow \quad\quad c = 75$$

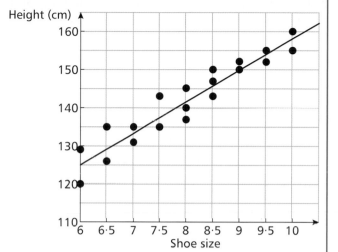

So the equation of the line of best fit is: $H = 8\frac{1}{3}S + 75$ where H is the height of the student in centimetres and S is their shoe size.

Note that the graph didn't show the y-intercept so we couldn't just read it off.

This equation can now be used to estimate height given the shoe size of the person.

Exercise 8.1

1 i Make scatter diagrams using the data in the tables.
  ii Add a line of best fit.
  iii Find the equation of your line of best fit.
  iv Use your equation to estimate the value of variable 2 for the given value of variable 1.

a Estimate variable 2 when variable 1 = 10·5.

	1	2	3	4	5	6	7	8	9	10	11	12	13	14	15
Variable 1	9	11	12	10	12	8	11	9	14	8	15	10	14	13	13
Variable 2	7	6	5	8	6	8	7	8	5	7	4	6	4	5	4

b Estimate variable 2 when variable 1 = 24.

	1	2	3	4	5	6	7	8	9	10	11	12	13	14	15
Variable 1	37	20	35	5	40	18	22	5	20	13	45	43	40	10	25
Variable 2	80	55	75	40	80	40	40	25	60	40	90	80	70	30	60

2 Lucy was investigating the burning times of candles. She lit candles of different lengths and then timed how long they lasted.

	1	2	3	4	5	6	7	8	9	10	11	12	13	14	15
Candle length (mm)	10	14	34	3	78	25	32	20	59	30	50	68	50	56	70
Lifetime (min)	20	17	40	14	86	35	44	24	71	32	58	79	53	60	75

   a Make a scatter diagram using the data.
   b Add a line of best fit.
   c Find the equation of your line of best fit, expressing the lifetime (T minutes) in terms of the candle length (L cm).
   d Use the equation to estimate the length of time a 50 mm candle would burn.

3 The table shows the times taken to do similar jobs with different widths of paintbrush.

	1	2	3	4	5	6	7	8	9	10	11	12	13	14	15
Brush width (cm)	8	6	3	8	9	4	2	7	10	4	7	5	4	6	3
Time taken (min)	6	7	10	7	5	9	12	7	5	11	6	9	10	9	11

   a Make a scatter diagram using the data.
   b Add a line of best fit.
   c Find the equation of your line of best fit.
   d Use the equation to estimate the time it will take to do the job using a 6 cm brush.

4 This scatter diagram shows a connection between the length of time an antique clock will run and the number of times its key is turned.

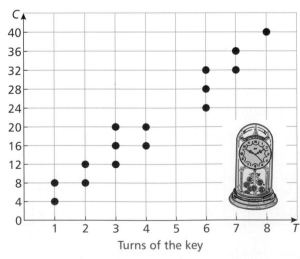

   a Describe the correlation.
   b Copy the diagram and add a line of best fit.
   c Work out the equation of the line of best fit.
   d Use your equation to estimate how long the clock will run if its key is given 5 turns.

◄◄ RECAP

Horizontal lines have an equation of the form $y = a$ where a is a constant.
The x axis has the equation $y = 0$.
Vertical lines have an equation of the form $x = a$ where a is a constant.
The y axis has the equation $x = 0$.

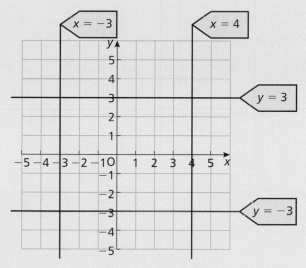

In mathematics the slope, or **gradient**, is defined as:

$$\text{gradient} = \frac{\text{vertical height}}{\text{horizontal distance}}$$

Horizontal lines have a gradient of 0.
The gradient of vertical lines is undefined.

In coordinate geometry the gradient of a line is defined by:

$$m_{AB} = \frac{y_2 - y_1}{x_2 - x_1}$$ where (x_1, y_1) and (x_2, y_2) are two points on the line.

Lines that slope down from left to right have a **negative** gradient.
Lines that slope up from left to right have a **positive** gradient.
Lines running in the x-direction have a gradient of 0.
The gradient for lines running in the y-direction is undefined.

With the exception of lines running in the y-direction,
all lines have an equation of the form $y = mx + c$
where m is the gradient of the line and c is the y-intercept
(where it cuts the y axis), e.g. $y = 3x + 2$.

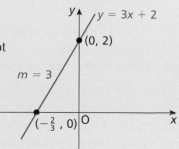

Given two points, we can always find the equation
of the line which passes through them both.

The technique can be used to find the equation
for the line of best fit in a scatter diagram.
The equation can then be used to estimate y values for given x values.

1 State the equation of each labelled line.

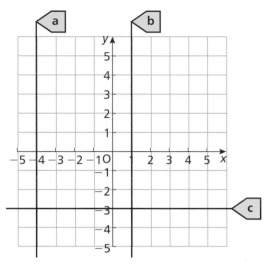

2 Describe the gradient of each line.

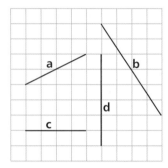

3 Calculate the gradient of the line passing through:
   a (2, 11) and (4, 19)   b (−1, 4) and (2, −5)   c (4, −1) and (12, 1)
   d (2, 3) and (6, 1)   e (4, 1) and (4, 8)   f (2, 1) and (10, 1)

4 a Copy and complete this table:

x	0	1	2	3	4
y = 2x + 1	1				

   b Draw the line with equation $y = 2x + 1$.
   c What is its gradient?
   d Where does it cut the y axis?

5 Write down the equation of the line with:
   a a gradient of 1, cutting the y axis at −1
   b a gradient of −3, passing through (0, 2)
   c a gradient of $\frac{1}{2}$, cutting the y axis at −3
   d a gradient of $-\frac{2}{3}$, passing through (0, 0)

REVISE

6 Draw a sketch of the following lines:

   a $y = 3x + 4$ **b** $y = -2x - 5$ **c** $y = \frac{1}{2}x + 1$ **d** $y = -\frac{1}{2}x$

7 The diagram shows how weight increases as more apples are put on a set of scales.

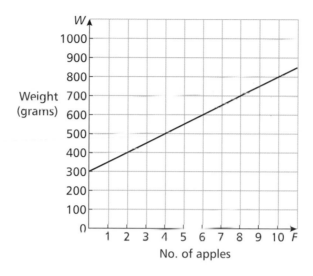

   a What is the reading when the pan is empty?

   b Work out the equation of the line, expressing the total weight in grams (W g) in terms of F, the number of apples.

8 Fifteen pupils were surveyed. They were asked how many dental fillings they had and how much they spent a week on sweets.

Number of fillings	7	1	8	1	9	3	4	8	2	3	9	10	7	1	10
Spent on sweets (£)	3·00	1·00	3·50	1·50	3·50	1·50	2·50	3·00	2·00	2·50	4·00	3·50	3·50	2·50	4·00

   a Make a scatter diagram to illustrate the data.

   b Add a line of best fit to the diagram.

   c Work out the equation of your line.

   d Use the equation to estimate how much is spent by a pupil with five fillings.

REVISE

8 Quadratic equations

Babylonian mathematics was based on the sexagesimal number system, using 60s rather than 10s.
Thousands of their clay tablets have survived to the present day.

Their mathematicians were, in many ways, more advanced than Egyptian mathematicians. They could work out square and cube roots, solve linear equations and some quadratic equations.

Quadratic equations are important in modern-day life where they are used in a variety of contexts, from the design of bridges and satellite TV aerials to working out problems on the flight of rockets and golf balls.

1 Review

A **linear** expression is one where the highest power of x is 1,
e.g. $2x$, $1 - 3x$, $4x + 3$.
A **quadratic** expression is one where the highest power of x is 2,
e.g. x^2, $x^2 + 2x + 1$, $3 - x^2$, $2x + x^2$.

◀◀ **Exercise 1.1**

1 Solve these equations:
  a $3x + 1 = 10$ **b** $2x = -8$ **c** $4 - 2x = 2x$
  d $4x - 1 = x$ **e** $5x - 3 = 2x + 7$ **f** $4 - 2x = 2 - 5x$

2 Expand these brackets:
  a $3(x - 4)$ **b** $-2(4 - 3x)$ **c** $3x(2 - x)$

3 Expand and then simplify:
  a $2(x + 1) + 3(x - 2)$ **b** $6(2x - 1) - 2(3x + 2)$ **c** $3x(2x - 1) + 4x(2 - x)$

4 Factorise each expression:
  a $6x + 8$ **b** $12y - 8$ **c** $8a^2 + a$ **d** $2t^2 + 6t$
  e $3x^2 - 9x$ **f** $6h - 2h^2$ **g** $15k^2 - 5k$ **h** $12ab^2 + 6a^2b$

5 Factorise:

a $a^2 - b^2$ **b** $k^2 - l^2$ **c** $25 - x^2$ **d** $y^2 - 36$

e $x^2 - 49y^2$ **f** $100a^2 - 64b^2$ **g** $49s^2 - 121t^2$ **h** $1 - 16x^2$

6 Now factorise these:

a $x^2 + 5x + 6$ **b** $x^2 + 7x + 12$ **c** $y^2 + 8y + 12$ **d** $x^2 + 10x + 21$

e $x^2 - 3x - 18$ **f** $y^2 - 3y - 10$ **g** $s^2 - 10s + 24$ **h** $x^2 - 8x + 12$

i $2x^2 + 5x + 2$ **j** $3y^2 + 2y - 8$ **k** $6x^2 - 13x + 6$ **l** $4s^2 - 12s - 27$

---

2 Introducing quadratic equations

John strikes his golf ball.
This graph shows how its height (h m) varies with time (t seconds).

Its height after t seconds is given by the formula $h = 20t - 5t^2$.
This is called a **quadratic function**.
(More of this in the next chapter.)

If we want to find when the height is zero (on the ground) then we have to solve the equation $20t - 5t^2 = 0$.
This is called a **quadratic equation**.

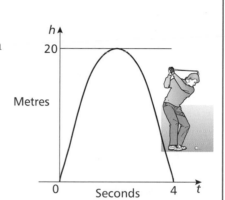

The golf ball takes off at $t = 0$ and lands at $t = 4$.
When $t = 0$ and $t = 4$, the height, h, of the ball is 0 metres.

So the **solutions** of $20t - 5t^2 = 0$ are $t = 0$ and $t = 4$.

$t = 0$ and $t = 4$ are called the **roots** of the equation.

Exercise 2.1

1 Here is the graph of another golf shot.
The height, h m, of the ball after t seconds is given by $h = 15t - 5t^2$.

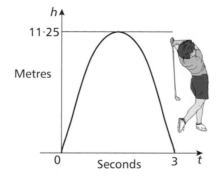

Such a curve is known as a **parabola**.

a For what values of t does $h = 0$?

b What are the solutions of the quadratic equation $15t - 5t^2 = 0$?

2 A cannon ball is fired into the air.
This is the graph of how its height varies with time.
The height of the cannon ball is given by $h = 40t - 5t^2$.
a What are the values of t when $h = 0$?
b What are the solutions of the quadratic equation
$40t - 5t^2 = 0$?

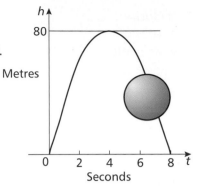

3 On the planet Org, gravity is weaker than on Earth.
The graph shows how high Hana hopped.
The height after t seconds is given by the
quadratic formula $h = 10t - 2t^2$.
a From the graph, what are the roots of the
quadratic equation $10t - 2t^2 = 0$?
b What is the height of Hana when
i $t = 1$ second   **ii** $t = 4$ seconds?
c What are the solutions of the equation
$10t - 2t^2 = 0$?
d What is the solution of the quadratic equation $10t - 2t^2 = 12.5$?

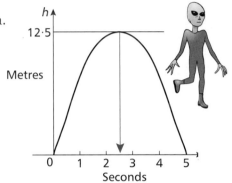

4 The height, h m, of a rocket t seconds after launch is given by the formula $h = 20t - 2t^2$.
a Copy and complete this table.

t (seconds)	0	1	2	3	4	5	6	7	8	9	10
h (metres)	0	18	32								

b What are the roots of the quadratic equation $20t - 2t^2 = 0$?
c What are the solutions of these quadratic equations?
i $20t - 2t^2 = 18$ **ii** $20t - 2t^2 = 32$ **iii** $20t - 2t^2 = 42$ **iv** $20t - 2t^2 = 50$

3 Using factors to solve quadratic equations

$A \times B = 6$ has various answers, for example:
$A = 1$ and $B = 6$; $A = 2$ and $B = 3$; $A = 3$ and $B = 2$; $A = 6$ and $B = 1$.

but **if $A \times B = 0$ then $A = 0$ or $B = 0$.**

If a product of factors is equal to zero then we can deduce that one of the factors must be zero.
If we wish to make a product equal to zero then we must make one of the factors equal to zero.

Puzzle: Expand $(x - a)(x - b)(x - c)(x - d)(x - e)(x - f) \ldots (x - y)(x - z)$

Answer: Zero!   … because one of the factors of the product is $(x - x)$

Example 1   Solve $x^2 - 4x = 0$.

Factorising gives $x(x - 4) = 0$.

If the expression equals zero, then either $x = 0$ or $(x - 4) = 0$ so the solution of $x^2 - 4x = 0$ is $x = 0$ or $x = 4$.

We can write: $x^2 - 4x = 0 \Rightarrow x = 0$ or $x = 4$.

Example 2   Solve the equation $16t - 6t^2 = 0$.

$$16t - 6t^2 = 0$$
$$\Rightarrow \qquad 2t(8 - 3t) = 0$$
$$\Rightarrow \qquad 2t = 0 \text{ or } 8 - 3t = 0$$
$$\Rightarrow \qquad t = 0 \text{ or } t = \tfrac{8}{3}$$

Example 3   Solve the equation $x^2 - 9 = 0$.

$$x^2 - 9 = 0$$
$$\Rightarrow \qquad (x - 3)(x + 3) = 0 \qquad \text{(factorising)}$$
$$\Rightarrow \qquad (x - 3) = 0 \text{ or } (x + 3) = 0$$
$$\Rightarrow \qquad x = 3 \text{ or } x = -3$$

Exercise 3.1

Solve the equations in questions **1** to **4**.

1 a $x(x + 1) = 0$
   b $4x(x + 2) = 0$
   c $x(x - 3) = 0$
   d $3x(x - 4) = 0$
   e $x(5 + x) = 0$
   f $3x(4 - x) = 0$
   g $x(5 - 2x) = 0$
   h $3x(7 + 2x) = 0$
   i $5x(3 - 4x) = 0$
   j $5x(2 - 3x) = 0$
   k $2x(11 - 3x) = 0$
   l $6x(5x - 7) = 0$

2 a $x^2 + 3x = 0$
   b $y^2 - 5y = 0$
   c $t^2 - 12t = 0$
   d $4m^2 + 8m = 0$
   e $6t^2 - 12t = 0$
   f $6l^2 - 18l = 0$
   g $4y^2 + 12y = 0$
   h $3t^2 - 12t = 0$
   i $4y^2 - 10y = 0$
   j $6k^2 + 12k = 0$
   k $12x^2 - 16x = 0$
   l $12x - 3x^2 = 0$
   m $15c - c^2 = 0$
   n $12s - 4s^2 = 0$
   o $5t + 6t^2 = 0$

3 a $(x - 4)(x + 4) = 0$
   b $(x - 5)(x + 5) = 0$
   c $(x - 10)(x + 10) = 0$
   d $(2x - 3)(2x + 3) = 0$
   e $(5 - 2x)(5 + 2x) = 0$
   f $(7 + 3x)(7 - 3x) = 0$
   g $(2 - 3x)(2 + 3x) = 0$
   h $3(5 - 6x)(5 + 6x) = 0$
   i $(4 + 7x)(4 - 7x) = 0$

4 a $x^2 - 4^2 = 0$
   b $x^2 - 5^2 = 0$
   c $d^2 - 7^2 = 0$
   d $f^2 - 64 = 0$
   e $100 - a^2 = 0$
   f $121 - b^2 = 0$
   g $144 - b^2 = 0$
   h $81 - s^2 = 0$
   i $(4x)^2 - 3^2 = 0$
   j $(9x)^2 - 2^2 = 0$
   k $25x^2 - 4^2 = 0$
   l $64t^2 - 4 = 0$
   m $100s^2 - 25 = 0$
   n $81c^2 - 25 = 0$
   o $x^2 = 9$
   p $4f^2 = 0$
   q $3y^2 = 12$
   r $49 - 36r^2 = 0$

5 If you wanted to find the height, h metres, of a cliff, you could drop a stone down it and count the number of seconds, t, before you heard the splash.

The height is calculated from the formula $h = 5t^2$.

a What is the cliff's height if it takes 3 seconds?

b How long will the stone take to fall 125 m? (Form an equation and solve it.)

Exercise 3.2

1 Solve these quadratic equations:

a $2x^2 - 8 = 0$	**b** $3x^2 - 12 = 0$	**c** $5x^2 - 45 = 0$
d $9y^2 - 36 = 0$	**e** $4x^2 - 64 = 0$	**f** $24t^2 - 24 = 0$
g $12b^2 - 48 = 0$	**h** $6c^2 - 54 = 0$	**i** $8t^2 - 72 = 0$
j $10k^2 - 1000 = 0$	**k** $9p^2 - 81 = 0$	**l** $7n^2 - 175 = 0$

2 Now solve these:

a $8x^2 - 18 = 0$	**b** $12t^2 - 27 = 0$	**c** $18t^2 - 50 = 0$
d $24k^2 - 54 = 0$	**e** $20y^2 - 125 = 0$	**f** $28b^2 - 63 = 0$
g $32d^2 - 72 = 0$	**h** $90e^2 - 250 = 0$	**i** $72f^2 - 32 = 0$
j $100g^2 - 196 = 0$	**k** $98h^2 - 50 = 0$	**l** $108t^2 - 75 = 0$
m $18 - 50s^2 = 0$	**n** $12 - 75d^2 = 0$	**o** $80 - 125e^2 = 0$

3 The height, h m, of a rollercoaster above a line AB can be calculated from the formula $h = 900 - x^2$ where x m is the distance from the axis of symmetry of the ride, as shown.

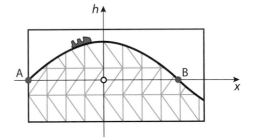

a Solve the equation $900 - x^2 = 0$ to find how far A and B are from the axis.

b How far apart are A and B?

4 More quadratic equations: trinomials

In many cases the quadratic equation can be expressed as a **trinomial**. The trinomial can then be factorised and the equation solved.

Example 1   Solve the quadratic equation $x^2 + 5x + 6 = 0$.

$$x^2 + 5x + 6 = 0$$
$$\Rightarrow \qquad (x + 3)(x + 2) = 0$$
$$\Rightarrow \qquad x + 3 = 0 \text{ or } x + 2 = 0$$
$$\Rightarrow \qquad x = -3 \text{ or } -2$$

Example 2   Solve the quadratic equation $2x^2 + 9x + 4 = 0$.

$$2x^2 + 9x + 4 = 0$$
$$\Rightarrow \qquad (2x + 1)(x + 4) = 0$$
$$\Rightarrow \qquad 2x + 1 = 0 \text{ or } x + 4 = 0$$
$$\Rightarrow \qquad x = -\tfrac{1}{2} \text{ or } -4$$

Exercise 4.1

1 Solve the quadratic equations:

a $(x + 2)(x + 1) = 0$ **b** $(2x + 1)(x - 2) = 0$ **c** $(x - 3)(x + 4) = 0$

d $(2 - x)(3 + x) = 0$ **e** $(2 - 5x)(1 + 2x) = 0$ **f** $(2 + 5x)(3 - 4x) = 0$

2 Factorise to help you solve each quadratic equation:

a $x^2 + 5x + 4 = 0$ **b** $x^2 + 6x + 8 = 0$ **c** $x^2 + 7x + 10 = 0$

d $x^2 + 9x + 20 = 0$ **e** $x^2 + 8x + 12 = 0$ **f** $x^2 + 11x + 28 = 0$

g $x^2 + 11x + 30 = 0$ **h** $x^2 + 9x + 14 = 0$ **i** $x^2 + 10x + 24 = 0$

j $x^2 + 12x + 35 = 0$ **k** $x^2 + 4x + 4 = 0$ **l** $x^2 + 9x + 18 = 0$

3 Solve these quadratic equations:

a $x^2 + x - 2 = 0$ **b** $x^2 + x - 6 = 0$ **c** $x^2 - 2x - 8 = 0$

d $x^2 - 2x - 15 = 0$ **e** $x^2 - 3x - 4 = 0$ **f** $x^2 + 3x - 18 = 0$

g $x^2 - 5x + 4 = 0$ **h** $x^2 - 6x + 8 = 0$ **i** $x^2 - 8x + 15 = 0$

j $x^2 - 7x + 12 = 0$ **k** $x^2 - 9x + 20 = 0$ **l** $x^2 - 9x + 18 = 0$

m $x^2 - 3x - 28 = 0$ **n** $x^2 - 4x - 45 = 0$ **o** $x^2 - 11x + 30 = 0$

4 Now solve these:

a $2x^2 + 5x + 2 = 0$ **b** $3x^2 + 13x + 4 = 0$ **c** $5x^2 + 11x + 2 = 0$

d $4x^2 + 13x + 3 = 0$ **e** $6x^2 + 13x + 2 = 0$ **f** $4x^2 + 14x + 6 = 0$

g $2x^2 - 5x + 2 = 0$ **h** $3x^2 - 7x + 2 = 0$ **i** $5x^2 - 16x + 3 = 0$

j $1 + x - 6x^2 = 0$ **k** $6 - 5x - 6x^2 = 0$ **l** $3 - 10x - 8x^2 = 0$

m $4 + 5x - 6x^2 = 0$ **n** $3 + 10x - 8x^2 = 0$ **o** $2 - 11x + 12x^2 = 0$

5 Engineers have designed the cross-section of a road to be curved to allow water to run off.

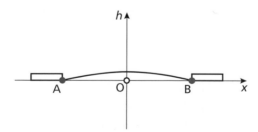

The height, h mm, above AB is calculated using $h = 16 + 6x - x^2$ where x m is the distance from a fixed point on the road as shown.
The road meets the kerb at A and B.

a Solve the equation $16 + 6x - x^2 = 0$ to find how far A and B are from O.

b How wide is the road?

Manipulation

Sometimes a bit of algebra is needed to get the trinomial.

Example 3 Solve $(x + 4)^2 = 36.$

$$(x + 4)^2 = 36$$
$$\Rightarrow \quad x^2 + 8x + 16 = 36$$
$$\Rightarrow \quad x^2 + 8x - 20 = 0$$
$$\Rightarrow \quad (x + 10)(x - 2) = 0$$
$$\Rightarrow \quad x = -10 \text{ or } x = 2$$

Example 4 Solve $5x(2x + 1) - 10 = x(7x + 6).$

$$5x(2x + 1) - 10 = x(7x + 6)$$
$$\Rightarrow \quad 10x^2 + 5x - 10 = 7x^2 + 6x$$
$$\Rightarrow \quad 3x^2 - x - 10 = 0$$
$$\Rightarrow \quad (3x + 5)(x - 2) = 0$$
$$\Rightarrow \quad x = 2 \text{ or } -\tfrac{5}{3}$$

Example 5 Solve $\dfrac{1}{x - 1} + \dfrac{1}{x + 2} = \dfrac{1}{2}.$

$$\dfrac{1}{x - 1} + \dfrac{1}{x + 2} = \dfrac{1}{2}$$

Multiplying throughout by $2(x - 1)(x + 2)$ removes denominators

$$\Rightarrow \quad 2(x + 2) + 2(x - 1) = (x - 1)(x + 2)$$
$$\Rightarrow \quad 2x + 4 + 2x - 2 = x^2 + x - 2$$
$$\Rightarrow \quad x^2 - 3x - 4 = 0$$
$$\Rightarrow \quad (x - 4)(x + 1) = 0$$
$$\Rightarrow \quad x = 4 \text{ or } x = -1$$

Exercise 4.2

1 Solve:

a $(x + 3)^2 = 1$ **b** $(x + 4)^2 = 9$ **c** $(x - 2)^2 = 16$

d $(2x - 1)^2 = 64$ **e** $(3x - 1)^2 = 49$ **f** $(4x - 1)^2 = 81$

g $(1 - 3x)^2 = 100$ **h** $(2 - 5x)^2 = 100$ **i** $(3 - 4x)^2 = 64$

2 Find the roots of these quadratic equations:

a $2x^2 = 5x + 3$ **b** $3x^2 = 10x - 8$ **c** $6t^2 = 48 - 2t$

d $7x - 5 = -6x^2$ **e** $2x(3x - 1) - 20 = 0$ **f** $y(6y - 17) = -5$

g $3x(x + 1) - 2(x + 1) = 0$ **h** $2x(4 - x) + 3(4 - x) = 0$ **i** $5x(2 - x) = 4(2 - x)$

j $x(2x + 5) - 12 = 0$ **k** $4b^2 = b(b + 2) + 16$ **l** $2(3x^2 - 2) = 23x$

m $3(x + 1)^2 = 3$ **n** $(x + 1)^2 = (2x + 3)^2 + (1 + 2x)$

3 Solve:

a $\dfrac{x}{2} = \dfrac{8}{x}$ [Hint: multiply throughout by $2x$.]

b $\dfrac{2x}{1} = \dfrac{1}{2x}$ [Hint: multiply throughout by $2x$.]

c $\dfrac{x+1}{4} = \dfrac{1}{x+1}$ [Hint: multiply throughout by $4(x+1)$.]

d $\dfrac{3x-1}{9} = \dfrac{1}{3x-1}$ [Hint: multiply throughout by $9(3x-1)$.]

e $\dfrac{2(2x-5)}{5} = \dfrac{10}{2x-5}$ **f** $\dfrac{6}{a-4} = \dfrac{a+4}{a}$

g $\dfrac{6}{x} - \dfrac{6}{x+1} = 1$ **h** $\dfrac{1}{x-2} - \dfrac{1}{x+3} = \dfrac{1}{10}$

i $x + \dfrac{3}{x} = 4$

Challenge

Find a quadratic equation whose roots are:
a 1, 3   **b** 4, 3   **c** 2, −3   **d** −2, 5

---

5 Problems involving quadratic equations

Example   A rectangular garden is twice as long as it is broad.
The area of the garden is 200 m².
What are the dimensions of the garden?

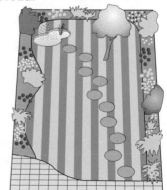

Let the breadth of the garden be x metres.

$\Rightarrow$ length $l = 2x$ metres

$\Rightarrow$ area $A = lb = 2x \times x$

$\Rightarrow$ $2x^2 = 200$

$\Rightarrow$ $x^2 - 100 = 0$

$\Rightarrow$ $(x-10)(x+10) = 0$

$\Rightarrow$ $x = 10$ or $x = -10$

The breadth is a positive quantity (eliminating $x = -10$).
Therefore, the breadth is 10 metres and the length is 20 metres.

Exercise 5.1

1 A 5 m ladder leans against a wall.
The distance from the top of the ladder to the ground is a
metre more than the distance from the foot of the ladder
to the wall.
The diagram illustrates the situation.
   a Use Pythagoras' theorem to show that $x^2 + x - 12 = 0$.
   b Solve this quadratic to help you find the distance from:
      i the foot of the ladder to the wall
      ii the top of the ladder to the ground.

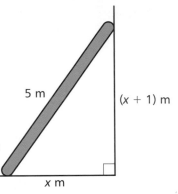

5 m

$(x + 1)$ m

x m

2 Two negative numbers differ by 4.
   a If the larger number is x, what is the smaller?
   b Their product is 60.
      Form a quadratic equation and solve it to find the two numbers.

3 A square plot of land is extended in one direction
by 3 metres and in the other by 1 metre.
The area of the rectangle created is 63 m².

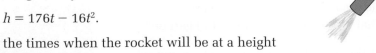

x m   3 m

1 m

   a If x m is the length of the square, write an
      expression for:
      i the length of the rectangle
      ii the breadth of the rectangle
      iii the area of the rectangle.
   b Form a quadratic equation and solve it to find the dimensions of:
      i the square   **ii** the rectangle.

4 The product of two positive consecutive numbers is 110.
   a Let x stand for the smaller number. Write an expression for the larger.
   b Form a quadratic equation and solve it to find the two numbers.

5 The height, in metres, of a rocket fired vertically
upwards is given by the formula:

$$h = 176t - 16t^2.$$

   a Find the times when the rocket will be at a height
      of 160 metres.
   b Is it possible for this rocket to reach a height of 188 metres? Explain your answer.

Exercise 5.2

1 The sum of the first n natural numbers is given by the formula $\frac{1}{2}n(n + 1)$.
   a Solve $\frac{1}{2}n(n + 1) = 210$ to find how many numbers have been added to give 210.
   b How many numbers have to be added to give 325?

2 The height, in metres, of a stone thrown vertically upwards is given by the formula
   $h = 25t - 5t^2$.
   a Find the times when the stone is 20 metres above the ground.
   b When does the stone land back on the ground?

3 Two cars travel 150 miles at different speeds.
The faster car travels 5 miles per hour faster than the slower car.
Let the speed of the slower car be s miles per hour.

  a Write down an expression for the speed of the faster car.

  b Using the formula $T = \dfrac{D}{S}$ write an expression for the time taken by

    i the faster car   **ii** the slower car. (Both answers are fractions.)

  c The time taken for the two cars to travel 150 miles differs by 1 hour.
    Form an equation in s. (Remember that the faster car has the shorter time.)

  d Show that $s^2 - 5s - 750 = 0$.

  e Solve this quadratic equation to find the speed of each car.

4 In a right-angled triangle, the hypotenuse is 30 cm in length.
The other two sides together add to give 42 cm.

  a Let one of the shorter sides be of length x cm.
    Give an expression for the length of the other one.

  b Use Pythagoras' theorem to form an equation in x and solve it for x.

  c Find the perimeter of the triangle.

5 A boat with an outboard motor can maintain a steady
speed in calm water.
A sailor takes it on a river which has a downstream
flow of 4 miles per hour.
He goes upstream for a mile and then downstream for a
mile back to the start.
This takes him 8 minutes ($\frac{8}{60}$ hour).

  a If his speed in calm water is x mph write an expression for:

    i his speed against the current

    ii his speed with the current

    iii his time for the upstream journey

    iv his time for the downstream journey

    v the total time for the journey in hours.

  b His whole journey takes him 8 minutes ($\frac{8}{60}$ hours).
    Form an equation in x and solve it to find his speed in calm water.

---

6 The standard form of the quadratic equation

Every quadratic equation can be written in the form $ax^2 + bx + c = 0$ where a, b and
c can be any number with the exception that a cannot be zero ($a \neq 0$).

a is the coefficient of x^2. b is the coefficient of x. c is the term independent of x.

Example 1   Consider each of the following quadratic equations in standard form,
         $ax^2 + bx + c = 0$, and identify a, b and c.

         a $x^2 + x - 12 = 0$ **b** $x^2 - 3x + 3 = 0$ **c** $2x^2 + 4 = 0$
         d $6 + 2x - 3x^2$ **e** $x^2 - 5x = 0$ **f** $3x^2 = 0$

         a $a = 1, b = 1, c = -12$ **b** $a = 1, b = -3, c = 3$ **c** $a = 2, b = 0, c = 4$
         d $a = -3, b = 2, c = 6$ **e** $a = 1, b = -5, c = 0$ **f** $a = 3, b = 0, c = 0$

Example 2   Arrange this equation in the form $ax^2 + bx + c = 0$ and give the values a, b and c:

$$x(x + 1) = 2(3x - 4)$$
$$\Rightarrow \qquad x(x + 1) = 2(3x - 4)$$
$$\Rightarrow \qquad x^2 + x = 6x - 8$$
$$\Rightarrow \qquad x^2 - 5x + 8 = 0$$
$$\Rightarrow \quad a = 1, b = -5 \text{ and } c = 8$$

Exercise 6.1

1 Each equation is in the form $ax^2 + bx + c = 0$. State the values of a, b and c.

a $x^2 + 2x + 2 = 0$ **b** $3x^2 - 5x + 1 = 0$

c $4x^2 + x - 3 = 0$ **d** $2x^2 - x - 1 = 0$

e $x^2 - 1 = 0$ **f** $2x^2 + x = 0$

g $2 + 3x + x^2 = 0$ **h** $4 - 3x + 5x^2 = 0$

i $1 + x - x^2 = 0$ **j** $6 - x - 2x^2 = 0$

k $9 - 2x^2 = 0$ **l** $x - x^2 = 0$

m $4 - 2x - x^2 = 0$ **n** $5x^2 = 0$

o $x^2 = 0$

2 i Rearrange each equation into the form $ax^2 + bx + c = 0$.

ii State the values of a, b and c.

a $x^2 + 2x + 2 = 4x - 3$ **b** $x^2 = 3x^2 + 5x - 1$

c $4x^2 + x - 3 = 3x^2 - 5x + 1$ **d** $5 - x - 3x^2 = 3 - 3x + 7x^2$

e $4 - 2x^2 = x - x^2$ **f** $2 - 3x + 4x^2 = 2x^2 - x + 1$

3 Why is $1 - 3x + 5x^2 = 5x^2 - x - 1$ *not* a quadratic equation?

4 i Multiply out the brackets.

ii Rearrange each equation into the form $ax^2 + bx + c = 0$.

iii State the values of a, b and c.

iv Identify the one that is not a quadratic equation.

a $x(x - 2) = 3(x + 4)$ **b** $(1 - 2x)x = x(5 + x)$

c $x(2 - 3x) = x(x + 2) + 6$ **d** $x(x + 4) - x(x - 2) = 5$

e $x(x + 2) + x(x - 2) = 0$ **f** $4(x^2 + 5) = 2(x^2 + 10) - (x - 3)$

5 Rearrange each equation into the form $ax^2 + bx + c = 0$.

a $\dfrac{1}{x + 1} + \dfrac{1}{x + 2} = 1$ **b** $\dfrac{x + 2}{x + 1} = \dfrac{1}{x}$

7 A formula for solving quadratics

In the early ninth century, Al-Khowarizmi devised many methods for solving quadratics but it wasn't till much later that a **formula** for doing the job was developed. It was found that if $ax^2 + bx + c = 0$ ($a \neq 0$), then we could find x using the formula:

$$x = \frac{-b \pm \sqrt{b^2 - 4ac}}{2a}$$

where the '$\pm$' means that $x = \dfrac{-b + \sqrt{b^2 - 4ac}}{2a}$ and $x = \dfrac{-b - \sqrt{b^2 - 4ac}}{2a}$ both work.

We can show this formula works by the following:

$$x = \frac{-b \pm \sqrt{b^2 - 4ac}}{2a}$$

$\Rightarrow \qquad 2ax = -b \pm \sqrt{(b^2 - 4ac)}$ ... multiplying both sides by $2a$

$\Rightarrow \qquad 2ax + b = \pm \sqrt{(b^2 - 4ac)}$ ... adding b to both sides

$\Rightarrow \qquad (2ax + b)^2 = b^2 - 4ac$ ... squaring both sides

$\Rightarrow \qquad 4a^2x^2 + 4abx + b^2 = b^2 - 4ac$

$\Rightarrow \qquad 4a^2x^2 + 4abx + 4ac = 0$ ... subtracting $(b^2 - 4ac)$ from both sides

$\Rightarrow \qquad ax^2 + bx + c = 0$ ... dividing both sides by $4a$ which is OK since $a \neq 0$

Example 1   Solve $3x^2 - 7x - 6 = 0$.

By inspection, $a = 3$, $b = -7$ and $c = -6$.

Substituting into $x = \dfrac{-b \pm \sqrt{b^2 - 4ac}}{2a}$

$\Rightarrow \qquad x = \dfrac{-(-7) \pm \sqrt{(-7)^2 - 4 \times 3 \times (-6)}}{2 \times 3}$

$\Rightarrow \qquad x = \dfrac{7 \pm \sqrt{49 + 72}}{6} = \dfrac{7 \pm \sqrt{121}}{6} = \dfrac{7 \pm 11}{6}$

$\Rightarrow \qquad x = \dfrac{7 + 11}{6} = \dfrac{18}{6} = 3$ or $x = \dfrac{7 - 11}{6} = \dfrac{-4}{6} = -\dfrac{2}{3}$

$\Rightarrow \qquad x = 3$ or $x = -\frac{2}{3}$.

Example 2   Solve $2x^2 + 5x + 1 = 0$, giving your answers correct to 1 decimal place.

By inspection, $a = 2$, $b = 5$ and $c = 1$.

Substituting into $x = \dfrac{-b \pm \sqrt{b^2 - 4ac}}{2a}$

$\Rightarrow \quad x = \dfrac{-5 \pm \sqrt{5^2 - 4 \times 2 \times 1}}{2 \times 2}$

$\Rightarrow \quad x = \dfrac{-5 \pm \sqrt{17}}{4}$

$\Rightarrow \quad x = \dfrac{-5 + \sqrt{17}}{4}$ or $x = \dfrac{-5 - \sqrt{17}}{4}$

$\Rightarrow \quad x = -0 \cdot 2$ or $x = -2 \cdot 3$ correct to 1 decimal place

Exercise 7.1

1 Use the formula to solve each equation.

a $x^2 + 3x + 2 = 0$ **b** $x^2 - 3x + 2 = 0$ **c** $x^2 + 7x + 12 = 0$

d $x^2 - 7x + 12 = 0$ **e** $x^2 + x - 2 = 0$ **f** $2x^2 + 3x + 1 = 0$

g $3x^2 - 5x - 2 = 0$ **h** $4x^2 - 13x + 3 = 0$

2 Solve these quadratic equations using   **i** factors   **ii** the quadratic formula.

a $x^2 + 9x + 20 = 0$ **b** $3x^2 - 14x + 8 = 0$ **c** $5x^2 - 9x - 2 = 0$

3 Solve each quadratic equation, giving your answers correct to 2 decimal places.

a $x^2 + 7x + 2 = 0$ **b** $2x^2 + 3x - 6 = 0$ **c** $3x^2 - 5x - 1 = 0$

d $4x^2 - x - 1 = 0$ **e** $1 + 6x - x^2 = 0$ **f** $8 - x - x^2 = 0$

g $7 + x - 3x^2 = 0$ **h** $6 - x^2 = 9x$ **i** $3 - 4x^2 = 10x$

j $-2x^2 = 8x - 4$ **k** $4x^2 + 20x + 1 = 0$ **l** $2x^2 + 7x + 2 = 0$

4 a What happens when you use the formula to try and solve $x^2 + 2x + 5 = 0$?
     Comment: *not every quadratic equation has a solution.*

     b i What happens when you use the formula to try and solve $x^2 - 6x + 9 = 0$?
     ii Make a comment.

5 i Sort the equation into the form $ax^2 + bx + c = 0$.

     ii Use the formula to try and solve it. Where no solution exists, say so.
     Work to 2 d.p. where appropriate.

     a $x^2 = 1 - 2x$ **b** $2 - x^2 = x(x + 1)$ **c** $2(x^2 - 2) = 3$

     d $x - 3 + \dfrac{1}{x} = 0$ **e** $\dfrac{x}{4} = \dfrac{3}{x + 1}$ **f** $\dfrac{1}{x} - \dfrac{1}{x + 1} = \dfrac{1}{6}$

6 The ancient Greeks were interested in a number which was
1 more than its own reciprocal.
They called it the **golden number**.

The reciprocal of x is $\dfrac{1}{x}$.

a Form an equation in x.

b Rearrange it in the form $ax^2 + bx + c = 0$.

c If the golden number is the bigger of the solutions, find the
golden number.

Exercise 7.2

1 The height, in metres, of a rocket fired vertically upwards is given by the equation
$h = 20t - 5t^2$.

     a Work out the times when the rocket is at a height of 8 metres.
     Give your answers correct to 2 decimal places.

     b How many seconds after taking off is the rocket back on the ground?

     c What is the height of the rocket after 2 seconds?

2 A flagpole is held up by a 9 m long guy rope.
The rope is tethered to the ground at a distance, x m, from the
foot of the pole.
The pole is 2 metres longer than this distance.
Calculate (correct to 1 decimal place):
  a the value of x
  b the height of the pole.

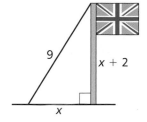

3 A small cuboid box with dimensions $2x$ cm, x cm and 3 cm is to be gift-wrapped.
The surface area of the box is 150 cm².
  a Find an expression for the surface area in terms of x.
  b Form an equation and solve it.
  c State the actual dimensions of the box, correct to 2 d.p.

4 A garden consists of a rectangular vegetable plot 3 metres by
5 metres surrounded by an area of grass x metres wide.
  a Write down an expression for the length and breadth of
  the whole garden.
  b Write down an expression for the area of the garden.
  c Given that the area of grass alone is 20 m², calculate
  the dimensions of the garden.

5 A ruler costs 5p more than a pen.
I have spent 300p on pens and 180p on rulers
  a Let xp represent the cost of a pen. Write
  expressions for:
    i the cost of a ruler
    ii the number of pens bought
    iii the number of rulers bought.
  b I have bought 18 items in total. Form
  an equation in x.
  c Solve the equation and find the
  unit cost of both items.

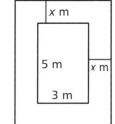

Investigation

Given that a formula exists for finding the solutions, it is possible to write a
spreadsheet to solve your equations. The illustration shows the solutions to
$x^2 + 2x - 1 = 0$.

	A	B	C	D	E	F
1	a =	1				
2	b =	2				
3	c =	−1				
4		8				
5			There are two answers	0·41421356237	−2·41421356	
6						

See if you can follow the logic of the sheet.
A1, A2 and A3 are as shown. The user puts appropriate values into B1, B2 and B3.

in C1: =IF(B1="","",IF(B1=0,"Can't use a=0","")) ... gives a warning if you use $a = 0$

in B4: =B2^2–4*B1*B3   ... what is it working out?

in C3: =IF(B4<0,"There are no answers","")   ... why would there be no answers?

in C4: =IF(B4=0,"There is one answer","")   ... why is there only one answer?

in D4: =IF(B4=0,–B2/(2*B1),"")   ... what is it working out?

in C5: =IF(B4>0,"There are two answers","")

in D5: =IF(B4>0,(–B2+SQRT(B4))/(2*B1),"")   ... what is it working out?

in E5: =IF(B4>0,(–B2-SQRT(B4))/(2*B1),"")   ... what is it working out?

Here the sheet has been used to solve $x^2 + 2x + 1 = 0$ and $x^2 + 2x + 2 = 0$.

	A	B	C	D	E	F
1	a =	1				
2	b =	2				
3	c =	1				
4		0	There is one answer	−1		

	A	B	C	D	E	F
1	a =	1				
2	b =	2				
3	c =	2	There are no answers			
4		−4				

8 Iteration – a guessing game

Look at the equation $2x^2 + 5x + 1 = 0$.
We are looking for the value of x which makes $2x^2 + 5x + 1$ worth nothing.

Guess what x is.

When $x = 0$ we have $2(0)^2 + 5(0) + 1 = 1$ ... too big
When $x = -1$ we have $2(-1)^2 + 5(-1) + 1 = -2$ ... too small

So the solution must be somewhere between $x = 0$ and $x = -1$ because zero lies between 1 and $-2 \Rightarrow -1 < x < 0$

Try $x = -0·5$: $2(-0·5)^2 + 5((-0·5) + 1 = -1$
                     ... zero lies between 1 and $-1 \Rightarrow -0·5 < x < 0$

Try $x = -0·25$: $2(-0·25)^2 + 5((-0·25) + 1 = -0·125$
                     ... zero lies between 1 and $-0·125 \Rightarrow -0·25 < x < 0$

Try $x = -0.1$: $\qquad$ $2(-0.1)^2 + 5((-0.1) + 1 = 0.52$

$\qquad\qquad\qquad$ ... zero lies between 0.52 and $-0.125 \Rightarrow -0.25 < x < -0.1$

Try $x = -0.15$: $\qquad$ $2(-0.15)^2 + 5((-0.15) + 1 = 0.295$

$\qquad\qquad\qquad$ ... zero lies between 0.295 and $-0.125 \Rightarrow -0.25 < x < -0.15$

So, to 1 decimal place, $x = -0.2$.

We could continue our search if more accuracy was required.

This sort of *trial and improvement* process is an example of **iteration**.

We can keep going to any degree of accuracy, e.g. $x = -0.219\,22$ to 5 decimal places.

The process can be used for any equation, not just quadratics.

Step 1	We find two values of x, one of which gives a positive answer and the other a negative answer. The switch of sign tells us we have the root trapped.
Step 2	We try different values for x. If we get a negative answer then we swap x for the x value that gave the previous negative answer. If we get a positive answer then we swap x for the x value that gave the previous positive answer.
Step 3	Eventually we trap the solution between two values which round to the same amount.
Example	Show that the equation $3x^2 - 5x - 4 = 0$ has a root that lies between $x = 2$ and $x = 3$. Find this root to 1 decimal place.

$$x = 2 \Rightarrow 3x^2 - 5x - 4 = -2$$

$$x = 3 \Rightarrow 3x^2 - 5x - 4 = 8$$

Since $-2 < 0 < 8$ then $2 < x < 3$

x	Quadratic	Root lies between	
		2	3
2.5	2.25	2	2.5
2.2	-0.48	2.2	2.5
2.3	0.37	2.2	2.3
2.25	-0.0625	2.25	2.3

A table is a good idea.

Orange values are those resulting in negative answers.

So, to 1 decimal place, the root is 2.3.

Exercise 8.1

1 a Prove that the quadratic equation $x^2 - 2x - 2 = 0$ has a solution between $x = 2$ and $x = 3$.

  b Using iteration, find this root to 1 decimal place.

  c Use the quadratic formula to check your answer.

2 $2x^2 + 2x - 5 = 0$ has a solution between $x = 1$ and $x = 2$.

  a Prove that this is true.

  b Find this root correct to 2 decimal places.

  c The other root lies between $x = -3$ and $x = -2$.
Find this root correct to 1 decimal place.

3 The equation $x^3 + 2x^2 - 3x - 6 = 0$ is called a cubic equation.

  a Show that it has a solution between $x = 1$ and $x = 2$.

  b Calculate this root correct to 2 decimal places.

4 $x^3 - 2x^2 - 2x + 3 = 0$

  a Show there is a solution between $x = -2$ and $x = -1$.

  b Find this root correct to 2 decimal places.

5 The equation $x^3 + 2x^2 - 3x - 3 = 0$ has a root at roughly $x = 1 \cdot 6$.

  a Improve this solution to 3 significant figures.

  b Show that another solution lies between $x = -2$ and $x = -3$.

  c Show that a third solution lies close to $x = -1$.

6 Here is another cubic equation: $x^3 - 6x + 3 = 0$.

  a Show that a solution lies in the region

    i $0 < x < 1$ **ii** $2 < x < 3$ **iii** $-3 < x < -2$

  b Find the negative solution to 2 significant figures.

7 a Show that the equation $x(x^2 + 1) - 3 = 0$ has a solution between 1 and 2.

  b Find the solution correct to 1 decimal place.

8 The equation $x^4 - 3 = 0$ has a solution that lies between $x = 1$ and $x = 2$.

  a Find this solution correct to 3 significant figures.

  b The equation has another solution. Can you estimate its value?

◀◀ RECAP

You should be able to recognise a **quadratic equation**.

These take many forms, e.g. $x^2 = 3x + 4$, $x(x + 1) = 5$, $\dfrac{1}{x + 1} = 2x - 3$, $3x^2 + 2x - 5 = 0$.

All quadratic equations can be reduced to a **standard form**: $ax^2 + bx + c = 0$. $(a \neq 0)$

Equations can be solved with the help of **factors**, making use of the fact that if two numbers multiply to make zero it means that one of them is zero.

$$A \times B = 0 \Rightarrow A = 0 \text{ or } B = 0$$

Common factors: $x^2 - 4x = 0 \Rightarrow x(x - 4) = 0 \Rightarrow x = 0$
     or $x - 4 = 0 \Rightarrow x = 0$ or $x = -4$

Difference of squares: $4x^2 - 49 = 0 \Rightarrow (2x - 7)(2x + 7) = 0 \Rightarrow (2x - 7) = 0$
     or $(2x + 7) = 0 \Rightarrow x = \pm \frac{7}{2}$

Trinomials: $x^2 + 3x + 2 = 0 \Rightarrow (x + 2)(x + 1) = 0 \Rightarrow x + 2 = 0$
     or $x + 1 = 0 \Rightarrow x = -2$ or -1

Quadratic equations can be solved by using the fact that

$$ax^2 + bx + c = 0 \; (a \neq 0) \Rightarrow x = \frac{-b \pm \sqrt{b^2 - 4ac}}{2a}$$

This cannot be used when: • $a = 0$ (you can't divide by zero)
                                  •   the value under the square root is negative.

We can use an **iterative** process to search for solutions in a systematic way. This works for other types of equation as well as quadratics.

When solving some expression in $x = 0$:

Step 1   We find two values of x, one of which gives a positive answer and the other a negative answer when substituted in the expression.
The switch of sign tells us we have the root *trapped*.

Step 2   We try different values for x.
If we get a negative answer, swap x for the x value that gave the previous negative answer.
If we get a positive answer, swap x for the x value that gave the previous positive answer.

Step 3   Eventually we trap the solution between two values which round to the same amount.

Quadratics can be used in many contexts in modern-day life, from the design of aerials for satellite TV or the reflectors in a car's main beam, to working out problems on the flight of rockets or golf balls.

1 Solve the following quadratic equations:

  a $16x^2 - 6x = 0$ **b** $14b - 20b^2 = 0$ **c** $8x^2 - 72 = 0$

  d $5ax + 25x^2 = 0$ **e** $12x^2 + 30x = 0$ **f** $16x - 36x^2 = 0$

2 Solve the following quadratic equations by first factorising:

  a $x^2 + 10x + 9 = 0$ **b** $x^2 - 10x - 24 = 0$ **c** $x^2 - 11x + 30 = 0$

  d $28 - 3x - x^2 = 0$ **e** $15 + 14x - 8x^2 = 0$ **f** $18 - 3x - 10x^2 = 0$

  g $35 + 11x - x^2 = 0$ **h** $15 - 26x + 8x^2 = 0$ **i** $6 - 29x + 35x^2 = 0$

3 Solve the following quadratic equations using the formula $x = \dfrac{-b \pm \sqrt{b^2 - 4ac}}{2a}$.
  Give your answers correct to 2 decimal places.

  a $x^2 + 11x + 1 = 0$ **b** $x^2 - 7x - 2 = 0$ **c** $3x^2 - 8x - 2 = 0$

  d $4 - 7x + x^2 = 0$ **e** $5 - 11x - 2x^2 = 0$ **f** $5 - x - 7x^2 = 0$

4 The cubic equation $x^3 + 5x^2 + 6x - 6 = 0$ has a solution between $x = 0$ and $x = 1$.

  a Prove that the solution lies between these two values.

  b Find the solution correct to 2 decimal places.

5 Solve these quadratic equations, giving your answers, where necessary, correct to 2 decimal places.

  a $\dfrac{4}{x} + x = 5$ **b** $\dfrac{3}{x - 1} + \dfrac{1}{x} = 2$

6 A farmer uses 12 m of fencing to form a rectangular pen against a wall.
  x m of fencing forms a side as shown.

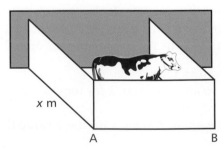

  a Express AB in terms of x.

  b Express the area enclosed in terms of x.

  c There are 14 m² enclosed.
      Form an equation and solve it to find x.

REVISE

9 Surds and indices

Jaina mathematicians were those who followed Jainism – a religion and philosophy which was founded in India around the sixth century BC.

One concept that the Jainas had begun to understand was that of roots. For example, one of their texts states: 'The first square root multiplied by the second square root is the cube of the second root.'

This can be written as $\sqrt{x}(\sqrt{\sqrt{x}}) = (\sqrt{\sqrt{x}})^3$. This will mean more to you later.

1 Review

◀◀ Exercise 1.1

1 Simplify each expression.
   a $x + x + x + x$ **b** $x \times x \times x$ **c** $x \times x \times x \times x \times x + x \times x \times x$
   d $5x \times 2x$ **e** $6x \div 2$ **f** $\dfrac{6x}{x}$

2 Given that $x = 3$, $y = -2$ and $z = -4$, evaluate these expressions:
   a x^2 **b** $3y^2$ **c** xyz **d** $z^2 + y^2$
   e x^2y^2 **f** $xz \div y$ **g** $y(x + z)$ **h** $z^2 - y^2$

3 Write down the values of:
   a 4^3 **b** 2^4 **c** 5^3 **d** $(-2)^3$
   e $\sqrt{9}$ **f** $\sqrt[3]{27}$ **g** $\sqrt[4]{16}$

4 Express each number in the form $a \times 10^n$ where $0 \leqslant a < 10$ and n is an integer.
   a 3210   **b** 543 670   **c** 6·78 million
   d 0·000 31   **e** 0·000 007 21

5 Calculate:
   a $(2·3 \times 10^2) \times (3·1 \times 10^6)$ **b** $(3·21 \times 10^{-2}) \times (4·1 \times 10^6)$
   c $(6·78 \times 10^8) \times (2·5 \times 10^2)$ **d** $(2·61 \times 10^3)^2$

6 Write down the first twelve square whole numbers.

7 Expand and simplify the following:
   a $(x - 2)(x + 2)$ **b** $(y + 3)(y - 3)$ **c** $(2x + 3)(2x - 3)$ **d** $(a + b)(a - b)$

2 Indices

From previous work we know that $x \times x \times x \times x$ can be simplified to x^4.
Likewise $x \times x \times x \times x \times x = x^5$.
Here the numbers 4 and 5 are called **index** numbers or **indices**; x is called the **base**.

Using the notation $x \times x = x.x$, we can say that, in general,
$x^n = x.x.x.x.x.x \ldots x.x$ (n factors of x)

In particular, $x^1 = x$

Example 1   Simplify $x^2 \times x^3$.
$$x^2 = x.x \text{ and } x^3 = x.x.x$$
$$\text{So } x^2 \times x^3 = (x.x) \times (x.x.x)$$
$$= x^{2+3} = x^5$$

In general $x^m \times x^n = \underbrace{x.x.x.x.x.x \ldots x.x}_{m \text{ factors}} \times \underbrace{x.x.x \ldots x.x.x}_{n \text{ factors}}$

$$= \underbrace{x.x.x.x.x.x \ldots x.x.x.x.x}_{m+n \text{ factors}}$$

$$= x^{m+n}$$

Rule 1 $x^m \times x^n = x^{m+n}$

To multiply numbers with the same base, you add the indices.

Example 2   Simplify $3x^5 \times 5x^3$.
$$3x^5 \times 5x^3 = 3 \times 5 \times x^5 \times x^3 = 15x^8$$

Exercise 2.1

1 Copy and complete:
   a $x^3 \times x^5 = (x \times \ldots\ldots\ldots\ldots) \times (x \ldots\ldots\ldots\ldots) = \ldots$
   b $x \times x^4 = \ldots\ldots\ldots\ldots \times (x \ldots\ldots\ldots\ldots) = \ldots$

2 Simplify:
   a $10^4 \times 10^2$ **b** $10^6 \times 10^2$ **c** 10×10^4 **d** $10^2 \times 10^5$

3 Simplify:
   a $x \times x^3$ **b** $x^2 \times x^6$ **c** $x^3 \times x^5$ **d** $x^7 \times x^8$
   e $x^4 \times x^5$ **f** $a^6 \times a^{10}$ **g** $b^7 \times b^5$ **h** $f^{12} \times f^{10}$
   i $x \times x^2 \times x^4$ **j** $x^2 \times x \times x^7$ **k** $x^4 \times x^5 \times x^6$ **l** $x^2 \times x^4 \times x^6 \times x^8$
   m $d^5 \times d^9 \times d$ **n** $y^6 \times y^7 \times y \times y^2$ **o** $a^{10} \times a^{12}$ **p** $f^{23} \times f^{10}$
   q $4^5 \times 4^8$ **r** $7^6 \times 7^{10}$ **s** $13^{10} \times 13^{11}$ **t** $60^{12} \times 60^{14}$

4 Simplify:

  a $2x \times 4x$ **b** $3x^2 \times 4x$ **c** $5x^4 \times 6x^2$ **d** $2x^6 \times 3x^4$

  e $x \times x^5$ **f** $\frac{2}{3}x^3 \times 12x^4$ **g** $2a^7 \times 3a^2 \times 4a$ **h** $4a^3 \times 3a^2 \times 5a$

  i $10x^2 \times 4x^5$ **j** $8a^4 \times 4a^5$ **k** $7b^5 \times 8b^4$ **l** $3c^{12} \times 4c^3$

  m $4r^6 \times 5r^7$ **n** $5t^7 \times 2t^4 \times 3t$ **o** $3s^2 \times 4s^4 \times (-2)s^6$ **p** $(-3)c^4 \times (-4)c^9$

 5 Multiply these numbers, leaving your answers in scientific notation.

  a $(8 \times 10^2) \times (4 \times 10^3)$ **b** $(5 \times 10^3) \times (6 \times 10^7)$

  c $(1 \cdot 1 \times 10^7) \times (8 \times 10^4)$ **d** $(2 \cdot 4 \times 10^6) \times (4 \times 10^5)$

  e $(8 \cdot 8 \times 10^5) \times (3 \times 10^7)$ **f** $(5 \cdot 25 \times 10^4) \times (5 \times 10^5)$

---

3 Division using indices

Example 1   Using the notation $x \times x = x.x$,

$$\frac{x^6}{x^2} = \frac{x.x.x.x.x.x}{x.x} = x.x.x.x = x^{6-2} = x^4$$

Example 2 $\dfrac{a^5}{a^3} = \dfrac{a.a.a.a.a}{a.a.a} = a.a = a^{5-3} = a^2$

Rule 2 $\dfrac{a^m}{a^n} = a^{m-n}$

> To divide numbers with the same base, subtract the indices.

In particular $\dfrac{a^m}{a^m} = a^{m-m} = a^0$ but $a^m \div a^m = 1 \rightarrow a^0 = 1$

Example 3 $\dfrac{12a^5}{3a^2} = \dfrac{12}{3} \times \dfrac{a^5}{a^2} = 4a^{5-2} = 4a^3$

Exercise 3.1

1 Copy and complete:

  a $\dfrac{a^7}{a^4} = \dfrac{a.a.a.a.a.a.a}{a.a.a.a} =$ **b** $\dfrac{x^8}{x^2} =$

2 Simplify the following, leaving your answers in index form.

  a $\dfrac{x^6}{x^1}$ **b** $\dfrac{x^8}{x^3}$ **c** $\dfrac{x^{10}}{x^3}$ **d** $\dfrac{x^{12}}{x^7}$

  e $\dfrac{x^{13}}{x^2}$ **f** $\dfrac{a^{11}}{a^2}$ **g** $\dfrac{b^{10}}{b^6}$ **h** $\dfrac{b^{13}}{b^3}$

  i $c^{12} \div c^4$ **j** $c^{15} \div c^5$ **k** $s^{14} \div s^6$ **l** $t^{20} \div t^3$

  m $10^8 \div 10^6$ **n** $10^{11} \div 10^3$ **o** $2^8 \div 2^5$ **p** $3^7 \div 3^3$

  q $\dfrac{8t^3}{2t}$ **r** $\dfrac{15x^5}{3x^2}$ **s** $\dfrac{20x^6}{4x^5}$ **t** $\dfrac{36x^7}{3x^4}$

  u $\dfrac{0 \cdot 6^8}{0 \cdot 6^2}$ **v** $\dfrac{1 \cdot 3^7}{1 \cdot 3^4}$ **w** $\dfrac{\left(\frac{3}{4}\right)^5}{\left(\frac{3}{4}\right)^2}$ **x** $\dfrac{(3x)^4}{3x}$

3 Simplify the following:

a $\dfrac{x^3 \times x^4}{x}$
b $\dfrac{x^2 \times x^6}{x^3}$
c $\dfrac{x^7}{x^4} \times x$
d $\dfrac{a^2 \times a^4}{a}$

e $\dfrac{3s^2 \times 2s^4}{s^3}$
f $\dfrac{5t^4 \times 4t^3}{2t^2}$
g $\dfrac{8s^5}{2s^4 \times 3s^3}$
h $\dfrac{24a^8}{2a^3 \times 3a^4}$

i $\dfrac{x^2 y^4}{x^2 y^3}$
j $\dfrac{a^5 b^8}{b^8 a^3}$
k $\dfrac{vw^5}{w^3}$
l $\dfrac{pq^9}{pq}$

---

4 Negative indices

We know that $1 \div x = \dfrac{1}{x}$ and that $x^0 = 1$ and $x^1 = x$

$$\Rightarrow \quad \frac{1}{x} = \frac{x^0}{x^1} = x^{0-1} = x^{-1}$$

In general $\quad \dfrac{1}{x^n} = \dfrac{x^0}{x^n} = x^{0-n} = x^{-n}$

Rule 3 $\qquad \dfrac{1}{a^n} = a^{-n}$ $\qquad \boxed{a^{-n} \text{ is the reciprocal of } a^n.}$

Example 1 $\qquad \dfrac{x^4}{x^6} = x^{4-6} = x^{-2} \left[= \dfrac{1}{x^2} \right]$

Example 2 $\qquad \dfrac{x^2}{x^8} = \dfrac{1}{x^6} = x^{-6}$

Example 3 $\qquad 4x^{-3} = \dfrac{4}{x^3}$

Exercise 4.1

1 Continue this number pattern for three more terms.

    a $1000, 100, 10, 1, \frac{1}{10}, \frac{1}{100}$ **b** $10^3, 10^2, 10^1, 10^0, 10^{-1}, 10^{-2}$

2 Express this pattern as a sequence of powers of 2.

    $8, 4, 2, 1, \frac{1}{2}, \frac{1}{4}, \frac{1}{8}$

3 Express each of the following using positive indices:

  a x^{-3} **b** x^{-6} **c** x^{-7} **d** a^{-8} **e** a^{-10}

  f $2s^{-3}$ **g** $7z^{-3}$ **h** $3d^{-4}$ **i** $\dfrac{a^{-9}}{2}$ **j** $\dfrac{e^{-5}}{6}$

  k $\dfrac{y^{-8}}{3}$ **l** $\dfrac{g^{-7}}{10}$ **m** $\dfrac{2x^{-4}}{3}$ **n** $\dfrac{4f^{-5}}{7}$ **o** $\dfrac{5x^{-8}}{9}$

  p $\dfrac{3d^{-7}}{5}$ **q** $\dfrac{1}{x^{-3}}$ **r** $\dfrac{1}{x^{-8}}$ **s** $\dfrac{1}{g^{-9}}$ **t** $\dfrac{1}{n^{-7}}$

  u $\dfrac{2}{c^{-9}}$ **v** $\dfrac{5}{m^{-10}}$ **w** $\dfrac{5}{x^{-6}}$ **x** $\dfrac{3}{5d^{-8}}$ **y** $\dfrac{7}{11t^{-12}}$

  z $\dfrac{13}{12x^{-12}}$

4 Express each of the following using negative indices:

a $\dfrac{1}{x^4}$ **b** $\dfrac{1}{x}$ **c** $\dfrac{1}{d^{10}}$ **d** $\dfrac{1}{b^{13}}$ **e** $\dfrac{3}{x^{13}}$

f $\dfrac{5}{x^9}$ **g** $\dfrac{7}{y^{10}}$ **h** $\dfrac{3}{x^{11}}$ **i** $\dfrac{1}{2x^4}$ **j** $\dfrac{1}{3x^7}$

k $\dfrac{1}{8a^5}$ **l** $\dfrac{1}{7b^6}$ **m** $\dfrac{3}{4x^5}$ **n** $\dfrac{2}{5x^{12}}$ **o** $\dfrac{4}{7x^9}$

p $\dfrac{3}{8a^7}$ **q** x^3 **r** y^8 **s** $5x^4$ **t** $8y^5$

u $\dfrac{s^8}{5}$ **v** $\dfrac{r^{13}}{9}$ **w** $\dfrac{4d^{14}}{9}$ **x** $\dfrac{6y^{15}}{5}$

5 Simplify each of the following expressions, leaving your answers with positive indices.

a $\dfrac{x^2}{x^3 \times x^4}$ **b** $\dfrac{x^5}{x \times x^7}$ **c** $\dfrac{x^3 \times x^6}{x^{11} \times x}$ **d** $\dfrac{x \times x^6}{x^2 \times x^3 \times x^4}$

e $\dfrac{4x^3}{2x^7 \times 6x^8}$ **f** $\dfrac{15x^4}{3x^3 \times 2x}$ **g** $\dfrac{12t^8}{3t^6 \times 6t^7}$ **h** $\dfrac{16s^6 \times 2s^5}{4s^7 \times 3s^8}$

6 Simplify each of the following expressions, leaving your answers with negative indices.

a $\dfrac{x^6 \times x^7}{x}$ **b** $\dfrac{x^9 \times x^{10}}{x^4}$ **c** $\dfrac{x \times x^{11}}{x^2 \times x^3}$ **d** $\dfrac{x^5 \times x^{11}}{x^4 \times x^5}$

7 Multiply out the expressions.
   a $x^4(x^3 + x^{-6})$ **b** $x^4(x^{-3} + x^{-5})$ **c** $(x^{-2} + x^3)(x^{-2} + x)$
   d $x^{-3}(x^4 + x^8)$ **e** $x^{-3}(x^{-6} + x^7)$ **f** $(x^{-3} + x^{-6})(x^7 + x^{-2})$

5 Powers of powers

Example 1 $(x^2)^3 = x^2 \times x^2 \times x^2 = x^{2+2+2} = x^{2\times3} = x^6$

Example 2 $(x^3)^4 = x^3 \times x^3 \times x^3 \times x^3 = x^{12} = x^{3\times4} = x^{12}$

In general we have

Rule 4 $\boxed{(a^m)^n = a^{mn}}$

Exercise 5.1

1 Simplify each expression:
   a $(x^4)^2$ **b** $(x^2)^6$ **c** $(x^4)^5$ **d** $(x^3)^5$ **e** $(x^7)^8$
   f $(x^{-2})^2$ **g** $(x^{-3})^4$ **h** $(x^{-5})^7$ **i** $(x^{-3})^4$ **j** $(x^{-4})^5$
   k $(x^2)^{-3}$ **l** $(x^4)^{-2}$ **m** $(x^6)^{-3}$ **n** $(x^7)^{-2}$ **o** $(x^8)^{-4}$

2 Simplify each expression, leaving only positive indices:

a $(x^4)^{-2}$ **b** $(x^{-3})^5$ **c** $(x^{-7})^6$ **d** $(x^8)^{-3}$ **e** $(x^9)^{-2}$

f $(x^6)^{-5}$ **g** $(s^{-5})^3$ **h** $(2x^3)^{-4}$ **i** $(3d^{-4})^2$ **j** $(4s^{-3})^{-2}$

3 Simplify each expression:

a $\left(\dfrac{1}{x^3}\right)^2$ **b** $\left(\dfrac{1}{x^4}\right)^5$ **c** $\left(\dfrac{1}{x^6}\right)^7$ **d** $\left(\dfrac{1}{x^4}\right)^7$ **e** $\left(\dfrac{1}{x^8}\right)^4$

f $\left(\dfrac{3}{x^4}\right)^{-5}$ **g** $\left(\dfrac{2}{x^6}\right)^{-3}$ **h** $\left(\dfrac{1}{x^{-4}}\right)^3$ **i** $\left(\dfrac{1}{x^{-5}}\right)^{-3}$ **j** $\left(\dfrac{4}{x^2}\right)^{-2}$

k $\left(\dfrac{1}{x^2}\right)^4$ **l** $\left(\dfrac{1}{x^4}\right)^2$ **m** $\left(\dfrac{5}{3x^4}\right)^3$ **n** $\left(\dfrac{5}{3x^3}\right)^4$ **o** $\left(\dfrac{2}{x^4}\right)^4$

4 Simplify:

a $(10^0)^5$ **b** $(3x^0)^3$ **c** $(x^3)^0$ **d** $\dfrac{x^2 \times x^4}{x^0}$

e $(2x^7)^0$ **f** $6(x^8)^0$ **g** $(8x^2)^0 \times (7x^0)^1$ **h** $((x^{-3})^2)^2$

5 Simplify, leaving your answers using positive indices only.

a $\left(\dfrac{1}{x^2} \times x^4\right)^2$ **b** $\dfrac{x}{(x^3 \times x^4)^2}$ **c** $\left(\dfrac{x^6}{x^{-1} \times x^4}\right)^{-1}$ **d** $\dfrac{(2x^5)^3}{3x^2 \times 2x^4}$

e $\dfrac{(3x)^4}{x^{-3} \times x^{-5}}$ **f** $\dfrac{2x^3}{(3x^{-1})^4 \times x^3}$ **g** $\dfrac{x^{-4}}{x^{-5} \times x^{-2}}$ **h** $\dfrac{x^{-8}}{(x^{-2} \times x^{-3})^{-2}}$

---

6 Fractional indices

Does $x^{\frac{1}{2}}$ have any meaning?

Note that $(\sqrt{x})^2 = x$.   Note also that $\left(x^{\frac{1}{2}}\right)^2 = x^{\frac{1}{2} \times 2} = x^1 = x$

$$\Rightarrow \quad (\sqrt{x})^2 = \left(x^{\frac{1}{2}}\right)^2$$

$$\Rightarrow \quad \sqrt{x} = x^{\frac{1}{2}} \quad \text{(assuming both are positive quantities)}$$

Note that $\sqrt{x}$, and hence $x^{\frac{1}{2}}$, is defined as the **positive root** of x.

Similarly $(\sqrt[3]{x})^3 = \left(x^{\frac{1}{3}}\right)^3 = x \quad \Rightarrow \quad \sqrt[3]{x} = x^{\frac{1}{3}}$

In general

Rule 5 $\boxed{\sqrt[b]{a} = a^{\frac{1}{b}}}$

Example 1   Find $16^{\frac{1}{4}}$.
$$16^{\frac{1}{4}} = \sqrt[4]{16} = 2$$

Note that the numerator of the power need not be 1.
$$x^{\frac{2}{3}} = x^{\frac{1}{3} \times 2} = \left(x^{\frac{1}{3}}\right)^2 = (\sqrt[3]{x})^2$$

We could equally have said $x^{\frac{2}{3}} = x^{2 \times \frac{1}{3}} = (x^2)^{\frac{1}{3}} = \sqrt[3]{(x^2)}$

Rule 6 $\boxed{a^{\frac{b}{c}} \quad \sqrt[c]{a^b} = (\sqrt[c]{a})^b}$

Example 1   Calculate $8^{\frac{2}{3}}$.
$$8^{\frac{2}{3}} = (\sqrt[3]{8})^2 = (2)^2 = 4$$

Note: we could calculate $\sqrt[3]{8^2} = \sqrt[3]{64} = 4$ but generally it is easier to take a root before raising to a power.

Example 2 $8^{-\frac{2}{3}} = \dfrac{1}{(\sqrt[3]{8})^2} = \dfrac{1}{(2)^2} = \dfrac{1}{4}$

Example 3   Using positive indices $\dfrac{x^{\frac{1}{4}}}{x^{\frac{2}{3}}} = x^{\frac{1}{4} - \frac{2}{3}} = x^{\frac{3}{12} - \frac{8}{12}} = x^{-\frac{5}{12}} = \dfrac{1}{x^{\frac{5}{12}}}$

Exercise 6.1

1 Simplify, leaving your answers with positive indices only.

a $\dfrac{x^{\frac{1}{4}}}{x^{\frac{1}{2}}}$
b $\dfrac{x^{\frac{2}{3}}}{x^{\frac{1}{4}}}$
c $\dfrac{x^{\frac{1}{6}}}{x^{\frac{1}{4}}}$
d $\dfrac{2x^{\frac{1}{2}}}{3x^{\frac{1}{5}}}$

e $\dfrac{x^{-\frac{1}{2}}}{x^{-\frac{1}{3}}}$
f $\dfrac{a^{-\frac{2}{3}}}{a^{\frac{1}{4}}}$
g $\dfrac{a^{-\frac{1}{5}}}{a^{\frac{2}{3}}}$
h $\dfrac{5x^{\frac{3}{4}}}{2x^{-\frac{1}{2}}}$

2 Write these in index form:

a $\sqrt{x}$
b $\sqrt[3]{x}$
c $\sqrt[2]{x^4}$
d $\sqrt{9x}$
e $\sqrt[3]{(8x)^4}$

f $(\sqrt{16x})^3$
g $(\sqrt[3]{27x})^4$
h $\sqrt[3]{x^{-3}}$
i $\sqrt[5]{(32x)^3}$
j $\dfrac{1}{\sqrt{25x^2}}$

k $\dfrac{3}{\sqrt{(36x^2)}}$
l $\dfrac{3}{2\sqrt{x^5}}$
m $\dfrac{3}{\sqrt[3]{125x^4}}$
n $\dfrac{11}{\sqrt[4]{81x^4}}$
o $\dfrac{2}{\sqrt[3]{(49x)^3}}$

3 Write these using the radical sign ($\sqrt{}$):

a $x^{\frac{3}{2}}$
b $x^{\frac{5}{3}}$
c $x^{-\frac{4}{3}}$
d $(3x)^{\frac{5}{2}}$

e $4x^{\frac{5}{4}}$
f $10x^{-\frac{1}{2}}$
g $12x^{-\frac{3}{4}}$
h $8x^{\frac{5}{7}}$

i $x^{\frac{5}{4}}$
j $x^{\frac{7}{2}}$
k $x^{\frac{11}{5}}$
l $(2x)^{\frac{0}{5}}$

m $7x^{\frac{9}{7}}$
n $8x^{-\frac{7}{6}}$
o $10x^{\frac{9}{8}}$
p $(4x)^{-\frac{6}{5}}$

4 Evaluate:

a $9^{\frac{1}{2}}$ **b** $27^{\frac{1}{3}}$ **c** $81^{\frac{1}{4}}$ **d** $100^{\frac{1}{2}}$ **e** $25^{\frac{1}{2}}$

f $32^{\frac{1}{5}}$ **g** $125^{\frac{1}{3}}$ **h** $49^{\frac{1}{2}}$ **i** $8^{-\frac{1}{3}}$ **j** $49^{-\frac{1}{2}}$

k $64^{-\frac{1}{2}}$ **l** $\left(\frac{1}{8}\right)^{\frac{2}{3}}$ **m** $4^{-\frac{5}{2}}$ **n** $8^{\frac{4}{3}}$ **o** $27^{\frac{5}{3}}$

p $32^{-\frac{3}{5}}$ **q** $16^{-\frac{3}{4}}$ **r** $\dfrac{7}{9^{-\frac{3}{2}}}$ **s** $\dfrac{8}{16^{-\frac{3}{4}}}$ **t** $\dfrac{4^{\frac{3}{2}}}{36^{-\frac{1}{2}}}$

5 Simplify, using positive indices:

a $x^2\left(x^{\frac{1}{2}} + x^{\frac{1}{3}}\right)$ **b** $x^{\frac{1}{2}}\left(x^4 + x^3\right)$ **c** $x^{-\frac{1}{4}}\left(x^8 + x^6\right)$

d $x^{\frac{2}{3}}\left(x^{\frac{1}{2}} + x^{\frac{1}{4}}\right)$ **e** $x^{-4}\left(x^{-\frac{1}{2}} - x^2\right)$ **f** $x^{-\frac{1}{3}}\left(x^{-2} - x^4\right)$

g $x^2\left(\dfrac{1}{x^{\frac{1}{2}}} + \dfrac{1}{x^{\frac{1}{3}}}\right)$ **h** $x^{\frac{1}{2}}\left(\dfrac{3}{x^{\frac{1}{4}}} - \dfrac{2}{x^{\frac{1}{5}}}\right)$ **i** $x^{-\frac{1}{2}}\left(\dfrac{1}{x^{-\frac{1}{4}}} - \dfrac{3}{x^{-\frac{1}{6}}}\right)$

6 Simplify, expressing your answer using positive indices only.

a $(2x^2)^2$ **b** $(8x^3)^{\frac{1}{3}}$ **c** $(16a^4)^{\frac{1}{4}}$ **d** $(2x^{\frac{1}{2}})^{-3}$

e $(16x^2)^{\frac{1}{4}}$ **f** $(27x^{-2})^{-\frac{1}{3}}$ **g** $\sqrt[4]{(625a^8)}$ **h** $\sqrt{\dfrac{64}{81x^{-\frac{1}{3}}}}$

Investigation

With real numbers we can't find $\sqrt{(-1)}$.
Try it on your calculator.
Explain the message.

Imagine there *was* a number that when you squared it you got -1.
Call the number i.

So $i^2 = -1$

$\Rightarrow$ $\sqrt{(-1)} = i$

So $\sqrt{(-4)} = \sqrt{(4 \times -1)} = \sqrt{4} \times \sqrt{(-1)} = 2i$

a Write down $\sqrt{(-9)}$, $\sqrt{(-16)}$, $\sqrt{(-25)}$.

b $i^2 = -1$, $i^3 = i^2 \times i = ?$ and $i^4 = (i^2)^2 = ?$

Now explore i^5, i^6, i^7, etc.

7 Surds

Class discussion

You have used many different sets of numbers.

N ... the set of natural, or counting numbers: 1, 2, 3, 4, ...

W ... the set of whole numbers: 0, 1, 2, 3, 4, ...

Z ... the set of integers: 0, ± 1, ± 2, ± 3, ...

Q ... the set of rational numbers. A number belongs to this set if it can be written as a ratio or division of two integers, for example:

4 is rational since it can be written as $4 \div 1$.
(So all integers are rational.)

0·6 is rational since it can be written as $6 \div 10$.
(Similarly with all terminating decimals:

0·15 = $15 \div 100$; 0·0123 = $123 \div 10\ 000$; etc.)

0·121 212... where the decimal is repeating is rational.

Consider $x = 0·121\ 212...$

Multiply both sides by 100: $100x = 12·121\ 212\ 12...$

$\Rightarrow$ $100x = 12 + 0·121\ 212\ 12...$

$\Rightarrow$ $100x = 12 + x$

$\Rightarrow$ $99x = 12$

$\Rightarrow$ $x = 12 \div 99$ (so x is rational)

The Pythagoreans believed that all numbers were rational.
However, they discovered a short proof that $\sqrt{2}$ was not.

Suppose $\sqrt{2} = a \div b$ where a and b are integers.
Squaring both sides gives $2 = a^2 \div b^2$

$\Rightarrow$ $2b^2 = a^2$

Now it is a known fact that all square numbers have an even number of prime factors, e.g. $4 = 2 \times 2$; $36 = 2 \times 2 \times 3 \times 3$.

So b^2 has an even number of factors and $2b^2$ has an odd number ... that extra factor of 2.

But a^2 has an even number ... and is equal to $2b^2$.

An odd number can't equal an even number, so $\sqrt{2} \neq a \div b$ for any pair of integers.

$\sqrt{2}$ is irrational.

$\sqrt{2} \approx 1·414$ but this is just a decimal approximation. $\dfrac{1414}{1000}$ is a rational approximation.

If we want to express the exact value we must write $\sqrt{2}$.

We now extend our set of numbers to:

R ... the set of real numbers, which includes all the irrational numbers, i.e. numbers that cannot be expressed as a ratio.

Surds are special irrational numbers. These are numbers such as square roots and cube roots which cannot be expressed as fractions.

For example, $\sqrt{2}$, $\sqrt[3]{7}$, $\sqrt{11}$ are examples of surds.

Simplifying surds

We know that $(ab)^n = a^n b^n \Rightarrow (ab)^{\frac{1}{2}} = a^{\frac{1}{2}} b^{\frac{1}{2}} \Rightarrow \sqrt{ab} = \sqrt{a} \times \sqrt{b}$

Example 1   Simplify $\sqrt{50}$.

Factorise 50 looking for square factors.

$\sqrt{50} = \sqrt{(25 \times 2)} = \sqrt{25} \times \sqrt{2} = 5\sqrt{2}$

This is its simplest form since 2 has no square factors.

Example 2   Simplify $\sqrt{80}$.

$\sqrt{80} = \sqrt{(16 \times 5)} = \sqrt{16} \times \sqrt{5} = 4\sqrt{5}$

(Note: if smaller, factors are easier to spot.

$\sqrt{80} = \sqrt{(4 \times 4 \times 5)} = \sqrt{4} \times \sqrt{4} \times \sqrt{5} = 2 \times 2 \times \sqrt{5} = 4\sqrt{5})$

Example 3   Simplify $\sqrt{147} - \sqrt{75}$.

We can only add or subtract like terms, so first simplify each surd.

$$\sqrt{147} - \sqrt{75} = \sqrt{(3 \times 49)} - \sqrt{(3 \times 25)}$$
$$= \sqrt{3} \times \sqrt{49} - \sqrt{3} \times \sqrt{25}$$
$$= 7\sqrt{3} - 5\sqrt{3}$$
$$= 2\sqrt{3}$$

Exercise 7.1

1 Which of these numbers are surds?

 a $\sqrt{5}$ **b** $\sqrt{8}$ **c** $\sqrt{9}$ **d** $\sqrt[3]{5}$ **e** $\sqrt[4]{16}$
 f $\sqrt{144}$ **g** $\sqrt[4]{81}$ **h** $\sqrt{64}$ **i** $\sqrt[3]{25}$ **j** $\sqrt{169}$

2 Why is $\sqrt{1 \cdot 21}$ not a surd?

3 Copy and complete the following:

 a $\sqrt{20} = \sqrt{(4 \times \ldots)} = \sqrt{\ldots} \times \sqrt{\ldots} = 2\sqrt{\ldots}$
 b $\sqrt{75} = \sqrt{(\ldots \times 3)} = \sqrt{\ldots} \times \sqrt{\ldots} = \ldots\sqrt{\ldots}$
 c $\sqrt{96} = \sqrt{(\ldots \times \ldots)} = \sqrt{\ldots} \times \sqrt{\ldots} = \ldots\sqrt{\ldots}$

4 Simplify the following surds:

 a $\sqrt{18}$ **b** $\sqrt{8}$ **c** $\sqrt{24}$ **d** $\sqrt{27}$ **e** $\sqrt{200}$
 f $\sqrt{32}$ **g** $\sqrt{72}$ **h** $\sqrt{147}$ **i** $\sqrt{128}$ **j** $\sqrt{45}$
 k $\sqrt{28}$ **l** $\sqrt{405}$ **m** $\sqrt{180}$ **n** $\sqrt{112}$ **o** $\sqrt{245}$

5 Now simplify these surds:

 a $63^{\frac{1}{2}}$ **b** $150^{\frac{1}{2}}$ **c** $75^{\frac{1}{2}}$ **d** $243^{\frac{1}{2}}$

6 **a** Simplify: **i** $\sqrt{108}$ **ii** $\sqrt{48}$

 b Show that $\sqrt{108} - \sqrt{48} = 2\sqrt{3}$.

7 Simplify:

  a $5\sqrt{2} + 3\sqrt{2}$ **b** $9\sqrt{11} - 3\sqrt{11}$

  c $8\sqrt{13} + 2\sqrt{13} - 9\sqrt{13}$ **d** $\sqrt{162} - 6\sqrt{2}$

  e $6\sqrt{5} - \sqrt{125}$ **f** $\sqrt{96} - \sqrt{24}$

  g $\sqrt{72} - \sqrt{32}$ **h** $\sqrt{24} + \sqrt{54}$

  i $\sqrt{12} + \sqrt{48}$ **j** $\sqrt{20} + \sqrt{80}$

  k $\sqrt{44} + \sqrt{99}$ **l** $\sqrt{40} + \sqrt{90}$

  m $\sqrt{252} - \sqrt{28} + \sqrt{175}$ **n** $\sqrt{50} + \sqrt{75} + \sqrt{98} + \sqrt{147}$

  o $\sqrt{300} + \sqrt{72} - \sqrt{108} + \sqrt{242}$

8 Simplify:

  a $\sqrt{3} \times \sqrt{3}$ **b** $\sqrt{7} \times \sqrt{7}$ **c** $2\sqrt{10} \times \sqrt{10}$ **d** $5\sqrt{11} \times 2\sqrt{11}$

  e $3\sqrt{5} \times 6\sqrt{5}$ **f** $\sqrt{8} \times \sqrt{2}$ **g** $\sqrt{5} \times \sqrt{20}$ **h** $\sqrt{20} \times \sqrt{8}$

9 Simplify the following, leaving your answer in surd form where applicable:

  a $3\sqrt{5} \times 2\sqrt{5}$ **b** $4\sqrt{6} \times 3\sqrt{8}$ **c** $2\sqrt{5} \times 3\sqrt{10}$

  d $2\sqrt{8} \times 3\sqrt{10}$ **e** $5\sqrt{8} \times 4\sqrt{2}$ **f** $3\sqrt{3} \times 4\sqrt{6}$

10 Expand the brackets and simplify where appropriate:

  a $8(2 + \sqrt{3})$ **b** $5(\sqrt{3} - 1)$ **c** $\sqrt{2}(6 + \sqrt{3})$

  d $\sqrt{3}(4 - \sqrt{3})$ **e** $\sqrt{5}(2 - 3\sqrt{5})$ **f** $\sqrt{5}(\sqrt{5} - 4)$

  g $\sqrt{(2x)}(\sqrt{(5x)} - 7)$ **h** $\sqrt{3x}(6 - \sqrt{8x})$ **i** $(4 + \sqrt{2x})(5 + \sqrt{3x})$

  j $(6 - \sqrt{7x})(5 + \sqrt{5x})$ **k** $(\sqrt{3} - \sqrt{2})(\sqrt{3} + \sqrt{2})$ **l** $(\sqrt{3} + \sqrt{5})(\sqrt{5} - \sqrt{3})$

11 Expand each set of brackets and simplify. Comment on your answers.

  a $(5 + \sqrt{2})(5 - \sqrt{2})$ **b** $(3 - \sqrt{5})(3 + \sqrt{5})$ **c** $(\sqrt{5} + 4)(\sqrt{5} - 4)$

  d $(\sqrt{2} + 1)(\sqrt{2} - 1)$ **e** $(\sqrt{3} - \sqrt{5})(\sqrt{3} + \sqrt{5})$ **f** $(\sqrt{7} + \sqrt{2})(\sqrt{7} - \sqrt{2})$

  g $(2\sqrt{5} + 1)(2\sqrt{5} - 1)$ **h** $(3\sqrt{7} - 2\sqrt{2})(3\sqrt{7} + 2\sqrt{2})$ **i** $(\sqrt{5} + 5\sqrt{3})(\sqrt{5} - 5\sqrt{3})$

12 Calculate the hypotenuse of this triangle, leaving your answer as a surd.

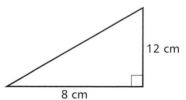

13 A square has a side that is 1 unit long.
Calculate the length of a diagonal, leaving your answer as a surd.

14 A box has sides 20 cm, 30 cm and 40 cm.

  a Calculate the length of the diagonal AC.

  b Calculate the length of the space diagonal, leaving your answer as a surd in its simplest form.

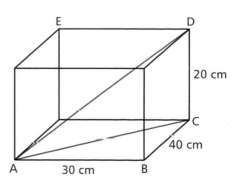

8 Further simplification

Rule 2 $\sqrt{\dfrac{a}{b}} = \dfrac{\sqrt{a}}{\sqrt{b}}$

Example 1 Simplify $\sqrt{\dfrac{4}{9}}$.

$$\sqrt{\dfrac{4}{9}} = \dfrac{\sqrt{4}}{\sqrt{9}} = \dfrac{2}{3}$$

Example 2 Find the value of $\sqrt{0\cdot09}$.

$$\sqrt{0\cdot09} = \sqrt{\dfrac{9}{100}} = \dfrac{\sqrt{9}}{\sqrt{100}} = \dfrac{3}{10} = 0\cdot3$$

Example 3 Evaluate $\dfrac{\sqrt{28}}{\sqrt{4}}$.

$$\dfrac{\sqrt{28}}{\sqrt{4}} = \sqrt{\dfrac{28}{4}} = \sqrt{7}$$

Example 4 $\sqrt{\dfrac{12}{27}} = \dfrac{\sqrt{12}}{\sqrt{27}} = \dfrac{\sqrt{4 \times 3}}{\sqrt{9 \times 3}} = \dfrac{2\sqrt{3}}{3\sqrt{3}} = \dfrac{2}{3}$

Exercise 8.1

1 Simplify the following:

    a $\dfrac{\sqrt{20}}{\sqrt{5}}$ **b** $\dfrac{\sqrt{24}}{\sqrt{3}}$ **c** $\dfrac{\sqrt{80}}{\sqrt{20}}$ **d** $\dfrac{\sqrt{72}}{\sqrt{8}}$ **e** $\dfrac{\sqrt{75}}{\sqrt{3}}$

    f $\dfrac{\sqrt{48}}{\sqrt{6}}$ **g** $\dfrac{\sqrt{48}}{\sqrt{3}}$ **h** $\dfrac{\sqrt{50}}{\sqrt{2}}$ **i** $\dfrac{\sqrt{84}}{\sqrt{7}}$ **j** $\dfrac{\sqrt{60}}{\sqrt{3}}$

    k $\dfrac{\sqrt{90}}{\sqrt{5}}$ **l** $\dfrac{\sqrt{242}}{\sqrt{22}}$ **m** $\dfrac{\sqrt{5}}{\sqrt{125}}$ **n** $\dfrac{\sqrt{6}}{\sqrt{96}}$ **o** $\dfrac{\sqrt{8}}{\sqrt{96}}$

    p $\dfrac{\sqrt{7}}{\sqrt{56}}$ **q** $\dfrac{\sqrt{50}}{\sqrt{18}}$ **r** $\dfrac{\sqrt{5}}{\sqrt{200}}$ **s** $\dfrac{\sqrt{6}}{\sqrt{120}}$ **t** $\dfrac{\sqrt{10}}{\sqrt{500}}$

2 Calculate the following:

    a $\sqrt{0\cdot16}$ **b** $\sqrt{0\cdot36}$ **c** $\sqrt{0\cdot64}$ **d** $\sqrt{0\cdot04}$

    e $\sqrt{0\cdot0025}$ **f** $\sqrt{0\cdot0081}$ **g** $\sqrt{0\cdot000\,049}$ **h** $\sqrt{0\cdot000\,000\,1}$

3 Simplify the following:

    a $\dfrac{2\sqrt{75}}{\sqrt{3}}$ **b** $\dfrac{6\sqrt{54}}{5\sqrt{6}}$ **c** $\dfrac{3\sqrt{96}}{4\sqrt{6}}$ **d** $\dfrac{5\sqrt{5}}{2\sqrt{20}}$ **e** $\dfrac{3\sqrt{8}}{4\sqrt{72}}$ **f** $\dfrac{2\sqrt{7}}{5\sqrt{112}}$

4 Simplify:

a $\sqrt{\dfrac{8}{18}}$ **b** $\sqrt{\dfrac{20}{45}}$ **c** $\sqrt{\dfrac{24}{54}}$ **d** $\sqrt{\dfrac{54}{96}}$

e $\sqrt{\dfrac{28}{63}}$ **f** $\sqrt{\dfrac{108}{147}}$ **g** $\sqrt{\dfrac{162}{98}}$ **h** $\sqrt{\dfrac{192}{75}}$

i $\sqrt{\dfrac{112}{63}}$ **j** $\sqrt{\dfrac{20}{125}}$ **k** $\sqrt{\dfrac{147}{75}}$ **l** $\sqrt{\dfrac{24}{96}}$

m $\sqrt{\dfrac{50}{98}}$ **n** $\sqrt{\dfrac{44}{99}}$ **o** $\sqrt{\dfrac{117}{52}}$ **p** $\sqrt{\dfrac{60}{375}}$

---

9 Rationalising the denominator

If we are left with a fractional form, it is customary to have the denominator as a rational number.

If a surd ends up in the denominator, then we rationalise it.

Example 1   Rationalise the denominator of $\dfrac{1}{\sqrt{3}}$.

$$\frac{1}{\sqrt{3}} = \frac{1}{\sqrt{3}} \times \frac{\sqrt{3}}{\sqrt{3}} = \frac{\sqrt{3}}{\sqrt{9}} = \frac{\sqrt{3}}{3}$$

Note $\dfrac{\sqrt{3}}{\sqrt{3}} = 1$, so in fact we have multiplied by 1.

We have created an equivalent fraction.

The fraction has the same value as $\dfrac{1}{\sqrt{3}}$, but now the denominator is rational.

Example 2   Rationalise the denominator of $\dfrac{3}{2\sqrt{5}}$.

$$\frac{3}{2\sqrt{5}} = \frac{3}{2\sqrt{5}} \times \frac{\sqrt{5}}{\sqrt{5}} = \frac{3\sqrt{5}}{2\sqrt{25}} = \frac{3\sqrt{5}}{2 \times 5} = \frac{3\sqrt{5}}{10}$$

Example 3   Rationalise the denominator of $\dfrac{4 - \sqrt{2}}{\sqrt{3}}$.

$$\frac{4 - \sqrt{2}}{\sqrt{3}} = \frac{4 - \sqrt{2}}{\sqrt{3}} \times \frac{\sqrt{3}}{\sqrt{3}} = \frac{\sqrt{3}(4 - \sqrt{2})}{3} = \frac{4\sqrt{3} - \sqrt{6}}{3} = \frac{4\sqrt{3}}{3} - \frac{\sqrt{6}}{3}$$

Exercise 9.1

1 Rationalise the denominators:

a $\dfrac{1}{\sqrt{5}}$ **b** $\dfrac{1}{\sqrt{2}}$ **c** $\dfrac{1}{\sqrt{7}}$ **d** $\dfrac{1}{\sqrt{8}}$

e $\dfrac{2}{\sqrt{3}}$ **f** $\dfrac{3}{\sqrt{5}}$ **g** $\dfrac{4}{\sqrt{8}}$ **h** $\dfrac{7}{\sqrt{7}}$

i $\dfrac{1}{3\sqrt{5}}$ **j** $\dfrac{1}{7\sqrt{3}}$ **k** $\dfrac{1}{8\sqrt{8}}$ **l** $\dfrac{1}{9\sqrt{7}}$

m $\dfrac{2}{3\sqrt{5}}$ **n** $\dfrac{5}{3\sqrt{7}}$ **o** $\dfrac{6}{5\sqrt{8}}$ **p** $\dfrac{2}{7\sqrt{5}}$

q $\dfrac{\sqrt{5}}{\sqrt{2}}$ **r** $\dfrac{\sqrt{7}}{\sqrt{3}}$ **s** $\dfrac{3\sqrt{2}}{\sqrt{5}}$ **t** $\dfrac{5\sqrt{3}}{3\sqrt{5}}$

2 Rationalise the denominator in each fraction:

a $\dfrac{1+\sqrt{3}}{\sqrt{3}}$ **b** $\dfrac{\sqrt{5}+2}{\sqrt{2}}$ **c** $\dfrac{\sqrt{3}-4}{\sqrt{3}}$ **d** $\dfrac{\sqrt{5}-2}{\sqrt{5}}$

e $\dfrac{\sqrt{2}+3}{\sqrt{5}}$ **f** $\dfrac{3\sqrt{3}+4}{\sqrt{2}}$ **g** $\dfrac{3\sqrt{3}+\sqrt{2}}{\sqrt{2}}$ **h** $\dfrac{2\sqrt{5}+\sqrt{3}}{2\sqrt{2}}$

Harder examples

Note that $(a+\sqrt{b})(a-\sqrt{b}) = a^2 - (\sqrt{b})^2 = a^2 - b$ (the difference of two squares)

$a+\sqrt{b}$ and $a-\sqrt{b}$ are known as **conjugates**.
When multiplied together their product is rational.
This can be made use of in the following cases.

Example 1   Rationalise the denominator of $\dfrac{1}{2+\sqrt{3}}$.

Here we multiply numerator and denominator by the conjugate of the denominator,

in this case by $2-\sqrt{3}$: $\dfrac{2-\sqrt{3}}{2-\sqrt{3}} = 1$, so we are multiplying $\dfrac{1}{2+\sqrt{3}}$ by 1.

$$\dfrac{1}{2+\sqrt{3}} \times \dfrac{2-\sqrt{3}}{2-\sqrt{3}} = \dfrac{1.(2-\sqrt{3})}{(2+\sqrt{3})(2-\sqrt{3})} = \dfrac{2-\sqrt{3}}{2^2-(\sqrt{3})^2} = \dfrac{2-\sqrt{3}}{4-3} = 2-\sqrt{3}$$

Example 2   Rationalise the denominator of $\dfrac{2}{3-\sqrt{5}}$.

$$\dfrac{2}{3-\sqrt{5}} = \dfrac{2}{(3-\sqrt{5})} \cdot \dfrac{(3+\sqrt{5})}{(3+\sqrt{5})} = \dfrac{6+2\sqrt{5}}{3^2-(\sqrt{5})^2} = \dfrac{6+2\sqrt{5}}{4} = \dfrac{3+\sqrt{5}}{2}$$

Exercise 9.2

1 Rationalise these denominators:

a $\dfrac{1}{2 + \sqrt{2}}$
    b $\dfrac{1}{1 + \sqrt{5}}$
    c $\dfrac{1}{3 + \sqrt{7}}$
    d $\dfrac{1}{5 + \sqrt{3}}$

e $\dfrac{2}{3 - \sqrt{5}}$
    f $\dfrac{4}{4 - \sqrt{2}}$
    g $\dfrac{6}{5 - \sqrt{3}}$
    h $\dfrac{7}{6 - \sqrt{2}}$

i $\dfrac{2}{\sqrt{2} - 1}$
    j $\dfrac{6}{\sqrt{5} + 2}$
    k $\dfrac{4}{\sqrt{3} - 1}$
    l $\dfrac{7}{\sqrt{2} - 2}$

2 Rationalise these denominators:

a $\dfrac{1}{\sqrt{3} + \sqrt{2}}$
    b $\dfrac{1}{\sqrt{5} - \sqrt{2}}$
    c $\dfrac{1}{\sqrt{8} + \sqrt{2}}$
    d $\dfrac{1}{\sqrt{7} - \sqrt{3}}$

e $\dfrac{3}{\sqrt{6} - \sqrt{2}}$
    f $\dfrac{5}{\sqrt{7} - \sqrt{3}}$
    g $\dfrac{2}{\sqrt{8} + \sqrt{3}}$
    h $\dfrac{5}{\sqrt{5} - \sqrt{3}}$

i $\dfrac{2}{\sqrt{5} - \sqrt{2}}$
    j $\dfrac{\sqrt{3}}{\sqrt{8} - \sqrt{2}}$
    k $\dfrac{4}{2\sqrt{5} - 1}$
    l $\dfrac{3}{3\sqrt{5} + 2}$

m $\dfrac{3}{3\sqrt{5} + 2\sqrt{2}}$
    n $\dfrac{4}{4\sqrt{2} - 2\sqrt{3}}$
    o $\dfrac{8}{3\sqrt{6} - 2\sqrt{2}}$
    p $\dfrac{5}{4\sqrt{7} - 2\sqrt{3}}$

Challenge

We have seen that surds such as $\sqrt{8}$ and $\sqrt{20}$ can be simplified by considering square factors.

Simplify these surds. (Hint: do not use squares.)

a $\sqrt[3]{24}$
    b $\sqrt[3]{81}$
    c $\sqrt[3]{320}$

d $\sqrt[4]{32}$
    e $\sqrt[4]{162}$
    f $\sqrt[5]{64}$

◀◀ RECAP

Definitions
  i $x^n = x.x.x.x.x \ldots .x.x.x$ (for n factors)
  ii x is known as the **base**.
  iii n is the **index**.

Basic facts
  i $x^1 = x$
  ii $x^0 = 1$
  iii $x^{-1} = \dfrac{1}{x}$

Basic rules **i** $x^a \times x^b = x^{a+b}$

ii $\dfrac{x^a}{x^b} = x^{a-b}$

iii $(x^a)^b = x^{ab}$

iv $x^{-a} = \dfrac{1}{x^a}$

v $\sqrt[b]{x} = x^{\frac{1}{b}}$

vi $\sqrt[b]{x^a} = x^{\frac{a}{b}} = (\sqrt[b]{x})^a$

Irrationals

An **irrational number** is one that cannot be expressed as the result of a division of one integer by another.

A surd is an irrational number which is expressed using the 'radical' sign ($\sqrt{\ }$).

Basic rules **i** $\sqrt{a \times b} = \sqrt{a} \times \sqrt{b}$

ii $\sqrt{\dfrac{a}{b}} = \dfrac{\sqrt{a}}{\sqrt{b}}$

Surds can be simplified by:

i extracting square factors, e.g. $\sqrt{80} = \sqrt{(16 \times 5)} = \sqrt{16} \times \sqrt{5} = 4\sqrt{5}$

ii reducing them, then adding or subtracting *like* terms, e.g.

$$\sqrt{98} - \sqrt{50} + \sqrt{8}$$
$$= \sqrt{2 \times 49} - \sqrt{2 \times 25} + \sqrt{2 \times 4}$$
$$= 7\sqrt{2} - 5\sqrt{2} + 2\sqrt{2}$$
$$= 4\sqrt{2}$$

iii combining them, e.g.

$$\frac{\sqrt{28}}{\sqrt{7}} = \sqrt{\frac{28}{7}} = \sqrt{4} = 2$$

Rationalising the denominator

i The denominator is a product, one of whose 'factors' is a surd.

In this case, multiply the numerator and denominator by the surd, e.g.

$$\frac{3}{2\sqrt{5}} = \frac{3}{2\sqrt{5}} \times \frac{\sqrt{5}}{\sqrt{5}} = \frac{3\sqrt{5}}{2 \times 5} = \frac{3\sqrt{5}}{10}$$

ii The denominator is of the form $a + b\sqrt{c}$.

Multiply the numerator and denominator by the conjugate surd, $a - b\sqrt{c}$, e.g.

$$\frac{4}{2 + \sqrt{3}} = \frac{4}{2 + \sqrt{3}} \times \frac{2 - \sqrt{3}}{2 - \sqrt{3}}$$

$$= \frac{4(2 - \sqrt{3})}{(2 + \sqrt{3})(2 - \sqrt{3})} = \frac{8 - 4\sqrt{3}}{2^2 - (\sqrt{3})^2}$$

$$= \frac{8 - 4\sqrt{3}}{4 - 3} = 8 - 4\sqrt{3}$$

1 Simplify, expressing your answers with positive indices:

a $x^9 \times x^6$ **b** $2x^8 \times 6x^5$ **c** $\dfrac{6x^8}{2x^5}$ **d** $\dfrac{x^7}{x^8}$ **e** $6x^{-8}$

f $\dfrac{5x}{x^6}$ **g** $(2x^2)^3$ **h** $(x^2)^{-3}$ **i** $\dfrac{3}{2x^{-3}}$ **j** $(3x^0)^4$

k $243^{-\frac{3}{5}}$ **l** $x^{\frac{2}{3}}(x^{\frac{1}{2}} + x^{\frac{1}{4}})$ **m** $2x^4(3x^{-3} + 2x^{\frac{1}{2}})$

2 Simplify, leaving your answers with negative indices:

a $x^3 \times x^8$ **b** $x^5 \div x^2$ **c** $(x^3)^{-5}$ **d** $(7x^0)^{-3}$

e $2 \div x^4$ **f** $5 \div (3x^5)$ **g** $6(x^5)^4$ **h** $3x \div 2x^{-3}$

3 Evaluate:

a $64^{\frac{5}{6}}$ **b** $128^{\frac{1}{7}}$ **c** $243^{-\frac{3}{5}}$ **d** $\dfrac{3}{125^{-\frac{2}{3}}}$

4 Calculate:

a $\sqrt{0{\cdot}0064}$ **b** $\sqrt{0{\cdot}000\,016}$

5 Simplify these surds:

a $\sqrt{250}$ **b** $\sqrt{32}$ **c** $\sqrt{147}$ **d** $\sqrt{150} + \sqrt{24}$

e $\sqrt{125} - 3\sqrt{50}$ **f** $\dfrac{\sqrt{50}}{\sqrt{94}}$ **g** $\sqrt{3}(\sqrt{8} - 2\sqrt{3})$ **h** $2\sqrt{5}(3\sqrt{5} - \sqrt{50})$

6 Rationalise the denominator in each case.

a $\dfrac{1}{\sqrt{7}}$ **b** $\dfrac{3}{2\sqrt{10}}$ **c** $\dfrac{\sqrt{5}}{3\sqrt{8}}$ **d** $\dfrac{1}{2 + \sqrt{3}}$

e $\dfrac{5}{\sqrt{5} - 1}$ **f** $\dfrac{\sqrt{3}}{7 + \sqrt{2}}$ **g** $\dfrac{\sqrt{7}}{\sqrt{8} + \sqrt{3}}$ **h** $\dfrac{\sqrt{5}}{3\sqrt{2} + 2\sqrt{3}}$

REVISE

10 Functions

The word **function** was first used by the German mathematician Leibniz in 1694.

The Belgian mathematician Dirichlet gave it its modern meaning in 1837.

Functions describe how one quantity relates to another.
Functions are a key part of mathematics.

A reconstruction of Leibniz's function machine

1 Review

◄◄ Exercise 1.1

1 Find the OUT entries in each of these:

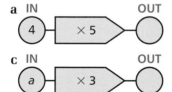

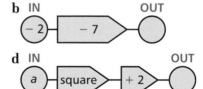

2 Find the value of $3x - 2$ when:

  a $x = 2$ **b** $x = 1$ **c** $x = 0$ **d** $x = -1$

3 Find the value of each expression when $x = 3$:

  a $2x - 7$ **b** $1 - 4x$ **c** 2^x **d** $(-5)^x$

4 Which of these equations represent a straight line? Which represent a quadratic? Which represent neither?

  a $y = 3x$ **b** $y = x^2 - 4$ **c** $y = 2 - 5x - 3x^2$

  d $x + y = 7$ **e** $x^2 y = 1$ **f** $x^2 - y = 6$

  g $xy = 5$ **h** $(x - 3)(x + 4) = y$ **i** $y + 4x = 0{\cdot}5$

5 **a** What is the gradient of the line $y = 3x - 2$?

  b What are the coordinates of the point where it cuts the y axis?

  c Where does it cut the x axis?

  d Sketch the line $y = 3x - 2$.

6 Sketch the graph of $y = -2x + 3$.

7 Solve, to 1 decimal place:

   a $x^2 + x - 6 = 0$ **b** $6x^2 - 7x - 3 = 0$ **c** $2x^2 + 5x - 1 = 0$

8 Find the value of $5 + 2x - 3x^2$ when:

   a $x = 1$ **b** $x = 2$ **c** $x = -3$

9 Heather drives off at the eighteenth.

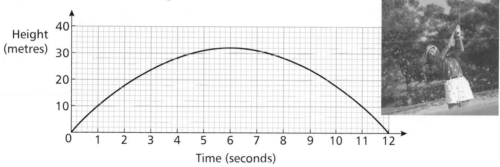

   a How long is the golf ball in the air?
   b How high does it go?
   c **i** When is the ball 20 metres above the ground?
      ii Why are there two answers?
   d What is the golf ball's height after 2 seconds?

10 Brad's parachute drifts down from a height of 1500 m.

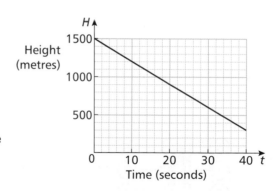

   a Write down:
      i the gradient of the line
      ii the point where it cuts the vertical axis
      iii the equation of the line.
   b Use the equation to calculate the time it takes Brad to reach the ground.

2 Illustrating a function

Class discussion

The **function** of the car assembly line is to take lots of different parts and put them together to make a motor car.

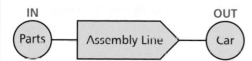

The **function** of a washing machine is to take dirty washing and turn it into clean washing.

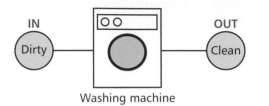

Washing machine

The chef's **function** is to take the milk, flour, eggs and sugar, and use the recipe to make pancakes.

Each process can be summed up as:

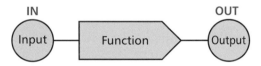

Each input has a definite output. For example, the chef would expect her ingredients and recipe to produce pancakes, not custard!

Example   Let *f* be a mathematical function that means 'add 6'.
The input set of numbers is {0, 1, 2, 3, 4}.
The output set is {6, 7, 8, 9, 10}.
Use diagrams to show how *f* links the input with the output.

There are three common ways to do this.

i function machines   **ii** arrow diagram   **iii** functional notation

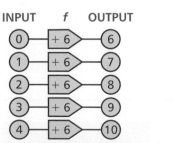

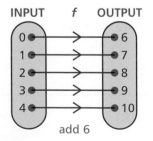

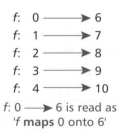

The input set is called the **domain**.
The output set is called the **range**.

The domain of this example is {0, 1, 2, 3, 4}. The range is {6, 7, 8, 9, 10}.

A **function** from a set A to a set B is a rule that links each member of A to **one** member of B, i.e. a rule which links each member of the domain to one member of the range.

Exercise 2.1

1 For each function diagram, list the set of numbers in each output:

a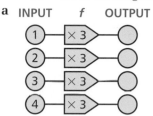

b
f: 0 ⟶
f: 2 ⟶
f: 4 ⟶
f: 6 ⟶
f: 8 ⟶

f means subtract 5

c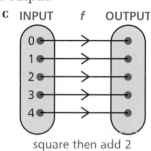

square then add 2

2 List the set of numbers in each input:

a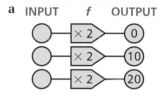

b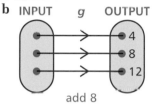

add 8

c
h: ⟶ 1
h: ⟶ 3
h: ⟶ 5

h means divide by 5

3 Find the function for each of these mapping diagrams.

a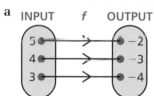

b
g: 48 ⟶ 6
g: 40 ⟶ 5
g: 24 ⟶ 3
g: 16 ⟶ 2

c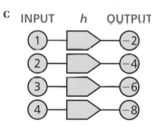

4 You pay 80 pence for each can of juice.
Copy and complete the diagram.

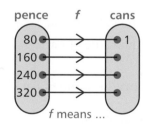 

f means ...

5 Copy and complete this mapping diagram:

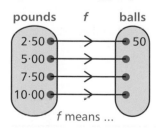 

f means ...

DOUGIE'S DRIVING RANGE

50 balls
for £2.50

6 Draw and complete mapping diagrams like the one in question **1c** for the following data:

  a Input {0, 1, 2, 3}, function f 'square'

  b Input {10, 5, 0}, function g 'add -3'

  c Domain {1, 3, 5, 7}, range {$-4, -12, -20, -28$}, function h

  d Domain {$-6, -4, -2$}, range {0, 2, 4}, function p

  e Function f 'add 2', output {$-4, -2, 0$}

  f Function h 'double and add 3', output {3, 9, 15, 21}

7 The function f is defined by: $f : 1 \rightarrow 26, 2 \rightarrow 27, 3 \rightarrow 28$.

  a List the set of numbers in   **i** the domain   **ii** the range.

  b Describe a function which links the two sets.

  c Illustrate the function using the three types of diagram in question **1**.

8 Describe a suitable rule for each function.

  a $f : 0 \rightarrow 0, 2 \rightarrow 4, 4 \rightarrow 16, 6 \rightarrow 36$

  b $g : 2 \rightarrow 1, 4 \rightarrow 2, 6 \rightarrow 3$

  c $h : 4 \rightarrow -3, 3 \rightarrow -4, 2 \rightarrow -5, 1 \rightarrow -6$

---

3 Notation

Using the function f 'add 4', $f : 1 \rightarrow 5$

                                $f : 2 \rightarrow 6$

                                  $f : 3 \rightarrow 7$

                                  ....

                                  ....

We can generalise this as $f : x \rightarrow x + 4$

f can be defined by the formula $f(x) = x + 4$.

We read this as 'f of x equals $x + 4$' or 'the function of x is $x + 4$' or 'the image of x under f is $x + 4$'.

          $f(1) = 1 + 4 = 5$ ... 5 is the value of f at 1

          $f(2) = 2 + 4 = 6$ ... 6 is the value of f at 2

          ......

          $f(a) = a + 4$ ... $a + 4$ is the value of f at a

Example 1   **a** $g(x) = x^2 - 3$. Find the value of g at

             i $x = 1$ **ii** $x = 2$.

          b $h(x) = 4x - 1$. For what value of x is $h(x) = -9$?

          a **i** $g(x) = x^2 - 3 \Rightarrow g(1) = 1^2 - 3 = 1 - 3 = -2$

              ii $g(2) = 2^2 - 3 = 4 - 3 = 1$

          b $h(x) = -9 \Rightarrow 4x - 1 = -9$

                     $\Rightarrow 4x = -8$

                     $\Rightarrow x = -2$

Exercise 3.1

1 **a** $f(x) = x - 5$. Find the value of:
   i $f(8)$ **ii** $f(5)$ **iii** $f(2)$
   b $g(x) = 3x$. Find the value of:
   i $g(4)$ **ii** $g(0)$ **iii** $g(-2)$

2 **a** $f(x) = 2x + 5$. Find the value of:
   i $f(1)$ **ii** $f(3)$ **iii** $f(-1)$
   b $g(x) = x^2 + 1$. Find the value of:
   i $g(1)$ **ii** $g(3)$ **iii** $g(5)$

3 **a** $t : x \to 10 - x^2$. Find the value of:
   i $t(1)$ **ii** $t(3)$ **iii** $t(-2)$
   b $h : x \to 6 - 3x$. Find the value of:
   i $h(1)$ **ii** $h(2)$ **iii** $h(-1)$

4 For every hour that passes, a plane travels 500 miles.
   The formula for this is $d(x) = 500x$ where d is
   the distance travelled in miles and x is the
   time in hours.
   a Calculate
   i $d(3)$ **ii** $d(5)$ **iii** $d(7)$
   b $d(3)$ means the distance travelled in 3 hours.
   What is the meaning of $d(7)$?

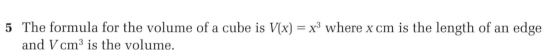

5 The formula for the volume of a cube is $V(x) = x^3$ where x cm is the length of an edge
   and V cm^3 is the volume.
   a Calculate
   i $V(2)$ **ii** $V(5)$
   b Explain the meaning of $V(3)$.

6 The formula for the area of a circle is $A(r) = \pi r^2$.
   a Approximate π with $\frac{22}{7}$ and calculate
   i $A(7)$ **ii** $A(1)$.
   b Explain the meaning of $A(7)$.

7 $f(x) = 4x$. For what value of x is $f(x) = 2$?

8 $g(x) = 1 + 3x$. For what value of x is $g(x) = 16$?

9 $h(x) = \dfrac{24}{x}$. For what value of x is $h(x) = 6$?

10 $f : x \to 3 - x$. If $f(a) = 9$, find the value of a.

11 $p : x \to 1 - 2x$. If $p(t) = 7$, find t.

Using the notation

Example 2 $f(x) = 4x^2$
  a Find the value of f at $x = 3$.
  b Find a formula for
    i $f(t)$ **ii** $f(2t)$ **iii** $f(t - 1)$.

  a $f(3) = 4 \times 3^2 = 4 \times 9 = 36$
  b i $f(t) = 4 \times t^2 = 4t^2$
    ii $f(2t) = 4 \times (2t)^2 = 4 \times 4t^2 = 16t^2$
    iii $f(t - 1) = 4(t - 1)^2 = 4(t^2 - 2t + 1) = 4t^2 - 8t + 4$

Example 3 $g(x) = 2x - 3$. Find a formula for $g(k) + g(-k)$.
    $g(k) + g(-k) = (2k - 3) + (2(-k) - 3)$
          $= 2k - 3 - 2k - 3 = -6$

Example 4 $f : x \to 3x - 1$
  a Find a formula for $f(a)$.
  b Find the value of f at $a =$
    i 0   **ii** 1.

  a $f(a) = 3a - 1$
  b i $f(0) = 3 \times 0 - 1 = 0 - 1 = -1$
    ii $f(1) = 3 \times 1 - 1 = 3 - 1 = 2$

Exercise 3.2

1 $f(x) = x^2 - 4x$. Find:
  a a formula for: **i** $f(t)$ **ii** $f(a)$
  b the value of f at $x =$ **i** 5   **ii** 1.

2 $f(x) = x^2 - 3x + 4$.
  Find the value of f at $x =$
  a 0   **b** 1   **c** -1   **d** 3.

3 $f : x \to 1 - x^2$.
  a Find a formula for $f(a)$.
  b Find the value of f at $x =$ **i** 1   **ii** 2   **iii** -2.

4 For every hour a swimming pool heating system
  is switched on, the water temperature rises 3 °C.
  Therefore the water temperature, T °C, is a function
  of the number of hours, x, the system is switched on.
  The formula is $T(x) = 12 + 3x$.
  a Calculate: **i** $T(2)$ **ii** $T(5)$.
  b For how many hours is the system switched
    on for the water temperature to be 33 °C?

5 The number of CDs, N, Eve can buy with her money is a function of their cost, £x.

The formula is $N(x) = \dfrac{72}{x}$.

a Calculate:

   i $N(6)$ **ii** $N(8)$ **iii** $N(12)$.

b i How much money does Eve have?

   ii Explain in a sentence the meaning of $N(8)$.

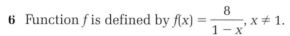

6 Function f is defined by $f(x) = \dfrac{8}{1 - x}$, $x \neq 1$.

a Why can x not be equal to 1?

b Calculate:

   i $f(0)$ **ii** $f(2)$ **iii** $f(-1)$ **iv** $f(\tfrac{1}{2})$ **v** $f(-\tfrac{1}{2})$.

c For what value of x is $f(x) = 2$?

7 $h(x) = \dfrac{2}{x^2}$. Find:

   a $h(2)$ **b** $h(\tfrac{1}{2})$ **c** $h(\tfrac{1}{3})$.

8 $f(x) = 3x$. Find a formula for:

   a $f(2a)$ **b** $f(a - 1)$.

9 $g: x \rightarrow 2x + 1$. Find a formula for:

   a $g(2a)$ **b** $g(a + 1)$.

10 $h: x \rightarrow x^2$. Find a formula for:

   a $h(2t)$ **b** $h(t + 3)$.

11 $f: x \rightarrow 1 - 3x$. Find $f(a) + f(-a)$ in its simplest form.

12 $f(x) = ax + b$. $f(1) = 8$ and $f(2) = 14$.

   a Write down two equations in a and b.

   b Solve the equations to find the values of a and b.

   c Write down the formula for f.

13 $g(x) = px + q$. $g(4) = 10$ and $g(6) = 16$.

   a Write down two equations in p and q.

   b Solve the equations to find the values of p and q.

   c Write down the formula for g.

14 $t: x \rightarrow ax + b$. $t(-3) = 8$ and $t(4) = -6$.

   a Write down two equations in a and b.

   b Solve the equations to find the values of a and b.

   c Write down the formula for t.

4 Graphs of linear and quadratic functions

A graph gives a picture of a function.

It shows the link between the numbers in the input (or **domain**) and the output (or **range**).

A Linear functions

The straight line shown here is the graph of the function f defined by $f(x) = \frac{1}{2}x$, for $0 \leqslant x \leqslant 6$.

$y = \frac{1}{2}x$ is called the equation of the graph of the function.

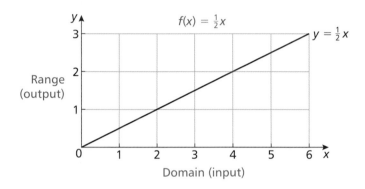

$$f(x) = \frac{1}{2}x$$

Range (output)

Domain (input)

$y = \frac{1}{2}x$

A function of the form $f(x) = ax + b$ is a **linear function**.
Its graph is a straight line with equation $y = ax + b$.

B Quadratic functions

The parabola shown here is the graph of the function f defined by $f(x) = x^2 - 1$, for $-4 \leqslant x \leqslant 4$. Its equation is $y = x^2 - 1$.

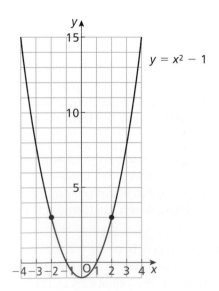

$y = x^2 - 1$

From the graph we see:
i   the minimum value of f is -1, at $x = 0$
ii   the minimum turning point (TP) is $(0, -1)$
iii $f(x) = 0$ at $x = -1$ and $x = 1$
iv   the axis of symmetry of the parabola is the line $x = 0$.

A function of the form $f(x) = ax^2 + bx + c$, $a \neq 0$, is a **quadratic function**, and its graph is a parabola with equation $y = ax^2 + bx + c$.

Exercise 4.1

1 Draw the graph of the function f, where $f(x) = 3 - x$, for $-4 \leqslant x \leqslant 4$.
Its equation is $y = 3 - x$. First copy and complete this table.

x	−4	−3	−2	−1	0	1	2	3	4
y									

2 Draw the graph of the function h, where $h(x) = 2x + 1$, for $-3 \leqslant x \leqslant 3$.
Its equation is $y = 2x + 1$. First copy and complete the table.

x	−3	−2	−1	0	1	2	3
y							

3 *Sketch* the graphs of these linear functions.
Remember that you only need to plot two points, and draw the straight line through them. However, to avoid mistakes, plot a third point and check that it lies on the line.
a $f(x) = 3x - 1$ **b** $g(x) = 8 - 2x$ **c** $h(x) = 4x - 3$

4 a Draw the graph of $y = x^2$, for $-3 \leqslant x \leqslant 3$, using this table of values:

x	−3	−2	−1	0	1	2	3
y							

    b Write down the coordinates of the parabola's turning point.
    c What is the equation of its axis of symmetry?

5 a Draw the graph of $y = x^2 + 3$, for $-4 \leqslant x \leqslant 4$, for the function $f(x) = x^2 + 3$.
Use a table like this:

x	−4	−3	−2	−1	0	1	2	3	4
y									

    b Write down:
      i the minimum value of f
      ii the coordinates of the minimum turning point
      iii the equation of the axis of symmetry of the parabola.
    c Is $f(x)$ ever equal to zero?

6 a Draw the graph of the function $f(x) = 5 - x^2$, for $-4 \leqslant x \leqslant 4$.
Use a table like the one in question **5**.
    b Write down
      i the maximum value of f
      ii the coordinates of the maximum turning point
      iii the equation of the axis of symmetry.
    c For what values of x is $f(x) = 0$?

7 Use the graph of the function
$f(x) = 4x^2 - 12x - 16$ to write down:
  a the coordinates of the minimum turning point
  b the minimum value of f
  c the equation of the line of symmetry
  d the values of x for which:
    i $f(x) = 0$
    ii $f(x) = -16$.

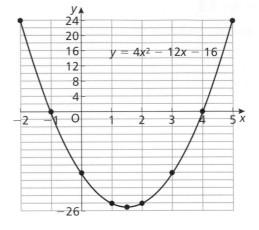

8 Use the graph of the function
$g(x) = 16 + 4x - 2x^2$ to write down:
  a the coordinates of the maximum turning point
  b the maximum value of g
  c the equation of the line of symmetry
  d the values of x for which
    i $g(x) = 0$
    ii $g(x) = 16$.

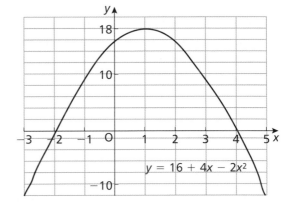

9 a Copy and complete the table for the graph of the function $f(x) = 10 - 3x - x^2$, for $-6 \leqslant x \leqslant 3$.
(By calculator, when $x = -6$, $y = 10 - 3 \times (-6) - (-6)^2 = -8$.)

x	−6	−5	−4	−3	−2	−1	0	1	2	3
y										

b Write down:
  i the maximum or minimum value of f and the coordinates of the turning point
  ii the values of x for which $f(x) = 0$.

For questions **10** to **13** draw the graphs of the functions and write down the same information as you gave for parts **b i** and **ii** of question **9**.

10 $h(x) = x^2 - 3x$,   for $-1 \leqslant x \leqslant 4$.

11 $p(x) = 4 - 4x^2$,   for $-2 \leqslant x \leqslant 2$.

12 $q(x) = x^2 - 6x + 9$,   for $0 \leqslant x \leqslant 6$.

13 $t(x) = 3 + 2x - x^2$,   for $-2 \leqslant x \leqslant 4$.

Exercise 4.2

1 The graph shows how height is related to time in the flight path of a golf ball.

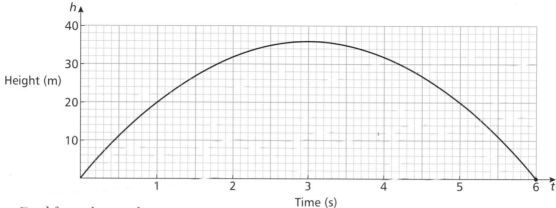

Read from the graph:
a the maximum height of the golf ball and the time taken to reach this height
b the time the ball was in the air
c the ball's height after
  i 4 seconds   **ii** 0·5 second
d after what times the height of the ball was at 20 metres.

2 Sal bungee jumps a distance of d metres in t seconds as shown in the table.

t	0	1	2	3	4
d	0	5	20	45	80

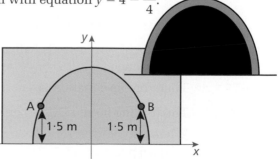

a Draw the graph of d against t for $0 \leqslant t \leqslant 4$, on 2 mm paper using the scales shown.
b Read from the graph:
  i d when $t = 2\cdot5$ **ii** t when $d = 30$.
c The curve is part of the parabola $d = 5t^2$.
  Calculate:
  i d when $t = 1\cdot5$ **ii** t when $d = 40$.

3 The mouth of a tunnel is a parabolic arch with equation $y = 4 - \dfrac{x^2}{4}$. The units are in metres.

a Draw the graph of $y = 4 - \dfrac{x^2}{4}$ in the interval $-4 \leqslant x \leqslant 4$.

b From your graph read the y value for x equal to
  i $-2\cdot5$ **ii** $3\cdot5$.
c What is the maximum height of the tunnel?
d What is the width of the tunnel at road level?
e Lights are placed at A and B.
  Find from the graph the distance between the lights.

4 The braking distance, D metres, for a new type of car travelling at s km/h is given by the formula $D = \dfrac{s^2}{200}$.

a Copy and complete this table:

s	0	20	40	60	80	100
D						

b Draw the graph of $D = \dfrac{s^2}{200}$, using scales of 20 km/h to 2 cm horizontally, and 5 m to 1 cm vertically.

c Use your graph to find:
    i the braking distance at 50 km/h and 90 km/h
    ii speeds for braking distances of 25 m and 40 m.

5 Sketching quadratic functions

Step 1	Find where the parabola cuts the x axis, i.e. the value(s) of x where $f(x) = 0$.
Step 2	Find the equation of its axis of symmetry.
Step 3	Find the turning point (TP) and determine if it is a maximum or minimum TP.
Step 4	Find where the curve cuts the y axis, i.e. the value of y for $x = 0$.

Mark these features on your sketch and they will act as a guide for drawing the curve.

Example   Sketch the quadratic $f(x) = 15 - 2x - x^2$.

Step 1
$$15 - 2x - x^2 = 0$$
$$\Rightarrow \quad (5 + x)(3 - x) = 0$$
$$\Rightarrow \quad 5 + x = 0 \text{ and } 3 - x = 0$$
$$\Rightarrow \quad x = -5 \text{ and } x = 3$$

So the curve cuts the x axis at $(-5, 0)$ and $(3, 0)$.

Step 2   The axis of symmetry is halfway between these two points so it must pass through $(-1, 0)$.

Therefore its equation is $x = -1$.

Step 3   The turning point (TP) lies on the line $x = -1$.

When $x = -1$, $y = 15 - 2 \times (-1) - (-1)^2 = 16$.

So the TP is $(-1, 16)$.

Also, as $(-5, 0)$ and $(3, 0)$ lie on the curve, and 0 is less than 16, $(-1, 16)$ must be a maximum TP.

Step 4   When $x = 0$, $y = 15$. So the y axis is cut at $(0, 15)$.

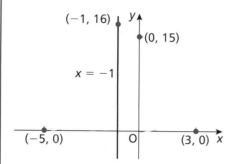

Put in your findings...

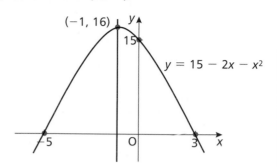

... and draw a smooth, symmetrical curve.

Exercise 5.1

1 Sketch the graphs of these quadratic functions:

a $f(x) = x^2 - 4$

b $f(x) = 9 - x^2$

c $f(x) = x^2 - 2x - 15$

d $f(x) = x^2 - 2x - 8$

e $f(x) = x^2 + 4x$

f $f(x) = 24 + 2x - x^2$

g $f(x) = x^2 - 7x + 6$

h $f(x) = x^2 - 3x$

i $f(x) = 5x - x^2$

j $f(x) = x^2 - 6x + 9$

k $f(x) = 4x^2 + 6x - 4$

l $f(x) = 4x^2 + 4x - 15$

m $f(x) = -x^2 + 4x - 4$

n $f(x) = -16 - 8x - x^2$

2 a Sketch the graphs of:

i $y = x^2$

ii $y = -x^2$

iii $y = x^2 - 1$

iv $y = 1 - x^2$

v $y = 2x^2$

vi $y = x^2 - 9$

b Explain how each can be obtained from the graph of $y = x^2$.

c Sketch the graphs of:

i $y = x^2 + 1$

ii $y = x^2 + 3$

iii $y = -x^2 - 1$

iv $y = 4 - x^2$

Brainstormer

Look at these two sketches.

a For what values of x on the parabolas is:

i $f(x) = 0$

ii $f(x) > 0$

iii $f(x) < 0$

iv $f(x)$ increasing as x increases

v $f(x)$ decreasing as x increases?

b Write down a *possible* equation for each graph.

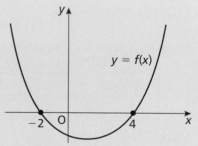

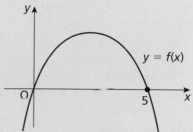

6 The function $f(x) = \dfrac{a}{x}$, $x \neq 0$

The function $f(x) = \dfrac{a}{x}$ is the simplest form of a **reciprocal** function.

This table of values is for the function $f(x) = \dfrac{24}{x}$ (i.e. for $a = 24$).

x	−24	−12	−8	−6	−4	−3	−2	−1	0	1	2	3	4	6	8	12	24
y	−1	−2	−3	−4	−6	−8	−12	−24	−	24	12	8	6	4	3	2	1

Note that x cannot be zero with this function as division by zero is undefined.

The graph of $y = \dfrac{24}{x}$ (called a **hyperbola**) is divided into two **branches**; one for $x < 0$ and the other for $x > 0$.

The further you go along the axes in either direction, the closer the curve approaches the axes.

The axes are called **asymptotes** to the curve.

(*Note:* the **reciprocal** of x is $\dfrac{1}{x}$. Hence the name of the function.)

$y = \dfrac{a}{x}$ or $xy = a$, where a is a constant, indicates that the two variables x and y are in inverse proportion or vary inversely with each other.

Note that the curve has two axes of symmetry at 45° to the coordinate axes.

Exercise 6.1

1 **a** Draw the graph with equation $y = \dfrac{36}{x}$ by first copying and completing this table.

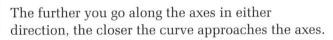

x	−36	−24	−12	−9	−6	−4	−3	−2	−1	1	2	3	4	6	9	12	24	36
y																		

  b Mark:
   i the axes of symmetry
   ii the centre of symmetry.
  c Calculate y when $x = 100, 1000, 1\,000\,000$.
   Copy and complete:
   'As x increases, y …, and the graph gets closer and closer to …'.
  d Calculate y when $x = 0.01, 0.001, 0.000\,001$.
   Copy and complete:
   'As x decreases, y …, and the graph gets …'.
  e Write down the equations of the asymptotes.

2 a This time the equation of the hyperbola is disguised as $xy = 16$.
Draw the graph by first completing this table.

x	−16	−8	−4	−2	−1	1	2	4	8	16
y										

b Describe any lines or points of symmetry.

3 Draw the graph of the hyperbola $xy = -16$ by making a table similar to the one in question **2**.

4 A new motorbike is test-driven round a circuit at different speeds and the times noted.

Speed (s m/s)	8	10	15	20	30	40	60
Time (t seconds)	90	72	48	36	24	18	12

a Draw the graph of t against s, for $8 \leqslant s \leqslant 60$.
(Make the scale 10 m/s to 2 cm on the s axis, 10 seconds to 1 cm on the t axis.)
b From your graph estimate the time for a speed of:
  i 35 m/s   **ii** 50 m/s
c From your graph estimate the speed for a time of 30 seconds.
d What is the length of the circuit?
e Write down the equation of the graph.

5 Terri has a fixed amount of money with which to buy garden tubs.
The table shows the number of tubs she can buy at the different prices.

Price per tub (£P)	6	10	12	15	20	30
Number of tubs (N)	30	18	15	12	9	6

a Draw the graph of N against P for $6 \leqslant P \leqslant 30$.
b From your graph estimate the price of the tubs if 25 of them can be bought with the money.
c How much money does Terri have?
d What is the equation of the hyperbola?

6 A new rectangular playpark has to be built. Its area is 1200 m².
Its length is L m and its breadth is B m.
The council planner drew up this table:

Length (L m)	15	20	25	30	40	48	50	60	80
Breadth (B m)	80	60							

a Copy and complete the table.
b Draw the graph of B against L.
c What is the equation of the hyperbola?

7 The function $f(x) = a^x$, for $a = 1, 2, 3, \ldots$

The function $f(x) = a^x$ where a is a constant and a
natural number (i.e. 1, 2, 3, 4, …) is called an
exponential function.

This table of values is for the function $f(x) = 2^x$.

x	-3	-2	-1	0	1	2	3
y	2^{-3}	2^{-2}	2^{-1}	2^0	2^1	2^2	2^3
y	$\frac{1}{8}$	$\frac{1}{4}$	$\frac{1}{2}$	1	2	4	8

The graph of $y = a^x$ is an **exponential curve**.

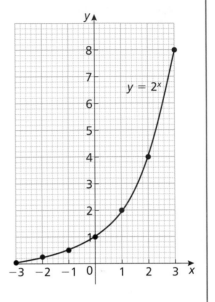

Reminder:

The y^x key on your calculator
calculates powers.
For example, to evaluate 3^5, key in

  to get 243.

Exercise 7.1

1 a Draw the graph with equation $y = 3^x$, by making a table of values with $-1 \leqslant x \leqslant 3$.
  b From your graph estimate the value of the function when x equals
    i 0·5   **ii** 2·6.

2 a Write down the value of:
    i 2^{10} **ii** 3^7 **iii** 5^5 **iv** 9^4
  b Check on your calculator that the value of each of these is 1:
    i 1^0 **ii** 3^0 **iii** 5^0 **iv** 16^0
  c Check on your calculator that these produce reciprocal values:
    i 1^{-1} **ii** 2^{-1} **iii** 3^{-1} **iv** 4^{-1}

3 a On the same graph as for $y = 3^x$ draw the graph of $y = 2^x$ with $-1 \leqslant x \leqslant 4$.
  b Estimate from your graph the value of:
    i $2^{1\cdot5}$ **ii** $2^{3\cdot5}$ **iii** $2^{3\cdot8}$
  c Check the accuracy of your answers to **b** with your calculator.

4 $f(x) = 4^x$. Use your calculator to evaluate:
    a $f(4)$ **b** $f(6)$ **c** $f(1\cdot2)$ **d** $f(-1\cdot2)$

5 $g : x \rightarrow 5^x$. Evaluate:
    a $g(0\cdot5)$ **b** $g(1\cdot3)$ **c** $g(2\cdot6)$ **d** $g(-1\cdot5)$.

6 A single bacterium is isolated. After 1 second it becomes two bacteria.
Each second the number of bacteria doubles.
  a How many of the bacteria are there after
    i 1 second   **ii** 2 seconds   **iii** 3 seconds   **iv** 4 seconds   **v** 10 seconds?
  b Express the situation as an exponential equation where N is the number of bacteria
    t seconds after the observations began.

7 A financial wizard trebles her money every year. After one year she has £3.
  a How much does she have after 5 years?
  b Express this situation as an exponential equation where £A is the amount y years
    after the investment began.
  c How much did she begin with ($y = 0$)?

Challenge

Alec's dad wants him to remove a large pile of boulders from their
garden. Alec agreed provided his dad paid him 2p for the first
boulder, 4p for the second boulder, 8p for the third, 16p for the
fourth, and so on, doubling the amount each time.

Express the situation as an exponential equation.
Should Alec's dad agree to this?

How much would it cost him if Alec removed
a 20 boulders   **b** 30 boulders   **c** 50 boulders?

8 Functions and graphs (miscellaneous)

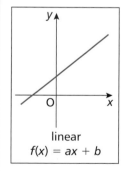

linear
$f(x) = ax + b$

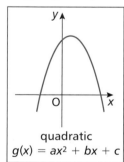

quadratic
$g(x) = ax^2 + bx + c$

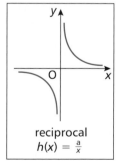

reciprocal
$h(x) = \frac{a}{x}$

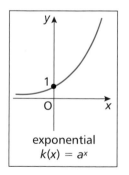

exponential
$k(x) = a^x$

Exercise 8.1

1 Examine the above graphs and check that each function is linked to the correct
graph.

2 Which function would you link with each of these?

a $y = \dfrac{5}{x}$ **b** $f : x \rightarrow x^2 - 6x + 5$ **c** $y = 1 + 2x$

d $g(x) = \frac{1}{2}x + 4$ **e** $h(x) = 6^x$ **f** $xy = 10$

g $f : x \rightarrow 10^x$ **h** $y = 3 - 4x + x^2$ **i** $m(x) = 2x^2 - 11x + 12$

j $3x - 4y + 1 = 0$ **k** $3y = 2x$ **l** $f(x) = (4x - 1)(x - 3)$

---

9 Mathematical models

Scientists looking for a connection between two or more quantities:
i collect data, using experiments, surveys, measurements, etc.
ii organise the data in tables and graphs
iii analyse the data, perhaps by matching it with graphs like the ones you have been studying
iv use the results to predict further results.

Example
i Fast Car Research are testing a new racing car. The table shows the data collected.

Time (x seconds)	0	0·5	1·0	1·5	2·0	2·5
Distance (y m)	0	2	8	18	32	50

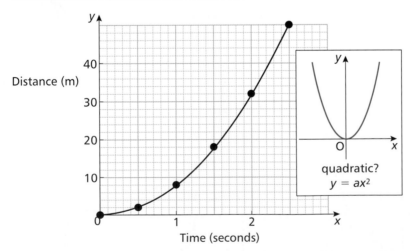

ii They draw a graph of the data and decide that it is like part of the parabola $y = ax^2$.
iii They then choose the point (1, 8) on the graph, and put it in the equation $y = ax^2$, giving $8 = a \times 1^2 \Rightarrow a = 8$.
So the equation of their graph is $y = 8x^2$.
They check this for other points on the graph, for example (2, 32): $32 = 8 \times 2^2$ ✓.
iv They are now able to *use* the equation.
At $t = 2·2$, $y = 8 \times 2·2^2 = 38·72$, so the car travels nearly 39 m in 2·2 seconds.

Exercise 9.1

1 For each sketch use the given data to find the equation of the graph.

a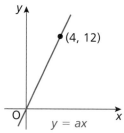

$y = ax$

b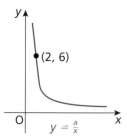

$y = \frac{a}{x}$

c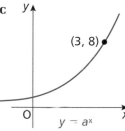

$y = a^x$

d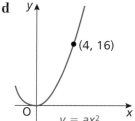

$y = ax^2$

e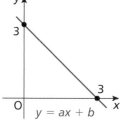

$y = ax + b$

f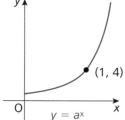

$y = a^x$

2 Chloe is experimenting to find the connection between *resistance* and *current* in an electrical circuit.

Resistance (x ohms)	0·5	1·0	1·5	2·0	2·5	5·0
Current (y amps)	24	12	8	6	4·8	2·4

a Draw a graph.
b Is it like one of the graphs in question **1**? Which one?
  Use the equation type of that graph to find the actual equation of Chloe's graph.
c From the equation calculate the current for a resistance of 4 ohms.
d Calculate the resistance for a current of 10 amps.

3 A marble is dropped into a liquid.

The results show the marble's speed against time.

Time (t seconds)	0	0·5	1·5	2·5	4·0
Speed (v m/s)	0	0·8	2·4	4·0	6·4

a Draw a graph.
b Which graph in question **1** is it like?
c Find the equation of the graph.
d Calculate the speed of the marble after
  i 1 second
  ii 3·5 seconds.
e Calculate how long it took the marble to reach a speed of 1 m/s.

4 In another experiment a stone is dropped from a height and it is timed as it falls to the ground.

Time (t seconds)	0	0·5	1·0	1·5	2·0	2·5	3·0
Distance travelled (d m)	0	2·5	10·0	22·5	40·0	62·5	90·0

a Draw the graph of d against t.
b Which graph in question **1** is it like?
c Determine the equation of the graph.
d Use the equation to calculate:
  i the distance fallen by the stone after 1·8 seconds
  ii the time taken to fall a distance of 50 metres.

5 Harbits are strange creatures that breed quickly.
Starting with one harbit, the population was counted every 24 hours.

Time (d days)	0	1	2	3	4
Number of harbits (n)	1	4	16	64	256

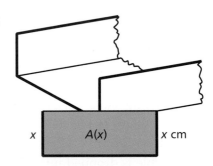

a Draw the graph of n against d.
b Determine the equation of the graph from its lookalike in question **1**.
c Use the equation to find the number of harbits there are likely to be after 6 days.

6 A sheet of metal 36 cm wide has its sides bent up at right angles to make a channel for carrying water. This channel has to be as big as possible to allow the maximum amount of water to be carried. The area of the cross-section depends on x, i.e. the area $A(x)$ is a function of x.
a A mathematical model of the situation is
$A(x) = x(36 - 2x)$.
Explain where this function comes from.
b Draw the graph of A against x, and find the dimensions of the channel with the greatest cross-sectional area.

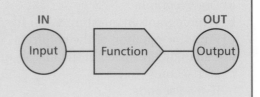

◀◀ RECAP

A **function** from a set A to a set B is a rule which links each member of A to one and only one member of B.

The input set is called the **domain**.
The output set is called the **range**.

IN   OUT

Input — Function ⟩ — Output

Example 1 The function 'add 4' maps the set {0, 3, 6, 9, 12} onto {4, 7, 10, 13, 16}. This can be represented in a diagram in several ways:

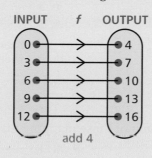

The function 'add 4' can be generalised as $f : x \rightarrow x + 4$, and f is defined by the formula $f(x) = x + 4$.

Example 2 $h(x) = 3x - 2$.

a What is the value of h at $x = 4$?

$h(4) = 3 \times 4 - 2 = 10$

b For what value of x is $h(x) = 7$?

$$h(x) = 3x - 2$$
$$\Rightarrow \quad 3x - 2 = 7$$
$$\Rightarrow \quad 3x = 9$$
$$\Rightarrow \quad x = 3$$

Graphs of some special functions

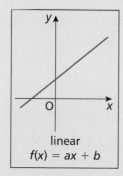

linear
$f(x) = ax + b$

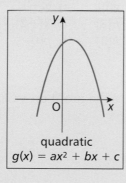

quadratic
$g(x) = ax^2 + bx + c$

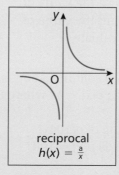

reciprocal
$h(x) = \frac{a}{x}$

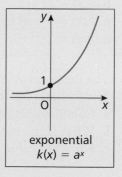

exponential
$k(x) = a^x$

Sketching quadratic functions

Step 1 Find where the parabola cuts the axes, i.e. the value(s) of x where $f(x) = 0$.

Step 2 Find the equation of the axis of symmetry.

Step 3 Find the turning point (TP) and determine if it is a maximum or minimum TP.

Step 4 Find where the curve cuts the y axis, i.e. the value of y for $x = 0$.

Example $y = x^2 - 2x - 3$

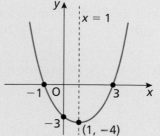

1 For each function diagram, list the set of numbers in each output.

a   INPUT f OUTPUT

b f: 0 ⟶

f: 3 ⟶

f: 5 ⟶

f means
double then add 1

2 Determine the function for each of these mapping diagrams.

a   INPUT g OUTPUT

g means

b h: 1 ⟶ 2

h: 2 ⟶ 5

h: 3 ⟶ 10

h: 4 ⟶ 17

3 Draw and complete a mapping diagram, like the one in question **1b**, for:
  a domain {0, 1, 2, 3}, function f 'square and subtract 1'
  b domain {1, 2, 3, 4, 5}, range {4, 7, 10, 13, 16}

4 **a** $f(x) = x - 4$.   Find the value of   **i** $f(2)$ **ii** $f(7)$.
   b $g(x) = 3x - 2$.   Find the value of   **i** $g(0)$ **ii** $g(-1)$.
   c $h(x) = 1 - x^2$.   Find the value of   **i** $h(2)$ **ii** $h(-2)$.
   d $p(x) = 3^x$.   Evaluate   **i** $p(3)$ **ii** $p(0)$ **iii** $p(1·5)$.

5 The profit on every CD sold in a shop is £4.
  The formula for this is $p(x) = 4x$ where £p is the profit on x CDs.
  a Calculate $p(100)$.
  b What is the meaning of $p(100)$?
  c For what value of x is $p(x) = 100$?

6 $f(x) = \dfrac{18}{x}$. For what value of x is $f(x) = 2$?

7 $g(x) = 2x$. Find a formula for
  a $g(a)$ **b** $g(2a)$ **c** $g(a - 3)$.

8 $h(x) = 4x - 3$. Determine $h(t + 1) + h(t - 1)$ in its simplest form.

9 $m(x) = \dfrac{3}{x} - 2$.
  a Find the value of $m(\frac{1}{2})$.
  b Given that $m(t) = 6$, find t.

10 From the graph of the function $f(x) = x^2 + 2x - 8$
write down:

a the coordinates of the minimum turning
point of the parabola

b the minimum *value* of f

c the equation of the axis of symmetry

d the values of x for which

   i $f(x) = 0$

   ii $f(x) = -8$.

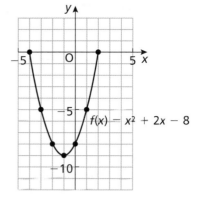

$f(x) = x^2 + 2x - 8$

11 a Sketch the graph of the quadratic function $f(x) = 15 - 2x - x^2$.

b For what values of x is $f(x) \geqslant 0$?

12 A rocket is tested at different speeds.
The times it takes to travel a distance of 4800 m are recorded:

Speed (v m/s)	20	30	60	100	120
Time (t seconds)	240	160	80		

a Copy and complete the table.

b Draw the graph of t against v.

c What is the shape of the graph?

d What is its equation?

e Use the equation to find the time taken by the rocket when its speed is 80 m/s.

13 Isa quadruples her savings every year. After one year she has saved £4.

a How much will she save over five years?

b The amount Isa has saved, $A(x)$, is a function of the number of years, x, she has
saved.

Write down the formula for the amount Isa has saved after x years.

14 Use the given data to find the actual equations of these graphs.

a
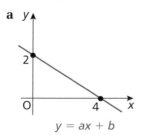

$y = ax + b$

b
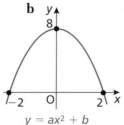

$y = ax^2 + b$

c
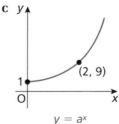

$(2, 9)$

$y = a^x$

d
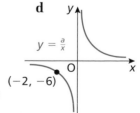

$y = \frac{a}{x}$

$(-2, -6)$

REVISE

11 Trigonometric graphs and equations

The Greek mathematicians and astronomers Hipparchus and Aristarchus are believed by many to be the inventors of trigonometry. They both lived in the second century BC.

The graph of sin x crops up in the representation of sound, light, radio and many other energy forms.

So today scientists and engineers make extensive use of the trigonometric functions.

1 Review

In right-angled triangle ABC:

$$\sin A = \frac{\text{opposite side}}{\text{hypotenuse}}$$

$$\cos A = \frac{\text{adjacent side}}{\text{hypotenuse}}$$

$$\tan A = \frac{\text{opposite side}}{\text{adjacent side}}$$

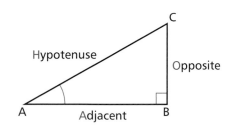

Throughout the chapter, unless otherwise instructed, give your answers correct to 3 significant figures where appropriate.

◄◄ Exercise 1.1

1 Write down the value of each of the following as a common fraction:
   a tan A   **b** sin C   **c** cos R   **d** tan P   **e** sin V   **f** cos W

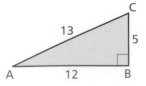
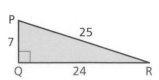
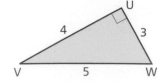

2 Calculate the value of *x* in each triangle.

a
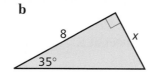
13
x
28°

b
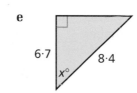
8
x
35°

c
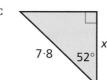
7·8
52°
x

d
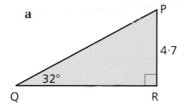
x°
9
7

e
6·7
8·4
x°

f
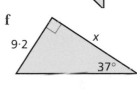
9·2
x
37°

3 Calculate the length of the hypotenuse in each right-angled triangle.

a
P
4·7
32°
Q
R

b
D
7·7
54°
E
F

4 A climber is lying injured on a hillside at C, 96 m above
sea level. The angle of elevation from the injured
climber to a rescue helicopter at H is 56° and the
distance between them is 145 m.

How high is the helicopter above sea level?

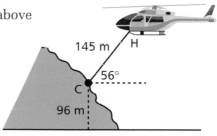

145 m
H
56°
C
96 m

5 In the kite ABCD, diagonal BD = 12 cm and AE = 5·0 cm.
  a Calculate the size of ∠ADB to the nearest degree.
  b ∠BDC = 60°. Calculate:
    i the lengths of the sides of the kite
    ii the length of diagonal AC.

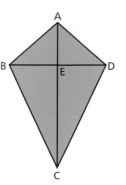

A
B
E
D
C

6 Two boats, A and B, are in line with D, the foot of a cliff.
The angle of depression from the top of the cliff at T to
boat A is 37° and to boat B is 58°.
The height of the cliff is 125 m.
Calculate the distance between the two boats.

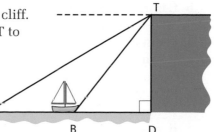

T
A
B
D

7 The *Sea Sprite* sets sail for 38 km on a course of 235° from a harbour at H. How far **a** west **b** south is it from the harbour, to the nearest kilometre?

8 A hole is drilled in the metal sheet at P. The coordinates of P are (2, 4). Another hole needs to be drilled at Q. Calculate the coordinates of Q, giving your answers to the nearest whole number.

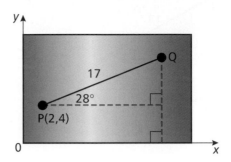

2 Exact values

Example   Given $\tan A = \frac{8}{15}$, and $0° < A < 90°$, calculate the *exact* value of sin A, cos A, sin B, cos B and tan B.

Since A is acute we can draw a right-angled triangle ABC, such that BC = 8 units and AC = 15 units.

By Pythagoras' theorem
$AB = \sqrt{(8^2 + 15^2)} = \sqrt{289} = 17$

Hence by inspection $\sin A = \frac{8}{17}$, $\cos A = \frac{15}{17}$, $\sin B = \frac{15}{17}$, $\cos B = \frac{8}{17}$ and $\tan B = \frac{15}{8}$.

Note that these are *exact* values and not decimal approximations.

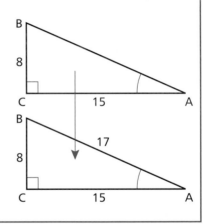

Exercise 2.1

1 **a** A is an acute angle such that $\tan A = \frac{12}{5}$.
   Find the exact value of
   i sin A
   ii cos A.
   b E is an acute angle such that $\cos E = \frac{3}{5}$.
   Find the exact value of
   i sin E
   ii tan E.
   c X is an acute angle such that $\sin X = \frac{21}{29}$.
   Find the exact value of
   i tan X
   ii cos X.

2 An angler raises and lowers his rod.
AB represents the rod; BC represents the line,
and AC the waterline.

Diagram 1

$$\sin A = \frac{BC}{AB}; \quad \cos A = \frac{AC}{AB}; \quad \tan A = \frac{BC}{AC}$$

a In diagram 1 as $\angle A$ gets smaller and smaller,
BC approaches zero and AB approaches the
length of AC.
Copy and complete these statements:

$$\sin 0° = ...; \quad \cos 0° = ...; \quad \tan 0° = ...$$

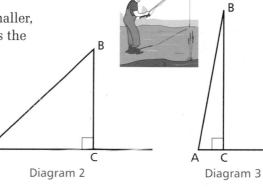

b In diagram 3 as $\angle A$ approaches 90°,
AC approaches zero and AB
approaches the length of BC.
Copy and complete these statements:

Diagram 2   Diagram 3

$$\sin 90° = ...; \quad \cos 90° = ...$$

c Comment on $\tan 90°$.

3 a An isosceles right-angled triangle can be made by halving a square as shown.
If the sides of the square are 1 unit long then, by Pythagoras' theorem, the diagonal
is $\sqrt{2}$ units.

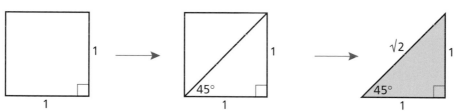

State the exact value of
i $\sin 45°$ **ii** $\cos 45°$ **iii** $\tan 45°$.
(To be exact, leave $\sqrt{2}$ in your answers. Don't use the decimal approximation.)

b An equilateral triangle of side 2 units is halved by an altitude.
By Pythagoras' theorem the altitude is $\sqrt{3}$ units long.
It forms a right-angled triangle with angles of 30° and 60°.

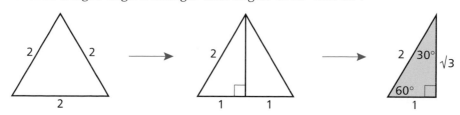

State the exact value of
i $\sin 30°$ **ii** $\cos 30°$ **iii** $\tan 30°$
and of
iv $\sin 60°$ **v** $\cos 60°$ **vi** $\tan 60°$
(To be exact, leave $\sqrt{3}$ in your answers. Don't use the decimal approximation.)

4 Copy and complete this table of results from questions **2** and **3**.
They are exact values and worth keeping for reference.

	0°	30°	45°	60°	90°
sin					
cos					
tan					

Notice that the cosine of an angle is the **sine** of its complement, i.e.
$\cos x° = \sin(90 - x)°$.

5 Work out the *exact* value of x in each of these triangles:

a

b

c

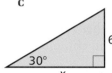

6 To reach the top of the building from the ground, two flights of stairs have to be climbed.
Calculate the exact height, h m, of the building.

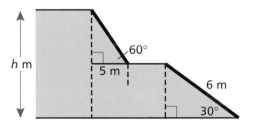

7 A cube has edges 8 cm long.
  a Sketch right-angled triangle EGH, and calculate the length of EG.
  b Sketch right-angled △AEG, and calculate ∠AGE.

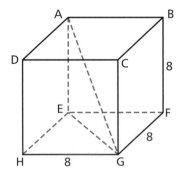

8 This packing case is in the shape of a cuboid.

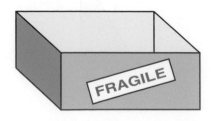

Calculate:
a AC   **b** AD   **c** ∠CAD.

9 The sun is 9.3×10^7 miles from the Earth, and the moon is 2.4×10^5 miles from the Earth. Calculate $\angle ESM$, correct to 2 decimal places, when $\angle SEM = 90°$.

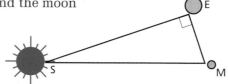

3 Sines, cosines and tangents of angles greater than 90°

So far trigonometry has helped us only to calculate sides and angles of right-angled triangles. With the use of coordinates, we can define the sine, cosine and tangent ratios for all angles, not just angles between 0° and 90°.

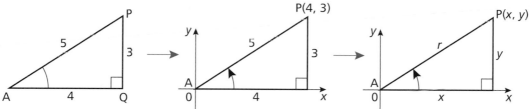

Consider the line OP, of length r units, rotating anticlockwise round the origin. The position of P will be unique for each angle A.

Using the coordinates of P, the trigonometric ratios for angle A are defined as:

$$\sin A = \frac{y}{r}, \quad \cos A = \frac{x}{r}, \quad \tan A = \frac{y}{x}.$$

Note that by convention:
- if OP = r rotates *anticlockwise* around O, angle A is **positive**
- if OP = r rotates *clockwise* around O, angle A is **negative**
- r is always positive.

Example

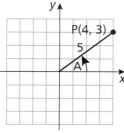
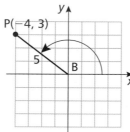
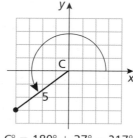
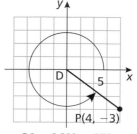

$A° = 37°$ to the nearest degree	$B° = 180° - 37°$ $= 143°$	$C° = 180° + 37° = 217°$	$D° = 360° - 37°$ $= 323°$

$\sin A° = \dfrac{3}{5}$ $\sin B° = \dfrac{3}{5}$ $\sin C° = -\dfrac{3}{5} = -\dfrac{3}{5}$ $\sin D° = \dfrac{-3}{5} = -\dfrac{3}{5}$

$\cos A° = \dfrac{4}{5}$ $\cos B° = \dfrac{-4}{5} = -\dfrac{4}{5}$ $\cos C° = \dfrac{-4}{5} = -\dfrac{4}{5}$ $\cos D° = \dfrac{4}{5}$

$\tan A° = \dfrac{3}{4}$ $\tan B° = \dfrac{3}{-4} = -\dfrac{3}{4}$ $\tan C° = \dfrac{-3}{-4} = \dfrac{3}{4}$ $\tan D = \dfrac{-3}{4} = -\dfrac{3}{4}$

Exercise 3.1

1 In the diagrams below, angles have been given to the nearest degree.
   In each case:
   i calculate the sine, cosine and tangent of each angle to 2 s.f.
   ii verify your answers using the trigonometric functions on your calculator.

a

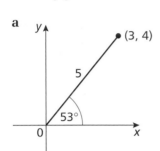

b

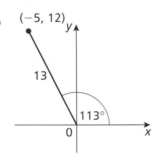

c

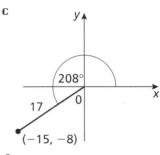

d

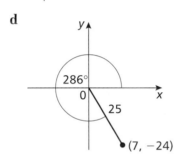

e

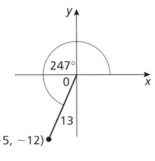

f

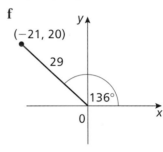

2 **a** Copy and complete this table, using a calculator to help you.

Angle (A°)	Quadrant	sin	cos	tan
65°	1	0·91	0·42	2·1
200°	3	−0·34	−0·94	0·36
285°	4	−0·97	0·26	−3·7
94°				
190°				
240°				
160°				
340°				
275°				
260°				

The quadrants

90 < A < 180 2nd quadrant	0 < A < 90 1st quadrant
3rd quadrant 180 < A < 270	4th quadrant 270 < A < 360

b In which quadrants are these positive?
   i sin A°
   ii cos A°
   iii tan A°
c In which quadrant are all three ratios positive?

3 Verify that your answers to question **2** agree with this chart.

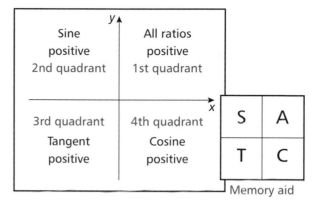

| Sine positive 2nd quadrant | All ratios positive 1st quadrant |
| 3rd quadrant Tangent positive | 4th quadrant Cosine positive |

| S | A |
| T | C |

Memory aid

4 Using the above chart, say which of these are positive and which are negative:

a sin 100°	**b** cos 100°	**c** tan 100°	**d** sin 200°	**e** cos 200°
f tan 200°	**g** sin 300°	**h** cos 300°	**i** tan 300°	**j** cos 150°
k sin 170°	**l** tan 190°	**m** sin 260°	**n** cos 280°	**o** tan 350°

(Check your decisions using a calculator.)

5 Write down the sign of each of these:

| **a** cos 85° | **b** tan 345° | **c** sin 172° | **d** tan 272° | **e** cos 95° |
| **f** sin 190° | **g** cos 269° | **h** tan 123° | **i** sin 285° | **j** cos 279° |

6 Use your calculator to find the values, correct to 3 decimal places, of the sines, cosines and tangents in question **5**.

7 Write down the sign of:

a the cosine of an acute angle

b the tangent of an obtuse angle

c the sine of a reflex angle.

4 Trig functions and their graphs

For each angle A, there is one and only one position of OP, so there is one and only one value of sin A.

As OP rotates, the largest y can be is r.

So the largest $\sin A = \dfrac{y}{r} = \dfrac{r}{r} = 1$, when A = 90.

Similarly, the smallest y can be is $-r$.

So the smallest $\sin A = \dfrac{y}{r} = \dfrac{-r}{r} = -1$, when A = 270.

So, there is a function which maps the set of all angles to the set of all real numbers between -1 and 1.

This function is called the **sine function**.

$$A° \rightarrow \sin A°$$

or

$$f(A°) = \sin A°$$

Similarly, the **cosine function** maps the set of all angles to the set of all real numbers between -1 and 1.

Exercise 4.1

1 The graph of the sine function

a Copy and complete this table (correct to 1 decimal place), using your calculator.

$x°$	0°	30°	60°	90°	120°		270°	300°	330°	360°
$\sin x°$										

b Plot all the points in the table on 2 mm graph paper.
Let 1 cm represent 30° on the x axis and 2 cm represent 0·5 on the y axis.

c Join the points to form a smooth curve.
This curve is called a **sine curve** or **sine wave**.

Note:

● For angles from 360° to 720° another complete sine wave is produced,
i.e. there is one wave or cycle for every 360°.
For this reason, the sine function is said to be a periodic function with a period
of 360°.

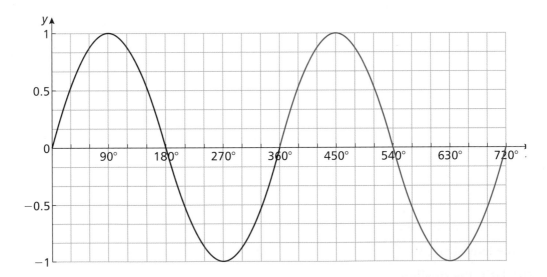

● $-1 \leqslant \sin x° \leqslant 1$

● A vertical axis of symmetry passes through each 'hill' or 'hollow' on the curve.
For the first hill it is $x = 90°$, and for the first hollow it is $x = 270°$.

2 The diagram above shows the graph of $y = \sin x°$ for $0 \leqslant x \leqslant 720$.

a For what values of x, $0 \leqslant x \leqslant 360°$, is:

i $\sin x° = 0$

ii $\sin x° = 1$

iii $\sin x° = -1$?

b The point A(30, 0·5) is shown.

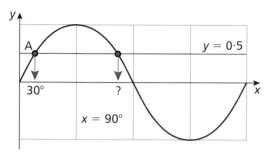

By symmetry, if 30 lies on the line then 180° − 30° also lies on the line.

Use the line of symmetry $x = 90$ to deduce the coordinates of another point where the line $y = 0.5$ cuts the curve.

c Use the periodicity of the curve to deduce another two points on this line.

d The point P(210, −0·5) lies on the curve.
Use the line of symmetry $x = 270$ to deduce the coordinates of another point where the line $y = -0.5$ cuts the curve.

e Use the periodicity of the curve to deduce another two points on this line if the graph is extended beyond 360.

3 The graph of the cosine function

a Make a table similar to the one in question **1** to help you draw the graph of
$y = \cos x°$ for $0 \leqslant x \leqslant 360$.

b Draw the graph of $y = \cos x°$ for $0 \leqslant x \leqslant 360$.

Note:

- The cosine function also has a period of 360°.
- Its graph is similar to that of the sine function, but 90° 'out of phase'.
  If you cover up the part of the sine graph from $x = 0$ to $x = 90$, you'll see the form of the cosine graph.
- The maximum value of $\cos x°$ is 1 and the minimum value is −1; so the range of $\cos x$ is: $-1 \leqslant \cos x° \leqslant 1$.

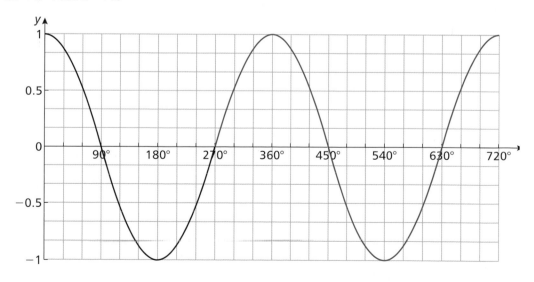

4 a For what values of x, $0 \leqslant x \leqslant 360$, is:

    i $\cos x° = 0$

    ii $\cos x° = 1$

    iii $\cos x° = -1$?

b What is the equation of the curve's axis of symmetry for the part from $x = 0$ to $x = 360$?

c The point $(60, 0·5)$ lies on the curve.

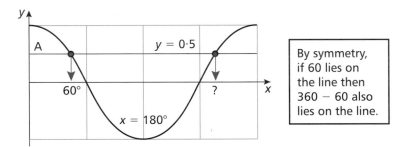

    i Use the line of symmetry $x = 180$ to help you find the coordinates of another point on the curve which lies on the line $y = 0·5$.

    ii Use the periodicity of the curve to deduce another two points on this line if the graph is extended beyond $x = 360$.

d The point $(120, -0·5)$ lies on the curve.

    i Use symmetry to find the coordinates of another point on the curve which also lies on the line $y = -0·5$.

    ii Use the periodicity of the curve to deduce another two points on this line.

5 The graph of the tangent function

a Copy and complete the table, with values of $\tan x°$ correct to 1 decimal place where necessary.

$x°$	0°	30°	45°	60°	75°	105°	120°	135°	150°	180°
$\tan x°$										

$x°$	210°	225°	240°	255°	285°	300°	315°	330°	360°
$\tan x°$									

b Plot the points in the table on 2 mm graph paper.
Let 1 cm represent 30° on the x axis and 2 cm represent 2 on the y axis.

c Join the points to form three separate parts of the tangent graph.

d Explore $\tan x$ when x is near 90, for example 89°, 89·9°, 89·999 999 9°, 90·000 001°.

e Explore $\tan x$ when x is near 270°, for example 269·9°, 269·9°, 270·1°.

f Draw vertical lines at $x = 90$ and $x = 270$. As x approaches 90 and 270 (from left and from right), the graph gets closer and closer to these lines without ever arriving. These lines are called **asymptotes** of the curve.

Note:
- The graph of $y = \tan x°$ consists of an endless number of branches. These occur at intervals of 180°. The period of the tangent function is 180°.
- Tan $x°$ has no value at $x = 90, 270$. The function is undefined at these points.

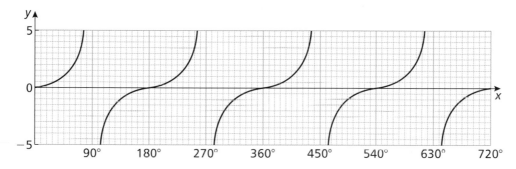

6 a For what values of x, $0 \leqslant x \leqslant 720$, is $\tan x° = 0$?

  b i The point (45, 1) lies on the graph above.
      Write down the coordinates of three other points that lie on the line $y = 1$.

    ii The point (135, −1) lies on the graph.
       Write down the coordinates of three other points that lie on the line $y = -1$.

---

5 Sketching trigonometric graphs

The graph of $y = 2 \sin x°$

For each value of x, the value of $2 \sin x°$ is twice the value of $\sin x°$.

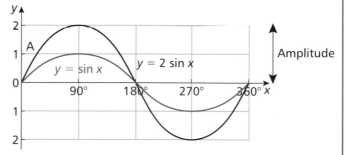

- The maximum value of $2 \sin x°$ is 2.
- The minimum value is −2.
- The **amplitude** is 2.
- The period of $y = 2 \sin x°$ is 360°.

Exercise 5.1

All the sketches in this exercise are for $0 \leqslant x \leqslant 360$, unless you are told otherwise.

1 a Write down the maximum and minimum values of $y = 3 \sin x°$.
  b What is the amplitude of
    i $y = \sin x°$ **ii** $y = 3 \sin x°$?
  c Sketch the graphs of $y = \sin x°$ and $y = 3 \sin x°$ on the same diagram.

2 a Write down the maximum and minimum values of $y = 2 \cos x°$.
  b Sketch the graphs of $y = \cos x°$ and $y = 2 \cos x°$ on the same diagram.
  c Add a sketch of $y = 3 \cos x°$ to your diagram.
  d Write down the period and the amplitude of:
    i $y = \cos x°$ **ii** $y = 2 \cos x°$ **iii** $y = 3 \cos x°$.

3 For $y = -\sin x°$, each value of y in $y = \sin x°$ is multiplied by -1.
  a Sketch the graph of $y = -\sin x°$.
  b On the same diagram sketch the graphs of $y = -2 \sin x°$ and $y = -3 \sin x°$.
  c Write down the period of each graph.
  d Write down the amplitude of each graph.

4 **a** Write down the maximum and minimum values of $y = n \sin x°$, for any integer, n.
  b What is the period of $y = n \sin x°$?
  c What is the amplitude of $y = n \sin x°$?
  d Note that $(90, n)$ lies on the curve $y = n \sin x°$.
    Sketch the graph of $y = n \sin x°$ for   **i** $n > 0$ **ii** $n < 0$.

5 Repeat question **4** for $y = n \cos x°$.

The graph of $y = \sin 2x°$

The maximum value of y is 1 when $2x = 90$, i.e. when $x = 45$.
The minimum value of y is 1 when $2x = 270$, i.e. when $x = 135$.
The amplitude is 1.
$y = 0$ when $2x = 0$ or 180 or 360, i.e. when $x = 0$, 90 or 180.
The period of the function $y = \sin x°$ is 360°, given by $0 \leqslant x \leqslant 360$.
The period of the function $y = \sin 2x°$ is given by $0 \leqslant 2x \leqslant 360$, i.e. by $0 \leqslant x \leqslant 180$, and so the period is 180°.
The graph will complete two cycles between 0 and 360.

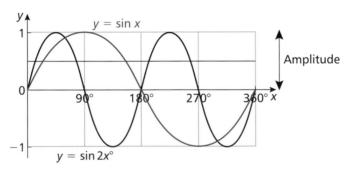

---

6 **a** Copy and complete this table.

$x°$	0	45	90	135	180	225	270	315	360
$\cos 2x°$									

  b Sketch the graph of $y = \cos 2x°$ on 2 mm graph paper, using the table as a guide.
  c You should complete two cycles for $0 \leqslant x \leqslant 360°$.
    What is the period of $y = \cos 2x°$?

7 **a** How many cycles has $y = \sin 3x°$ for $0 \leqslant x \leqslant 360$?
  b Sketch the graph.

8 How many complete cycles has $y = \cos 3x°$ for $0 \leqslant x \leqslant 360$? Sketch the graph.

9 **a** Copy and complete this table.

$x°$	0	60	90	180	270	300	360	420	450	540	630	660	720
$\cos \frac{1}{2}x°$													

 b Sketch the graph of $y = \sin \frac{1}{2}x°$ for $0 \leqslant x \leqslant 720°$.
 c How many complete cycles has $y = \sin \frac{1}{2}x$ for $0 \leqslant x \leqslant 360°$?
 d What is the period of $y = \sin \frac{1}{2}x°$?

Summary

For the graphs of $y = a \sin nx°$ and $y = a \cos nx°$:

- a gives the maximum/minimum values
- the amplitude is the same as the maximum value
- n gives the number of cycles for $0 \leqslant x \leqslant 360$
- the period of the function is $\dfrac{360}{n}$.

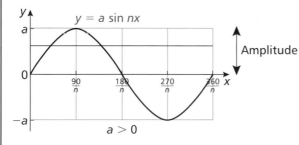

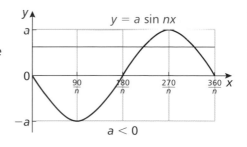

---

10 For each of the following:
 i write down the maximum and minimum values of y, and the number of cycles in the graph for $0 \leqslant x \leqslant 360$
 ii sketch the graph.

 a $y = \sin x°$ **b** $y = \cos x°$
 c $y = 4 \sin x°$ **d** $y = 4 \cos x°$
 e $y = \sin 2x°$ **f** $y = 4 \sin 2x°$
 g $y = 4 \cos 2x°$ **h** $y = \cos \frac{1}{2}x°$
 i $y = \frac{1}{2}\cos \frac{1}{2}x°$ **j** $y = -\sin 2x°$
 k $y = -6 \sin 3x°$ **l** $y = -3 \cos 4x°$
 m $y = \sin \frac{1}{3}x°$ **n** $y = 2 \cos \frac{1}{2}x°$

11 Find the equation of each graph. They are of the form $y = a \sin nx°$ or $y = a \cos nx°$.

a i

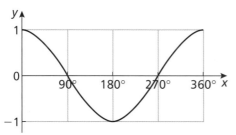

ii

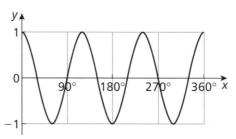

b i

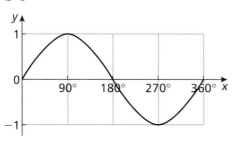

ii

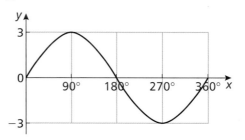

c i

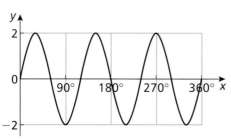

ii

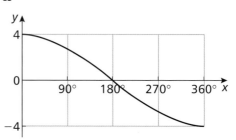

d i

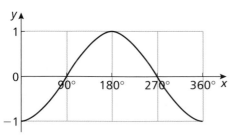

ii

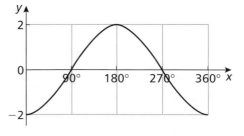

e i

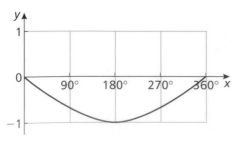

ii

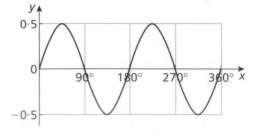

The graph of $y = \sin x° + 1$

Compared to $y = \sin x°$, each y value is increased by 1.
- The maximum value is $1 + 1 = 2$.
- Its minimum value is $-1 + 1 = 0$.
- The period is $360°$.
- The amplitude is 1.

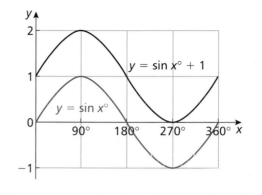

Exercise 5.2

1 For each of these functions, write down:
   i the maximum and minimum values   **ii** the amplitude   **iii** the period
   before sketching its graph.

   a $y = \sin x° + 2$

   b $y = \sin x° - 1$

   c $y = \cos x° + 3$

   d $y = \cos x° - 2$

   e $y = 2 \sin x° + 1$

   f $y = 2 \cos x° - 1$

   g $y = \sin 2x° + 1$

   h $y = 1 - \cos x°$

   i $y = \sin \frac{1}{2}x° - 1$

   j $y = 2 \sin 3x° + 3$

   k $y = 3 \cos 2x° - 1$

   l $y = \frac{1}{2} \cos \frac{1}{2}x° + 1$

2 Find the equation of each graph.
   They are of the form $y = a \sin nx° + b$ or $y = a \cos nx° + b$.

   a
   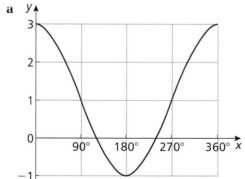

   b
   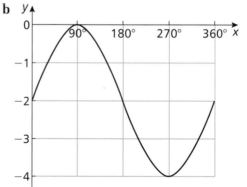

   c
   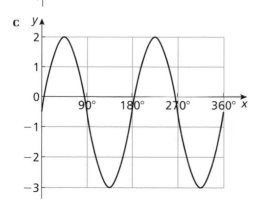

   d
   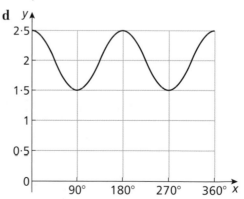

Challenge

Consider the function $y = \sin x° + \cos x°$ over the interval $0 \leqslant x \leqslant 360$.
a Prepare a table of values of x and y and sketch the curve. What is its
   i maximum value   **ii** minimum value   **iii** amplitude   **iv** period?
b Check your sketch with the curve obtained using a graphics calculator.

Investigation

Investigate the graphs of $y = a \sin nx° + b$ and $y = a \cos nx° + b$ with the aid of a computer or graphics calculator. Write a report on your findings.

6 Solving trigonometric equations

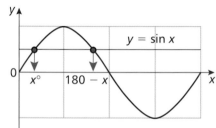

By symmetry x and $180 - x$
have the same sine.

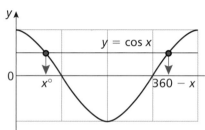

By symmetry x and $360 - x$
have the same cosine.

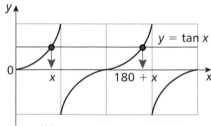

By glide symmetry x and $180 + x$
have the same tangent.

S	A
$180 - x$	$360 + x$
T	**C**
$180 + x$	$360 - x$

Memory aid

Example 1   Solve $\cos x° = 0{\cdot}7$, for $0 \leqslant x \leqslant 360$, to the nearest degree.

Step 1

Step 2   We are solving a problem involving 'cosine'.
         So, $x = 46$ or $360 - 46 = 314$.

Example 2 Solve $3 \tan x° + 2 = 0$, for $0 \leqslant x \leqslant 360$.

Step 1 $\tan x° = -\frac{2}{3}$

Step 2

Step 3 We are solving a problem involving 'tangent'.
So, $x = -34$ or $180 + (-34) = 146$ or $360 + (-34) = 326$.
Using only the answers in the range $0 \leqslant x \leqslant 360$, $x = 146$ or 326.

Example 3 Solve $\sin x° + 0.8 = 0$, for $0 \leqslant x \leqslant 360$.

Step 1 $\sin x° = -0.8$

Step 2

Step 3 We are solving a problem involving 'sine'.
So, $x = -53$ or $180 - (-53) = 233$ or $360 + (-53) = 307$.
Using only the answers in the range $0 \leqslant x \leqslant 360$, $x = 233$ or 307.

Exercise 6.1

1 Solve these equations for $0 \leqslant x \leqslant 360$, giving your answers to the nearest degree.

a $\sin x° = 0.2$	**b** $\cos x° = 0.2$
c $\tan x° = 0.2$	**d** $\sin x° = 0.4$
e $\cos x° = 0.8$	**f** $\tan x° = 0.75$
g $\cos x° = -0.3$	**h** $\tan x° = -3.2$
i $\sin x° = -0.85$	**j** $\tan x° = -1.7$
k $\sin x° = -0.25$	**l** $\cos x° = -0.47$
m $\sin x° = 0.5$	**n** $\cos x° = -0.5$
o $\tan x° = -1$	**p** $\sin x° = -1$
q $\cos x° = 1$	**r** $\tan x° = 0$

2 Solve these equations for $0 \leqslant x \leqslant 180$.

a $\sin x° = 0.15$

b $\cos x° = -0.55$

c $\tan x° = 7.8$

3 Solve these equations for $0 \leqslant x \leqslant 360$.

a $2 \sin x° - 1 = 0$	**b** $2 \tan x° + 3 = 0$
c $3 \cos x° + 4 = 5$	**d** $2 \cos x° + 3 = 1$
e $4 + \tan x° = 0$	**f** $1 + 5 \sin x° = 2$

4 Remember:

	0°	30°	45°	60°	90°
sin	0	$\dfrac{1}{2}$	$\dfrac{1}{\sqrt{2}}$	$\dfrac{\sqrt{3}}{2}$	1
cos	1	$\dfrac{\sqrt{3}}{2}$	$\dfrac{1}{\sqrt{2}}$	$\dfrac{1}{2}$	0
tan	0	$\dfrac{1}{\sqrt{3}}$	1	$\sqrt{3}$	–

Solve these equations for $0 \leqslant x \leqslant 360$.

a $\sin x° = \dfrac{\sqrt{3}}{2}$ **b** $\sqrt{3} \tan x° = 1$ **c** $2 \cos x° + \sqrt{3} = 0$

d $\sqrt{2} \cos x° - 1 = 0$ **e** $\sqrt{2} \sin x° = -1$ **f** $\sqrt{3} \tan x° + \sqrt{3} = 0$

5 $(\sin x°)^2$ is usually written as $\sin^2 x°$.
Solve these equations for $0 \leqslant x \leqslant 360$.
(Remember: $x^2 = 4 \Rightarrow x = 2$ or -2)
a $\sin^2 x° = 0{\cdot}38$ (Hint: $\sin x = \sqrt{0{\cdot}38}$ or $\sin x = -\sqrt{0{\cdot}38}$)
b $\cos^2 x° = 0{\cdot}64$
c $2 \tan^2 x° = 1$
d $4 \sin^2 x° - 3 = 0$

Exercise 6.2

1 The diagram shows part of the graph
of $y = \sin x°$.
  a The line $y = 0{\cdot}3$ cuts the graph at A
  and B.
  Calculate the coordinates of A and B.
  (Hint: solve $\sin x = 0{\cdot}3$.)
  b The line $y = -0{\cdot}6$ cuts the graph at C
  and D.
  Find the coordinates of C and D.
  c What is the value of y when $x = 75$?

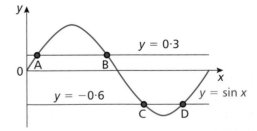

2 This is a sketch of $y = 2 \cos x°$,
for $0 \leqslant x \leqslant 360$.
  a Find the coordinates of C and D,
  the points where the curve cuts the
  x axis.
  b Find the coordinates of A and B, the
  points where the line $y = 1$ cuts the
  curve.
  c What is the value of y when x equals 100?

3 The sketch is part of the graph of $y = 2 + \cos x°$.
  a Find the coordinates of P, Q, R and S.
  b Write down:
    i the maximum value of y
    ii the minimum value of y.
  c Find the value of y when $x = 140$.

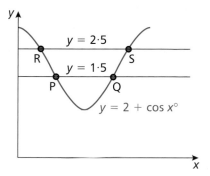

4 a Find the coordinates of the points where:
    i the line $y = -0.4$ cuts the curve $y = \sin x° - 1$
    ii the line $y = -1$ cuts the curve.
  b Find the coordinates of the maximum and minimum turning points of $y = \sin x° - 1$.

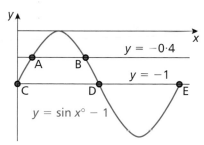

5 The sketch shows part of the graph $y = \sin 2x°$.
  a Find the coordinates of the points where:
    i the line $y = 0.2$ cuts the curve (hint: first find values for $2x$)
    ii the line $y = -0.8$ cuts the curve.
  b Calculate the value of y for:
    i $x = 30$
    ii $x = 160$.

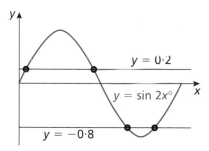

Challenge

Solve these quadratic equations for $0 \leqslant x \leqslant 360$.
a $\tan^2 x° - 2\tan x° + 1 = 0$ **b** $2\sin^2 x° + \sin x° - 1 = 0$ **c** $2\cos^2 x° = \cos x°$

7 Evaluating trigonometric functions

Example   The height, h km, of a satellite above the Earth after x hours is given by:

$$h(x) = 800 + 40\sin (10x)°.$$

  a What are the maximum and minimum heights of the satellite?
  b When do the maximum and minimum heights occur?
  c What is the period of h?
  d Draw a sketch of $h(x)$ for $0 \leqslant x \leqslant 180$.

a Maximum height = 800 + 40 × 1 = 840 km
Minimum height = 800 + 40 × (−1) = 760 km

b Max. height when sin (10x)° = 1
  ⟹ 10x = 90, 450, 810, ... ⟹ x = 9, 45, 81, ... hours
Min. height when sin (10x)° = −1
  ⟹ 10x = 270, 630, 990, ... ⟹ x = 27, 63, 99, ... hours

c Period = 360° ÷ 10 = 36°

d

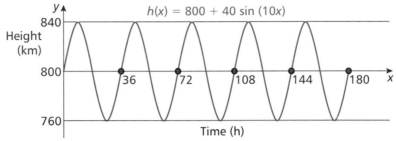

Exercise 7.1

1 Calculate the value of $f(t) = 2 \sin (6t)°$ to 1 decimal place for:
  a $t = 5$ **b** $t = 10$ **c** $t = 12$ **d** $t = 25$ **e** $t = 45$

2 a Calculate the value of $g(x) = 5 + 3 \cos (4x)°$ for
  i $x = 5$ **ii** $x = 20$ **iii** $x = 50$.
  b i What is the maximum value of $g(x)$?
  ii Give two values of x when $g(x)$ is at its maximum value.

3 The depth of the water in a harbour is given approximately by the formula
$d(t) = 12 + 8 \sin(30t)°$, where $d(t)$ is the depth in metres and t is the number of hours
after 6 am.
  a What is the depth of water at
   i 10 am   **ii** 1 pm?
  b When is the first high tide after 6 am?
  c What is the depth of the water at high tide?
  d When is the first low tide?
  e What is the depth of water at low tide?

4 The height, h m, of the tip of the blade of a
wind turbine above the ground is given on
one day by the formula $h(t) = 12 − 10$
$\cos(45t)°$, where t is the time in seconds.
  a What are the maximum and minimum
   heights of the tip of the blade above the
   ground?
  b When do these maximum and
   minimum heights first occur (for $t > 0$)?
  c How long does it take the blade to make
   one complete turn?

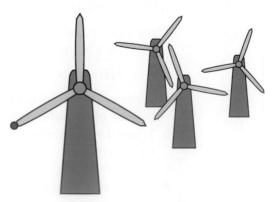

5 On this wall clock, the height, h cm, of the end of the minute hand above the floor is given by the formula
$h(t) = 180 + 12 \sin 6t°$, where t is measured in minutes.

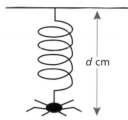

a What are the maximum and minimum heights of the end of the minute hand above the floor?

b After how many minutes do the first maximum and minimum heights occur?

c Calculate h after
   i 10 minutes   **ii** 20 minutes   **iii** 40 minutes.

d What is the period of h?   **e** Sketch the graph of h.

h cm

6 A child's toy is fixed to the ceiling and is on the end of a spring. It is made to bounce up and down.
Its distance from the ceiling is given by the formula
$d(t) = 25 - 20 \cos 45t°$, with d the distance in centimetres and the time in tenths of a second.

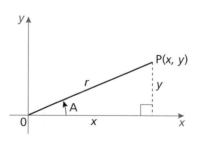

d cm

a How far from the ceiling is the toy at $t = 0$?

b What is its greatest distance from the ceiling?

c **i** What is d after 0·1 second?   **ii** When next does d have this value?

d What is the toy's distance from the ceiling after 1 second?

e What is the period of the graph of d against t?

f Sketch the graph of d against t.

7 A piston is moved vertically up and down under water so that its depth, D metres, is given by $D(n) = 2 - 2 \cos 30n°$, with n the time in hundredths of a second.

a When is the piston on the surface of the water for the first time?

b How long is it before the piston is again on the surface of the water?

c How deep does the piston go? When does it reach this depth for the first time?

d How deep is the piston after
   i 5 hundredths of a second   **ii** one tenth of a second?

e What is the period of the graph of D against n?

---

8 Related ratios

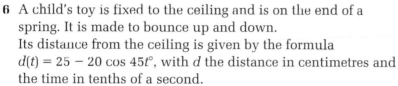

● $\dfrac{\sin A}{\cos A} = \dfrac{\frac{y}{r}}{\frac{x}{r}} = \dfrac{y}{r} \div \dfrac{x}{r} = \dfrac{y}{r} \times \dfrac{r}{x} = \dfrac{y}{x} = \tan A$

● By Pythagoras' theorem, $y^2 + x^2 = r^2$
Dividing throughout by r^2

$\Rightarrow \qquad\qquad \dfrac{y^2}{r^2} + \dfrac{x^2}{r^2} = \dfrac{r^2}{r^2}$

$\Rightarrow \qquad\qquad \left(\dfrac{y}{r}\right)^2 + \left(\dfrac{x}{r}\right)^2 = 1$

$\Rightarrow \quad (\sin A)^2 + (\cos A)^2 = 1$

$\Rightarrow \qquad \sin^2 A + \cos A^2 = 1$

y

$P(x, y)$

r

A

x

y

0 x x

Exercise 8.1

1 Check these with your calculator:

    a $\dfrac{\sin 70°}{\cos 70°} = \tan 70°$ **b** $\dfrac{\sin 28°}{\cos 28°} = \tan 28°$ **c** $\dfrac{\sin 89°}{\cos 89°} = \tan 89°$

2 Express the following in terms of sines and cosines:
    a $\tan 100°$ **b** $\tan 200°$ **c** $\tan 340°$

3 Both $\sin 90°$ and $\cos 90°$ are well defined. We know that $\tan 90°$ is not.

    Explain $\dfrac{\sin 90°}{\cos 90°} = \tan 90°$.

4 Check with your calculator that:
    a $\sin^2 30° + \cos^2 30° = 1$ **b** $\sin^2 140° + \cos^2 140° = 1$
    c $\sin^2 230° + \cos^2 230° = 1$ **d** $\sin^2 310° + \cos^2 310° = 1$

5 Verify, using *exact* values, that $\sin^2 A° + \cos^2 A° = 1$ when A is:
    a 45   **b** 60   **c** 90

6 If $0 \leqslant A \leqslant 90$ then $\sin A°$ is positive.

    $\sin^2 A° + \cos^2 A° = 1$

    $\Rightarrow$ $\sin^2 A° = 1 - \cos^2 A°$

    $\Rightarrow$ $\sin A° = \sqrt{(1 - \cos^2 A°)}$ ... taking the positive root since $\sin A°$ is positive

    Use this fact to help you calculate $\sin x°$, for $0 \leqslant x \leqslant 90$, when:
    a $\cos x° = 0·6$ **b** $\cos x° = 0·5$ **c** $\cos x° = 0·9$

7 Find a formula for $\cos x$ in terms of $\sin x$ and use it to help you calculate $\cos x°$ for $0 \leqslant x \leqslant 90$, when:
    a $\sin x° = 0·3$ **b** $\sin x° = 0·7$ **c** $\sin x° = 0·8$

8 Calculate both possible values of $\sin x°$, for $0 \leqslant x \leqslant 360$, when:
    a $\cos x° = 0·7$ **b** $\cos x° = -0·5$ **c** $\cos x° = -0·2$

9 $\sin^2 x° = 0·49$. Calculate: **a** $\cos^2 x°$ **b** $\tan^2 x°$.

10 Malcolm said that he had calculated $\sin x°$ as 0·4 and $\cos x°$ as 0·6.
    How did the teacher know he was wrong?

11 To solve the equation $4 \sin x° + 5 \cos x° = 0$, for $0 \leqslant x \leqslant 360$, copy and complete:
    $4 \sin x° + 5 \cos x° = 0$

    $\Rightarrow$ $4 \sin x° = -5 \cos x°$

    $\Rightarrow$ $4 \dfrac{\sin x}{\cos x} = -5$ (provided $\cos x° \neq 0$ since you cannot divide by zero)

    $\Rightarrow$ $4 \tan x° = -5$

    $\Rightarrow$ $\tan x° = \dots$

    $\Rightarrow$ $x° = \dots$ or $180° + \dots$ or $360° + \dots$

    $\Rightarrow$ $= \dots$ or $\dots$

12 Solve for $0 \leqslant x \leqslant 360$:

a $\sin x° - 3 \cos x° = 0$ **b** $2 \sin x° - \cos x° = 0$

c $3 \sin x° + 2 \cos x° = 0$ **d** $3 \cos x° - 4 \sin x° = 0$

Challenges

1 Use the formula $\sin^2 A + \cos^2 A = 1$ to prove that:

a $3 \sin^2 A + 3 \cos^2 A = 3$

b $(\sin x + \cos x)^2 = 1 + 2 \sin x \cos x$

c $(\sin x - \cos x)^2 + (\sin x + \cos x)^2 = 2$

d $\cos^2 y - \sin^2 y = 2 \cos^2 y - 1$

2 Using trigonometric formulae, prove that:

a $- \tan^2 x = 1 - \dfrac{1}{\cos^2 x}$ **b** $(\cos A \tan A)^2 + \left(\dfrac{\sin A}{\tan A}\right)^2 = 1$

◀◀ RECAP

Exact values

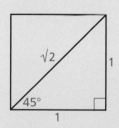

	0°	30°	45°	60°	90°
sin	0	$\dfrac{1}{2}$	$\dfrac{1}{\sqrt{2}}$	$\dfrac{\sqrt{3}}{2}$	1
cos	1	$\dfrac{\sqrt{3}}{2}$	$\dfrac{1}{\sqrt{2}}$	$\dfrac{1}{2}$	0
tan	0	$\dfrac{1}{\sqrt{3}}$	1	$\sqrt{3}$	–

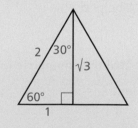

Definitions

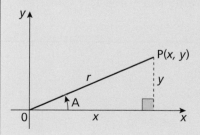

$\sin A° = \dfrac{y}{r}$ $\cos A° = \dfrac{x}{r}$ $\tan A° = \dfrac{y}{x}$

Sine, cosine and tangent for angles greater than 90°

S	A
$180 - x$	$360 + x$
T	C
$180 + x$	$360 - x$

Examples $\sin 130° = \sin (180 - 50)° = \sin 50°$
$\cos 200° = \cos (180 + 20)° = -\cos 20°$
$\tan 320° = \tan (360 - 40)° = -\tan 40°$

Graphs of the trigonometric functions

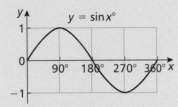
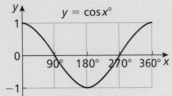
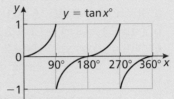

The **maximum** value of the sine and cosine functions is 1.
The **minimum** value of the sine and cosine functions is -1.
The **period** of the sine and cosine functions is 360°.
The period of the tangent function is 180°.

Graphs of $y = b + a \sin nx°$ and $y = b + a \cos nx°$

When $a > 0$, the maximum value is $b + a$, the minimum value is $b - a$.
When $a < 0$, the maximum value is $b - a$, the minimum value is $b + a$.
The period is $\dfrac{360}{n}$.

Two trigonometric formulae

$\tan A = \dfrac{\sin A}{\cos A}$ $\sin^2 A + \cos^2 A = 1$

Interpreting trigonometric functions

Example $h(t) = 5 + 4 \sin 15t°$
Maximum value is $5 + 4 = 9$; minimum value is $5 - 4 = 1$.
$h(2) = 5 + 4 \times \sin 30° = 5 + 4 \times \frac{1}{2} = 7$

1 Calculate x correct to 1 decimal place.

a

b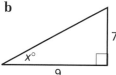

c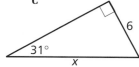

2 ABCDEFGH is a room in the shape of a cuboid.
A closed circuit TV camera at A points directly
at the opposite bottom corner G.
Use right-angled triangle ACG to calculate the
angle of depression, $\angle$CAG, of the camera.

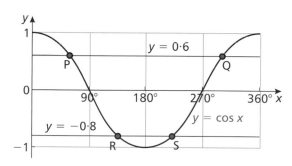

3 If angle A is acute and $\sin A = \frac{8}{17}$, calculate the *exact* value of:
  a $\cos A$ b $\tan A$.

4 Express each of these in terms of an acute angle, e.g. $\sin 110° = \sin 70°$.
  a $\sin 210°$ b $\cos 130°$ c $\tan 230°$ d $\cos 320°$

5 Find the coordinates of P, Q, R and S.

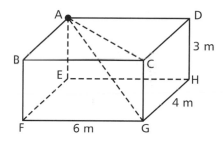

6 a Write down the maximum and minimum values of $y = 7 \cos x°$, for $0 \leqslant x \leqslant 360$.
    b What is the period of $y = 7 \cos x°$?
    c For what values of x is $7 \cos x° = 0$, $0 \leqslant x \leqslant 360$?
    d Sketch the graph of $y = 7 \cos x°$ for $0 \leqslant x \leqslant 360$.

7 a Write down the maximum and minimum values of $y = \sin 3x°$, $0 \leqslant x \leqslant 360$.
    b What is the period of $y = \sin 3x°$?
    c For what values of x is $\sin 3x° = 0$, $0 \leqslant x \leqslant 360$?
    d Sketch the graph of $y = \sin 3x°$ for $0 \leqslant x \leqslant 360$.

8 a Write down the maximum and minimum values of $y = 2 + 3 \sin \frac{1}{2}x°$.
    b What is its period?
    c For what values of x is $2 + 3 \sin \frac{1}{2}x° = 0$, $0 \leqslant x \leqslant 360$?
    d Sketch the graph of $y = 2 + 3 \sin \frac{1}{2}x°$ for $0 \leqslant x \leqslant 360$.

REVISE

9 Write down the equation of each of these graphs.
Each is of the form $y = a + b \sin nx°$ or $y = a + b \cos nx°$.

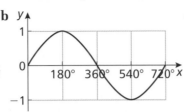

10 As the Ferris wheel rotates, the height, h m, of P
above the ground at time t seconds is given by
the formula $h(t) = 10 - 8 \cos 2t°$.

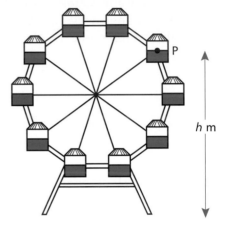

  a Write down the maximum and minimum
  heights of P.

  b What was the height of P when the wheel
  started, i.e. at $t = 0$?

  c What was the height of P after 1 minute?

  d When was 10 m the height of P for the first
  time?

  e How long did it take the wheel to revolve
  once?

  f Sketch the graph of $h(t) = 10 - 8 \cos 2t°$.

11 Solve these equations for $0 \leqslant x \leqslant 360$.

  a $\sin x° = 0.85$ **b** $\cos x° = -0.34$ **c** $\tan x° = 4.8$

  d $2 \cos x° = 0.5$ **e** $3 \tan x° + 1 = 0$ **f** $\sin 2x° = -0.6$

12 **a** Calculate $\sin x°$ for $0 \leqslant x \leqslant 360$ when $\cos x° = 0.4$.

  b Write down the value of $\tan x°$ when $\sin x° = 2 \cos x°$, $0 \leqslant x \leqslant 90$.

REVISE

12 Trigonometry and triangle calculations

Hipparchus and Aristarchus, the inventors of trigonometry in the second century BC, looked at the triangles formed by tangents to the sun, moon and Earth during eclipses and used trigonometry to calculate the sizes of these bodies and their distances from the Earth.

Today trigonometry is used by surveyors, architects and engineers, as well as navigators and astronomers.

1 Review

Reminders:

SOH – CAH – TOA

Sine, cosine and tangent for angles greater than 90°

Examples sin 130° = sin (180 − 50)° = sin 50°
cos 200° = cos (180 + 20)° = − cos 20°
tan 320° = tan (360 − 40)° = − tan 40°

S	A
$180 - x$	$360 + x$
T	C
$180 + x$	$360 - x$

Throughout the chapter give your answers correct to 3 significant figures where appropriate.

◀◀ **Exercise 1.1**

1 Use trigonometry to calculate x in each triangle.

a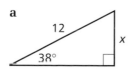

b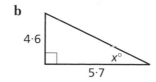

c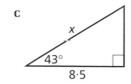

2 Calculate:
  a the height of the kite
  b the angle of elevation from A to the kite.

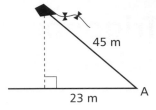

45 m

23 m

A

3 Calculate:
  a the height of the helicopter
  b the length FG
  c the angle of depression from H to G.

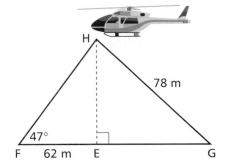

H

78 m

47°

F   62 m   E   G

4 The cairn is 580 m due west of the camp.
   The cave is 920 m due south of the camp.
   Calculate the bearing of the cave from the cairn.

Cairn   580 m   Camp

920 m

Cave

5 At A, the *Bounty* is 420 m due east of the rocks at R.
   It is sailing on a course of 222°.
   a Calculate how close the *Bounty* gets to the rocks.
   b The *Bounty* is due south of the rocks at P.
      Calculate the distance it travels in going from A to P.

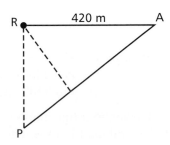

R   420 m   A

P

6 What is the bearing of:
   a B from A   **b** A from B?

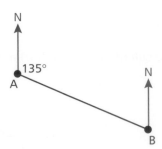

N

135°

A

N

B

7 Find the area of each triangle:

a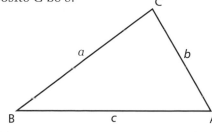

b

c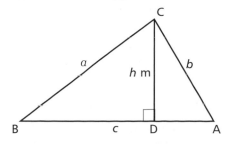

8 Relate each ratio to one involving an acute angle. (Hint: see reminders.)
   a sin 170°   **b** cos 220°   **c** tan 325°   **d** sin 255°   **e** cos 345°

9 Solve each equation for A. (Remember there may be more than one solution.)
   a sin A = 0·2, 0 ⩽ A ⩽ 180   **b** cos A = 0·35, 0 ⩽ A ⩽ 360
   c tan A = −0·9, 0 ⩽ A ⩽ 360   **d** cos A = −0·28, 0 ⩽ A ⩽ 360

10 The village of Bren is 35 km from the village of Cairns on a bearing of 158°.
   The village of Dent is 40 km from Cairns on a bearing of 210°.
   Choosing a suitable scale, make a scale drawing to find:
   a the distance between Bren and Dent   **b** the bearing of Bren from Dent.

11 Make x the subject of each of these:

   a $\dfrac{x}{a} = b$
                b $\dfrac{x}{a} = \dfrac{b}{c}$
                c $\dfrac{h}{x} = \dfrac{m}{n}$

   d $\dfrac{x}{\sin X} = y$
              e $\dfrac{x}{\sin A} = \dfrac{y}{\sin B}$
          f $\dfrac{\sin X}{x} = \dfrac{\sin Y}{y}$

---

2 The sine rule

Class discussion

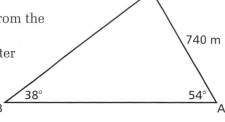

A helicopter at C is 740 m from a lifeboat at A.
The angle of elevation of the helicopter is 54° from the
lifeboat and 38° from the ship at B.
How can we find the distance from the helicopter
to the ship, i.e. the distance BC?
a Could we make a scale drawing?
b Could we use trigonometry?
   But we don't have a right-angled triangle.

Let's look at the problem in general. Sketch the situation.
Label the vertices A, B and C. Let the side opposite A be *a*, opposite B be *b* and
opposite C be *c*.

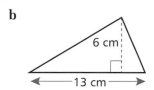

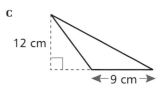

Draw CD perpendicular to AB. You now have two right-angled triangles.
Let CD = h units.

In triangle BCD, sin B = $\dfrac{h}{a}$ $\Rightarrow h = a$ sin B

In triangle ACD, sin A = $\dfrac{h}{b}$ $\Rightarrow h = b$ sin A

$$\Rightarrow a \sin B = b \sin A$$

$$\Rightarrow \frac{a}{\sin A} = \frac{b}{\sin B}$$

If, instead, we had drawn BE perpendicular to AC we would have obtained:

$$\frac{a}{\sin A} = \frac{c}{\sin C}$$

> Thus in any triangle $\dfrac{a}{\sin A} = \dfrac{b}{\sin B} = \dfrac{c}{\sin C}$. This is known as **the sine rule.**

The above argument works even if $\angle$A is obtuse.

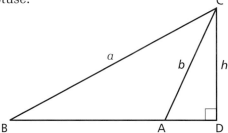

Note that for us to use the sine rule the problem should involve two pairs of 'opposites'.

Example 1   What is the distance from the helicopter to the ship as described above?

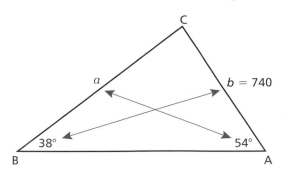

Take stock

$$\frac{\textcircled{a}}{\sin A} = \frac{b\checkmark}{\sin B} = \frac{c}{\sin C}$$

Circle what is wanted.

Tick what is given.

$$\frac{a}{\sin A} = \frac{b}{\sin B} \qquad \Rightarrow \qquad \frac{a}{\sin 54°} = \frac{740}{\sin 38°}$$

$$\Rightarrow \qquad a = \frac{740 \sin 54°}{\sin 38°} \qquad \Rightarrow \quad a = 972 \text{ (to 3 s.f.)}$$

Thus BC is a distance of 972 metres.

Exercise 2.1

1 In $\triangle ABC$, $\dfrac{a}{\sin 50°} = \dfrac{9}{\sin 60°}$

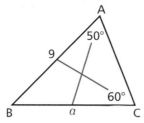

    a Calculate the value of a.

    b Write down the part of the sine rule that will help you calculate the labelled side
    (remember 'opposites') and calculate it.

i 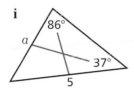   **ii** 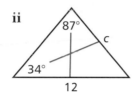   **iii** 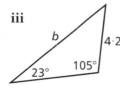   **iv**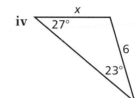

2 Calculate a in each triangle.

a 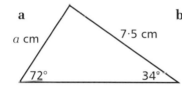   **b** 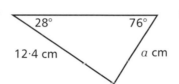   **c**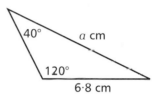

3 In $\triangle DEF$, calculate:
    a $\angle DEF$
    b e

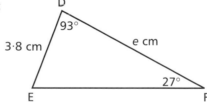

4 In $\triangle RST$, calculate:
    a r
    b $\angle RST$
    c s
    (All lengths are in centimetres.)

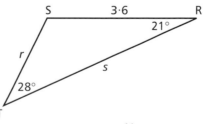

5 A tunnel is to be made along the line XY.
    To help in planning, a point Z is chosen from
    where X and Y can both be seen.
    a Sketch $\triangle XYZ$, marking $\angle X = 35°$, $\angle Z = 68°$
    and $XZ = 540$ m.
    b Calculate XY, to the nearest metre.

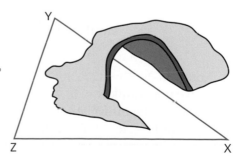

6 Sara sails 16 km in a south-easterly direction from Westquay (W) until she is hit by a storm at S. She changes course and sails on a bearing of 040°, dropping anchor at A, which is due east of Westquay.
Calculate:
a how far Sara is from Westquay now
b the distance she sails on the bearing of 040°.

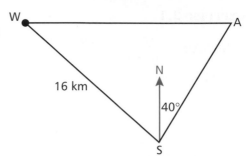

7 A and B are two points 600 m apart on an airport runway. From A the angle of elevation to the plane at P is 37°. From B the angle of elevation is 62°.
a Calculate, to the nearest metre, the distance the plane is from
i A **ii** B.
b Calculate the height of the plane.

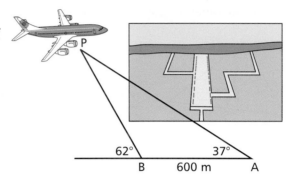

---

3 Calculating an angle

Example Calculate:
a ∠R
b ∠S, in △RST.

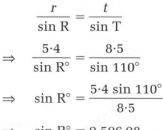

a Identify two pairs of 'opposites': one containing the unknown; one containing 'complete information'.

Identify opposites

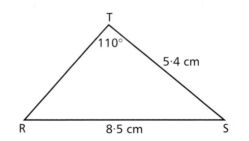

$$\frac{r}{\sin R} = \frac{t}{\sin T}$$

$$\Rightarrow \quad \frac{5 \cdot 4}{\sin R°} = \frac{8 \cdot 5}{\sin 110°}$$

$$\Rightarrow \quad \sin R° = \frac{5 \cdot 4 \sin 110°}{8 \cdot 5}$$

$$\Rightarrow \quad \sin R° = 0.596\ 98\ \dots$$

$$\Rightarrow \quad R° = 36 \cdot 7°\ \text{(3 s.f.)}$$

Note: sin 36·7 = 0·596 98… but sin (180 − 36·7) = 0·596 98… too.
So we may think R = 36·7 or R = 143·3 can be solutions.
However, the information given about △RST tells us that ∠R must be 37°. Explain.
b ∠S = 180 − ∠R − ∠T = 180° − 36·7° − 110° = 33·3° (to 3 s.f.)

Exercise 3.1

1 Calculate ∠A in each triangle:

a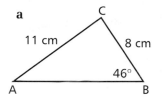

b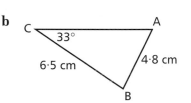

c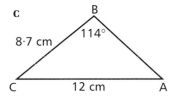

2 Calculate the sizes of the unmarked angles in each triangle.

a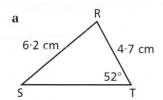

b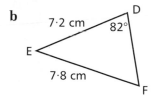

c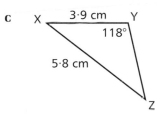

3 Sketch these triangles and in each case calculate the sizes of the unmarked angles.
  a △PQR, with ∠R = 65°, *p* = 14 and *r* = 17
  b △XYZ, with *x* = 34, *y* = 28 and ∠X = 64°

4 KLMN is a parallelogram with
  KL = 8 cm and LM = 5 cm. ∠LMK = 34°.
  Calculate:
  a ∠LKM (hint: consider a triangle)
  b the length of diagonal KM.

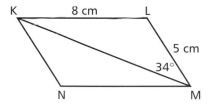

5 Three roads meet as shown in the sketch.
  Calculate:
  a the angles at junctions Y and Z
  b the length of Blossom Drive
    between X and Y.

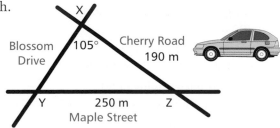

6 The sketch represents a triangular lawn.
  a Calculate the unknown angles of the lawn.
  b Calculate the lawn's perimeter.

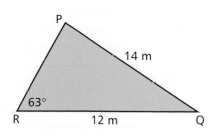

7 ABCD represents a field at Farmer Reid's farm.
He divides it into two triangular fields as shown.
Calculate:

a $\angle ACD$

b CD

c $\angle BCA$

d $\angle BAD$.

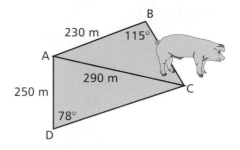

The ambiguous case

'Ambiguous' means open to more than one interpretation.

Consider the triangle ABC such that AB = 6 cm, BC = 3 cm and $\angle A = 20°$. Calculate $\angle C$.

An initial sketch gives:

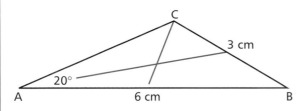

$$\frac{a}{\sin A} = \frac{c}{\sin C} \Rightarrow \frac{3}{\sin 20°} = \frac{6}{\sin C} \Rightarrow \sin C = \frac{6 \sin 20°}{3}$$

$\Rightarrow \qquad C = 43·2°$ (3 s.f.) by calculator, or C = 180 − 43·2 = 136·8°

There are two possible answers!
Both have to be checked to make
sure they are possible solutions
inside a triangle.

If $\angle C = 43·2°$ then
$\angle B = 180 − 43·2 − 20 = 116·8$.

A triangle with angles of 20°,
116·8° and 43·2° is possible.

If $\angle C = 136·8°$ then
$\angle B = 180 − 136·8 − 20 = 23·2$.

A triangle with angles of 20°, 23·2° and 136·8° is also possible.

The diagram shows how the given data
can describe two different triangles.

> When finding an angle using the sine rule, always check for a second possible answer.

Exercise 3.2

1 Calculate the sizes of the unknown angles in each triangle.
Consider whether more than one answer is possible.
  a In △XYZ, YZ = 9·8 cm, XY = 7·6 cm and ∠XZY = 39°
  b In △PQR, ∠PQR = 36°, QR = 14 cm and PR = 8·8 cm

2 **a** Sketch △EFG with EF = 8 cm, EG = 9 cm and ∠EFG = 55°.
  b Find the size of the largest possible angle in the triangle.

3 The diagram shows a tree growing on a hillside.
Calculate the height of the tree, to the nearest metre.

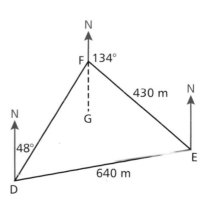

4 Boat B is 35 m further from the
foot of the cliff than boat A.
The angle of elevation from A
to the top of the cliff is 40°, and
from B it is 32°.
  a Calculate the distance
    i AC
    ii BC.
  b Calculate the height of the cliff to the nearest metre.
  c Calculate the distance of each boat from the foot of the cliff.

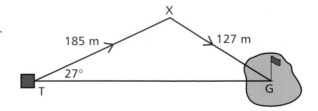

5 The sketch shows how Luke played
the first hole of his local golf course.
What is the length of the hole, i.e.
what is the length of TG?

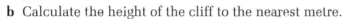

6 This diagram shows the positions of three buildings
at D, E and F.
  a Write down the size of:
    i ∠DFG
    ii ∠GFE
    iii ∠DFE.
  b Calculate the size of ∠EDF, using the sine rule.
  c What is the bearing of E from D?
  d What is the distance between D and F?

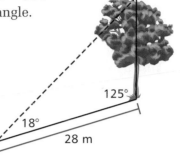

Brainstormer

Given $\angle A = 35°$, AB = 9 cm and BC = 6 cm, make an accurate construction for two different positions of C on the base-line. Measure, then calculate $\angle ACB$ in each case.

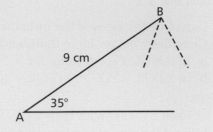

Challenges

1 Chloe is a surveyor. She needs to find the lengths of all the sides and the sizes of all the angles of a piece of land in the shape of a pentagon.

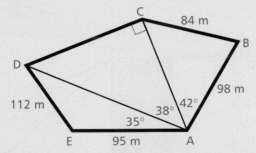

Chloe divides the land into triangles and takes the measurements given in the diagram. Calculate:
   a the length of side CD
   b the five angles of the pentagon.

2 To find the height, h metres, of the pyramid, a surveyor measures the distance d metres and angles $x°$ and $y°$.
   By investigating triangles ABP and PBM, can you find a formula involving h, d, x and y that he could use?

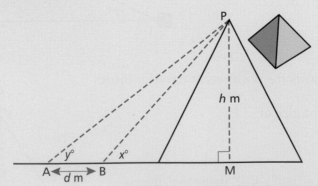

Use your formula to calculate h, taking $d = 200$, $x = 30$ and $y = 20$.
Use your formula to calculate the height of a local landmark.

4 The cosine rule

In the triangle opposite, we cannot establish pairs of 'opposites', so we can't use the sine rule.

We're back to *making* right-angled triangles and using Pythagoras' theorem.

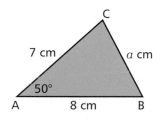

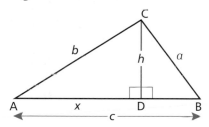

Draw altitude CD to make two right-angled triangles ADC and BDC.
Let $CD = h$ cm and $AD = x$ cm. So $DB = (c - x)$ cm.
Using Pythagoras in $\triangle BCD$:

$$a^2 = h^2 + (c - x)^2$$
$$= h^2 + c^2 + x^2 - 2cx$$
$$= (h^2 + x^2) + c^2 - 2cx$$
$$= b^2 + c^2 - 2cx \qquad \text{(using Pythagoras' theorem on triangle ADC)}$$
$$= b^2 + c^2 - 2c(b \cos A) \qquad \text{(from triangle ADC, } x = b \cos A)$$
$$= b^2 + c^2 - 2bc \cos A$$

A similar argument applies when A is obtuse.

> This result is known as the **cosine rule**: in any $\triangle ABC$, $a^2 = b^2 + c^2 - 2bc \cos A$

Similarly we can show that:

$$b^2 = a^2 + c^2 - 2ac \cos B$$
$$c^2 = a^2 + b^2 - 2ab \cos C$$

Note that, when two sides and the angle between them are known, the cosine rule can be used to find the third side.

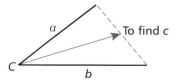

Example In $\triangle ABC$, $b = 3$, $c = 5$ and $\angle A = 125°$. Calculate a.

$$a^2 = b^2 + c^2 - 2bc \cos A$$
$$= 3^2 + 5^2 - 2 \times 3 \times 5 \times \cos 125°$$
$$= 9 + 25 - (-17 \cdot 207 \ldots) \ldots \text{using a calculator}$$
$$\Rightarrow \quad a^2 = 51 \cdot 207 \ldots$$
$$\Rightarrow \quad a = \sqrt{(51 \cdot 207 \ldots)} = 7 \cdot 16 \text{ cm, correct to 3 significant figures}$$

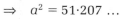

Exercise 4.1

1 Use the formula $a^2 = b^2 + c^2 - 2bc \cos A$ to calculate a, correct to 1 decimal place, in each triangle.

a
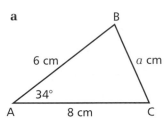

b
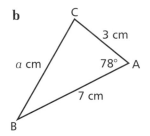

c
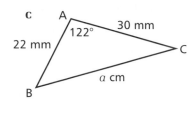

2 Use the cosine rule to find the unknown side in each triangle.

a
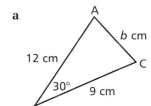

b
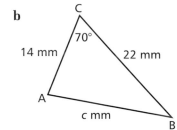

c
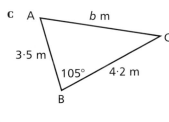

3 For $\triangle XYZ$, write down the cosine rule to find:
   a x^2
   b y^2
   c z^2

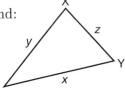

4 a For $\triangle PQR$, write down the cosine rule to find:
   i p^2
   ii q^2
   iii r^2
   (Hint: draw a sketch of triangle PQR.)
   b Calculate the length of the third side, given:
   i $q = 7$, $r = 5$ and $\angle P = 53°$
   ii $p = 5.6$, $q = 6.4$ and $\angle R = 130°$
   c What does the formula for p^2 become when $\angle P = 90°$?

5 A surveyor needs to find the width of a quarry face. He takes the measurements shown in the diagram. Calculate the width of the quarry face (BC).

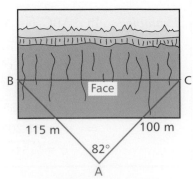

6 Calculate the length of the traffic jam.

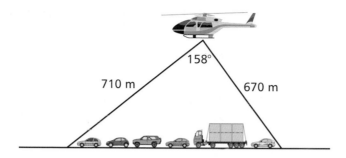

7 Calculate the width of each house.

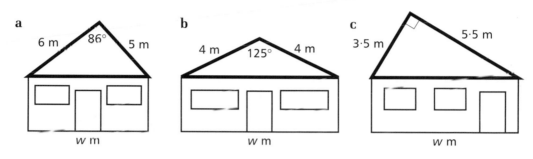

a 6 m 86° 5 m *w* m

b 4 m 125° 4 m *w* m

c 3·5 m 5·5 m *w* m

8 PQRS is a parallelogram.
Calculate the length of its shorter diagonal.

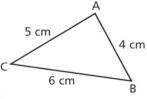

12 cm
8 cm
48°

---

5 Calculating an angle, given the lengths of the three sides

In $\triangle$ABC, $a^2 = b^2 + c^2 - 2bc \cos A$ (the cosine rule)

$\Rightarrow$ $2bc \cos A = b^2 + c^2 - a^2$

$\Rightarrow$ $\cos A = \dfrac{b^2 + c^2 - a^2}{2bc}$

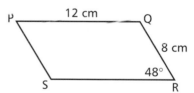

5 cm
4 cm
6 cm

Example   Calculate the *largest* angle in $\triangle$ABC.

The *largest* angle is opposite the *largest* side,
so $\angle$A is the one we want.

$$\cos A = \frac{b^2 + c^2 - a^2}{2bc} = \frac{5^2 + 4^2 - 6^2}{2.5.4} = \frac{5}{40} = 0\cdot125$$

$\Rightarrow$ $\angle$A = 82$\cdot$8°, correct to 3 s.f.

Exercise 5.1

1 From $a^2 = b^2 + c^2 - 2\,bc\cos A$, we can arrive at: $\cos A = \dfrac{b^2 + c^2 - a^2}{2bc}$.

Write down the expression for:

a $\cos B$, using $b^2 = a^2 + c^2 - 2ac\cos B$

b $\cos C$, using $c^2 = a^2 + b^2 - 2\,ab\cos C$.

2 Use the cosine rule to find the smallest angle in $\triangle PQR$.

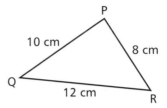

3 In $\triangle ABC$, calculate:

a $\angle A$, when $a = 5$ cm, $b = 3$ cm and $c = 7$ cm

b $\angle B$, when $a = 8$ cm, $b = 11$ cm and $c = 5$ cm

c $\angle C$, when $a = 4\cdot2$ cm, $b = 3\cdot8$ cm and $c = 4\cdot8$ cm.

4 In $\triangle PQR$, $p = 7$, $q = 5$ and $r = 9$.

a Write down formulae for $\cos P$, $\cos Q$ and $\cos R$.

b In the triangle find and calculate the size of

i the largest angle

ii the smallest angle.

5 The frame of the swing has legs $3\cdot2$ m long and $3\cdot0$ m long. At the bottom the legs are $2\cdot8$ m apart.

a Calculate the angles of the triangle formed.

b If the legs were changed to $2\cdot6$ m apart at the bottom, how would that affect the angles of the triangle?

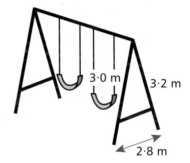

6 Calculate angle X in each diagram.

a

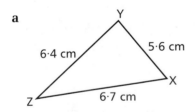

b

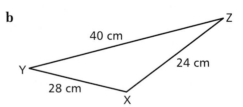

7 The diagram represents a playground.

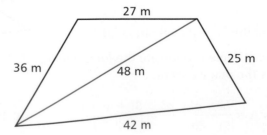

Calculate:

a the sizes of the angles at the corners of the playground

b the length of the other diagonal.

Exercise 5.2

1 Calculate the angles of △KLM.

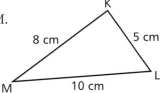

2 Calculate the length of the longer diagonal of parallelogram DEFG.

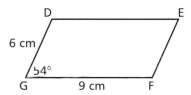

3 The planet Mars (M) has two moons. Phobos (P) is 5900 miles and Deimos (D) is 14 600 miles from the planet's centre, to the nearest 100 miles.
How far apart are the moons, to the nearest 100 miles, when:
a ∠PMD = 70°
b ∠PMD = 140°?

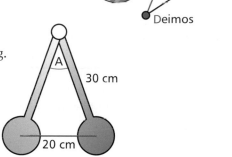

4 The arm of the pendulum is 30 cm long. The centre of the pendulum bob moves a horizontal distance of 20 cm.
Use trigonometry to calculate the size of ∠A in two ways.

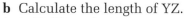

5 The diagram gives some wing dimensions of an aircraft.
a Z is the midpoint of AB. ZX = 18 m. Angle ZXY = 38°. Sketch triangle XYZ
b Calculate the length of YZ.
c Calculate ∠XYZ.

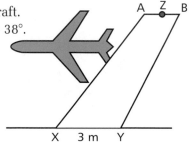

6 The diagram shows the distances between three towns as the crow flies. The bearing of Stirling from Callander is 130°.
a Calculate the bearing of Kilsyth from Callander.
b Calculate the bearing of Kilsyth from Stirling.

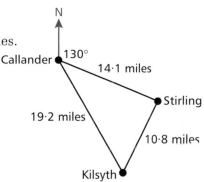

Brainstormer

The hour hand of a clock is 6 cm long, and the minute hand is 10 cm long. Calculate:
a the angle between the hands at ten minutes to two
b the distance between their tips at that time.

6 The area of a triangle

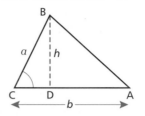

Triangle ABC has an altitude BD of length h units.

$$\text{Area} = \tfrac{1}{2}bh$$

From triangle CDB, $h = a \sin C$

$\Rightarrow \qquad \text{Area} = \tfrac{1}{2}b \times a \sin C$

$$\text{Area of a triangle} = \tfrac{1}{2}ab \sin C$$

It can be shown that the formula works when C is obtuse.

By considering the other altitudes of the triangle it can be shown that:

$$\text{Area} = \tfrac{1}{2}ab \sin C = \tfrac{1}{2}bc \sin A = \tfrac{1}{2}ac \sin B$$

> The area of a triangle is the product of two sides times the sine of the angle between them.

Example Calculate the area of $\triangle ABC$ in which $a = 8$, $b = 5$ and $\angle C = 110°$.
The lengths are in centimetres.

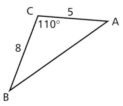

$\text{Area} = \tfrac{1}{2}ab \sin C$

$\qquad = \tfrac{1}{2} \times 8 \times 5 \times \sin 110°$

$\qquad = 18·8$, correct to 3 s.f.

The area is 18·8 cm² correct to 3 s.f.

Exercise 6.1

1 Calculate the area of each triangle below.

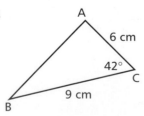

a

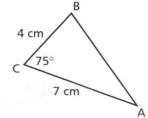

b

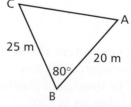

c

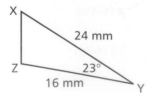

d

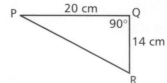

e

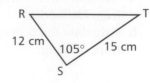

f

2 Calculate the area of:
   a △ABC, with ∠BAC = 26°, b = 9·5 cm and c = 8·6 cm
   b △KLM, with ∠MKL = 95°, l = 6·7 m and m = 9·3 m.

3 The measurements of the sails are given in metres.

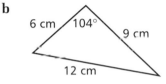
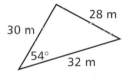

Which of the four sails has:
   a the largest area   **b** the smallest area?

4 Calculate the areas of these triangles.

   a
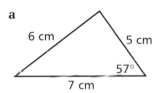

   b

   c

5 Calculate the area of:
   a parallelogram DEFG   **b** rhombus KLMN

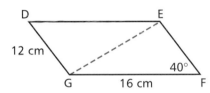
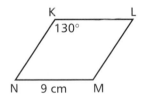

6 The area of △ABC is 108 cm². AC = 24 cm and BC = 18 cm.
   a Calculate two possible sizes of ∠ACB.
   b Sketch the two possible triangles.

7 The side of the factory building is triangular.
   a Calculate the size of ∠A.
   b Calculate the area of the side of the factory.

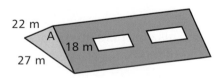

8 The diagram shows the area of ground for a
   new shopping mall.
   Calculate its area.

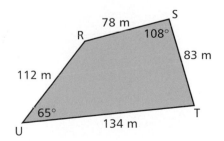

9 Metal Plant make aluminium window boxes.

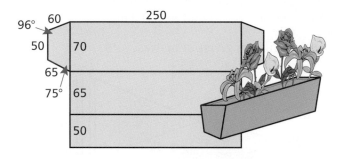

The front, back and base of the window boxes are rectangles.

The two sides are congruent quadrilaterals.

All lengths are in centimetres.

Calculate the area of aluminium needed to make a window box.

Challenges

1 Calculate the area of the field of maze.

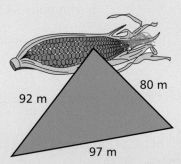

2 This cow field, in the shape of an acute-angled triangle, has an area of 8170 m². Calculate its perimeter, correct to 3 significant figures.

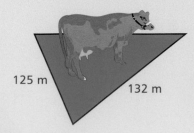

Investigation

Every regular polygon can be divided into congruent isosceles triangles.
For example:

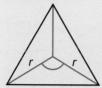 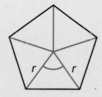 

a Find a formula in terms of r for the area of each of the four polygons.
b Find a formula in terms of n and r for a regular n-sided polygon.
c i As n increases (more and more sides), the polygon's shape approaches a circle.

Use this fact to prove that $\dfrac{n}{2} \sin \left(\dfrac{360}{n} \right)$ is a good approximation for π.

ii What approximation does your calculator give for π?
d Using your calculator, investigate how many sides a polygon must have before it gives an approximation as good as this.
e Construct a spreadsheet that calculates approximations for π for $n = 50, 100, 150, 200, \ldots$, and their differences from π.
Use your spreadsheet to draw a graph of the differences.

7 Which formula?

To find the area of a triangle, use

either $\boxed{\text{Area} = \tfrac{1}{2} \times \text{base} \times \text{height}}$ *or* $\boxed{\text{Area} = \tfrac{1}{2} ab \sin C}$

The table below summarises the correct formula to use when starting to solve a triangle.

Given	Sketch	Use
Three sides		$\cos A = \dfrac{b^2 + c^2 - a^2}{2bc}$
Two sides and the angle between them		$a^2 = b^2 + c^2 - 2bc \cos A$
Two sides and the angle *not* between them		$\dfrac{a}{\sin A} = \dfrac{b}{\sin B}$ Be on guard for a second answer
One side and two angles		Find the third angle then $\dfrac{a}{\sin A} = \dfrac{b}{\sin B}$

Exercise 7.1

1 Which would you use (the sine rule or the cosine rule) to calculate x in each diagram? Do not do the calculations.

a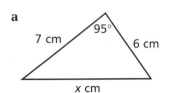

b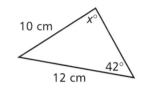

c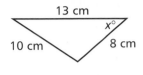

2 Calculate BC in each triangle.

a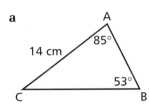

b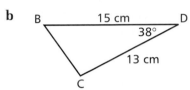

3 Calculate the size of $\angle RST$ in each triangle.

a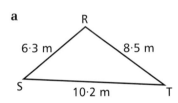

b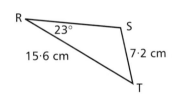

4 Calculate the area of $\triangle DEF$.

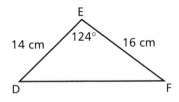

5 In $\triangle XYZ$, calculate:
   a $\angle ZXY$
   b the triangle's area.

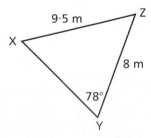

6 Sketch a triangle with sides 7 cm, 8 cm and 11 cm long. Calculate:
   a the sizes of all the angles in the triangle
   b the area of the triangle.

7 Three roads meet at P, Q and R.
Hazel Road is 165 m between P and R.
Adam's Gate is 142 m between Q and R.
Calculate the length of Blossom Drive
between P and Q.

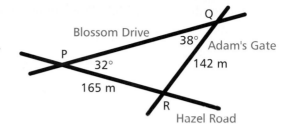

8 Two explorers are lost together in the desert.
One leaves camp and walks 5 km on a bearing of 310°.
The other walks 7 km on a bearing of 065°.
How far apart are the two explorers?

9 Two observers 80 m apart see a balloon in the air at C.
The angle of elevation from A to the balloon is 63°,
and from B it is 48°. C is directly above the line AB.
a Calculate the distance from A to the balloon.
b Calculate the height of the balloon.
(Hint: copy the triangle, and draw CD perpendicular
to AB.)

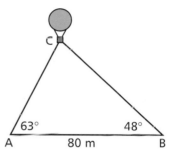

10 Calculate *x* in each part of question **1**.

Exercise 7.2

1 Calculate:
a the length of both diagonals of parallelogram
RSTU
b the area of the parallelogram.

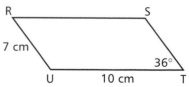

2 Calculate the area of △PQR.

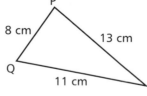

3 The *Sea Sprite* sails 34 km from Aberdeen harbour on a
bearing of 154°.
It then changes course to 075° and sails a further
46 km.
a How far is the *Sea Sprite* now from Aberdeen
harbour?
b What is its bearing from the harbour?

4 A paraglider is 65 m high. From an observer at A the angle of elevation to the paraglider is 59°.

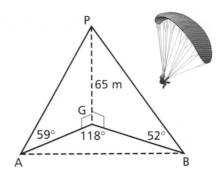

From an observer at B the angle of elevation of the paraglider is 52°.

A, B and G are on horizontal ground.

Calculate:

a AG

b BG

c the distance between the observers at A and B.

5 A piece of land, ABCD, is to be divided evenly between two brothers. They measure the diagonal AC and because the perimeter of △ADC is the same as the perimeter of △ABC, brother Dom is given △ADC and brother Tom is given △ABC.

Which brother is in fact given the bigger piece of land? How much bigger is it?

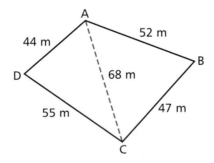

6 Calculate:

a the total length of the picture cord (A to B to C)

b ∠BAC when B is in the middle of the cord (and the picture is horizontal).

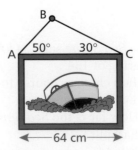

◀◀ RECAP

Solving right-angled triangles

$\sin A° = \dfrac{\text{opposite side}}{\text{hypotenuse}}$

$\cos A° = \dfrac{\text{adjacent side}}{\text{hypotenuse}}$

$\tan A° = \dfrac{\text{opposite side}}{\text{adjacent side}}$

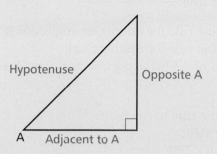

Solving scalene triangles

Given	Sketch	Use
Three sides		$\cos A = \dfrac{b^2 + c^2 - a^2}{2bc}$
Two sides and the angle between them		$a^2 = b^2 + c^2 - 2bc \cos A$
Two sides and the angle *not* between them		$\dfrac{a}{\sin A} = \dfrac{b}{\sin B}$ Be on guard for a second answer
One side and two angles		Find the third angle then $\dfrac{a}{\sin A} = \dfrac{b}{\sin B}$

Finding the area of a triangle

$\quad \text{Area} = \frac{1}{2} \times \text{base} \times \text{height}$

or $\quad \text{Area} = \frac{1}{2} ab \sin C$

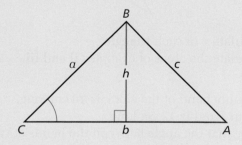

1 Copy and complete the following for △ABC.

  a the sine rule: $\dfrac{\ldots}{\ldots} = \dfrac{b}{\ldots} = \dfrac{c}{\sin C}$

  b the cosine rule for finding an angle: $\cos B° = \dfrac{\ldots + \ldots - b^2}{2ac}$

  c the cosine rule for finding a side: $c^2 = a^2 + \ldots$

  d the area of △ABC using trigonometry: Area $= \frac{1}{2} \ldots \sin B°$

2 Use the sine rule to calculate:

  a AB in △ABC

  b ∠DEF in △DEF

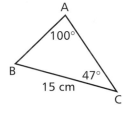
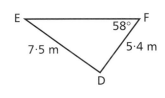

3 Use the cosine rule to calculate:

  a AC in △ABC

  b ∠XZY in △XYZ.

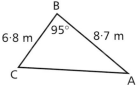
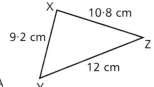

4 Sine rule, cosine rule or Pythagoras' theorem?

  a Which would you use to calculate x in each of these triangles?

i
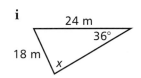

ii
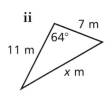

iii
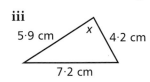

iv
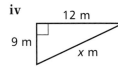

  b Calculate x in each triangle.

  c Calculate the areas of triangles **ii** and **iii**.

5 The minute hand of the clock is 18 cm long.
The hour hand is 12 cm long.

  a Show that the angle between the hands at twenty past eight is 130°.

  b Calculate the distance between the tips of the hands at that time.

6 The two triangles have the same area.
Calculate the value of x.

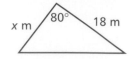
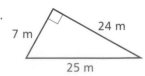

REVISE

7 Three friends, Abe, Bev and Clare, live in the same town.
Bev's house is on a bearing of 042° from Abe's.
From Bev's house, Clare lives 760 m, as the crow flies,
on a bearing of 155°.
From Abe's house, Clare's house is on a bearing of 138°.
Calculate the distance as the crow flies between:
a Abe's house and Clare's house
b Abe's house and Bev's house.

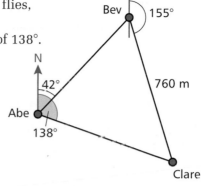

8 An airport runway is 3·2 km long.
The distance of a plane at P, after take-off, is
3·8 km from one end of the runway, and
4·7 km from the other end.
a Calculate the angle of elevation to the
plane from
i A   ii B.
b Calculate the height of the plane, h m.

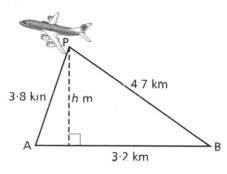

REVISE

13 Chapter revision

Revising Chapter 1   Area and volume

1 Calculate the area of each shape.

a

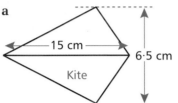

15 cm
6·5 cm
Kite

b

11·7 cm
4·9 cm
Parallelogram

c

23 cm
12 cm
Triangle

2 This is the sign for the 'Off the Square' café.
The diagonals of the rhombus measure 68 cm
and 1·14 m.
Calculate the area of the sign.

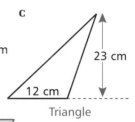

OFF THE
SQUARE

3 An old farm building is being converted
into a house. Three arches which were
entrances to the 'cart-shed' have been made
into windows.
Calculate the total area of glass required.

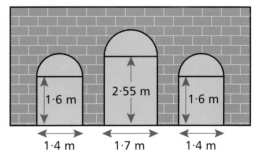

1·6 m   2·55 m   1·6 m
1·4 m   1·7 m   1·4 m

4 Calculate the volume
and surface area of
each object.

a

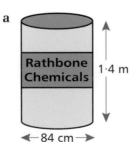

Rathbone
Chemicals
1·4 m
84 cm

b

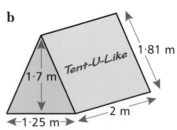

1·7 m   Tent-U-Like   1·81 m
1·25 m   2 m

5 Calculate the volume and surface area of
this cardboard box.
All measurements are in centimetres.

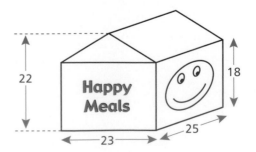

22
Happy
Meals
18
23   25

Revising Chapter 2 Money

1 Mr Forest works a 40 hour week. He is paid £8·65 per hour.
Calculate his:

   a weekly wage   **b** total annual pay.

2 Mrs Wood's annual salary is £29 616.

   a Calculate her monthly salary.

   b She receives a pay rise of 4%.
      Calculate her new
      i monthly   **ii** annual salary.

3 Jill starts work at 07 45. From Monday to Thursday she finishes at 16 15.
On Friday she finishes at 16 00. She has 45 minutes for lunch each day.

   a How many hours does she work in a week?

   b Her basic rate of pay is £6·76 for a 35 hour week.
      Overtime (above 35 hours) is paid at time and three-quarters.
      Calculate her total weekly pay.

4 **a** After a 6% pay rise Ged's weekly wage has risen to £676·28.
      What was his wage before the rise?

   b Geraldine earns £2445·30 a month. She has just had a 4·5% rise.
      Calculate her monthly salary before the rise.

5 This is Paul's payslip for March.

Name P Prentice	Employee number 53	NI number ZA98765A	Week number 12
Basic pay £2318·43	Overtime £74·42	Bonus £39·38	Gross pay
Income tax £434·38	NI £227·47	Pension £145·93	Total deductions
			Net pay

Calculate his:

   a gross pay   **b** total deductions   **c** net pay.

6 The table gives the National Insurance Contributions for 2004–05

Rate	Weekly	Annual
0%	<£91	<£4745
11%	£91–£610	£4745–£31 720
1%	>£610	>£31 720

Calculate the amount these workers pay in National Insurance Contributions.

   a Arnold, annual wage £3642

   b Benny, annual wage £25 543

   c Charlie, annual wage £39 600

REVISE

7 The table shows the income tax rates for different taxable incomes for 2004–05.

Taxable income	Rate of tax
£0–£2020	10%
£2020–£31 400	22%
over £31 400	40%

Calculate the total income tax paid by:
a Doreen, taxable income £2800
b Emma, taxable income £28 490
c Frances, taxable income £58 530

8 Calculate the total tax payable for these people:
a Mr Wright, total annual income £24 086. Tax allowance £5390.
b Ms Hall, total annual income £36 960. Tax allowance £4965.

9 Mr Singh invests £60 000 at 4·8% p.a. interest.
He leaves the interest in the account.
Calculate the balance of the account after 5 years.

10 a The value of Mr Short's investment portfolio has grown from £7400 to £7881.
Calculate the rate of growth as a percentage.
b Two years ago Mrs Long's portfolio was worth £12 600.
It depreciated by 1·6% in the first year and appreciated by 1·6% in the second.
What is its value now?
c Ms Broad's portfolio is now valued at £19 152.
It depreciated by 5% last year but appreciated by 12% the year before.
What was its value 2 years ago?

Revising Chapter 3   Similarity

1 On a map, the distance between two hills is 2·4 cm.
The representative fraction of the map is 1 : 40 000.
What is the actual distance between the hills?

2 In a hill race, the actual distance between two checkpoints was 3 km.
How long would this distance be on a map with representative fraction 1 : 20 000?

3 Joe sees Sam on a bearing of 220° and heads off on a bearing of 140° at a steady pace of 6 km/h.
Half an hour later he sees Sam again, this time on a bearing of 300°.
a Make an accurate drawing of Joe's route and identify Sam's position.
b How far was Sam away from Joe at the start?
c At the start, what was Joe's bearing from Sam?

4 A tree standing 32 metres tall casts a shadow which is 24 metres long.
At the same time, another tree casts a shadow which is 16 metres in length.
What is the height of the second tree? (Use similar triangles.)

REVISE

5 In New York they sell Statue of Liberty souvenirs.
Though different in size, the models are similar.

  a What is the reduction factor for the two models
shown?

  b The larger model uses up 42 cm³ of metal.
What volume of metal is needed for the
smaller model?

  c The base of the larger model is 6 cm².
Calculate the area of the base of the smaller
statuette.

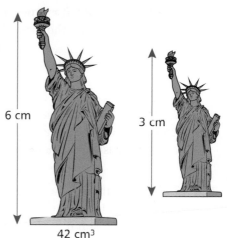

6 cm   3 cm

42 cm³

6 Two cuboids, A and B, are similar.
Their volumes are 720 000 cm³ and 368 640 cm³ respectively.
The length of box A is 150 cm.
What is the length of box B?

7 In each diagram there is a pair of similar triangles.
Calculate the value of the labelled length in each case.

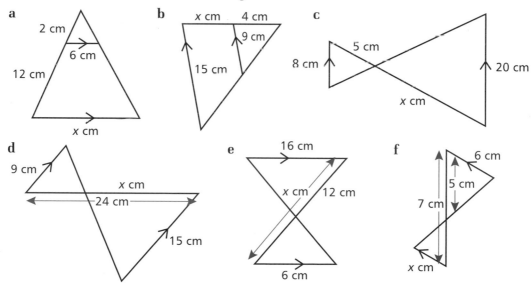

a 2 cm, 6 cm, 12 cm, x cm

b x cm, 4 cm, 9 cm, 15 cm

c 5 cm, 8 cm, 20 cm, x cm

d 9 cm, x cm, 24 cm, 15 cm

e 16 cm, x cm, 12 cm, 6 cm

f 6 cm, 5 cm, 7 cm, x cm

Revising Chapter 4 Formulae

1 Use the given formula to calculate the indicated quantity.
Give answers to 3 significant figures where necessary

  a The distance travelled when driving at 55 km/h for 2·5 hours $(D = ST)$

  b The area of a circle with radius 7 mm $(A = \pi r^2)$

  c The diameter of a circle with circumference 85 cm $\left(D = \dfrac{C}{\pi}\right)$

  d The perimeter of an 8·2 m × 3·6 m rectangle $(P = 2(l + b))$

  e The 'height' (altitude) of a triangle with area 45 cm² and base length 6 cm $\left(h = \dfrac{2A}{b}\right)$

REVISE

2 When £P is invested at r% per month (compound interest), the amount, £A, after m months is given by the formula:

$$A = P\left(1 + \frac{r}{100}\right)^m$$

a Calculate A when $P = 150$, $r = 10$ and $m = 3$.
b Which will produce the largest amount:
£250 invested at 6% per month for 4 months
or £200 invested at 5% per month for 8 months?

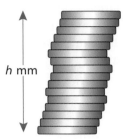

h mm

3 a Find a formula for the value, £V, of a pile of £1 coins
that is h mm tall if each coin is 3 mm thick.
b Calculate V when $h = 39$.

4 a Find a formula for the number of matches, M, needed to make design m.
(The designs continue in the same pattern.)

Design 1

Design 2

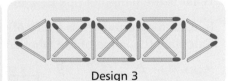

Design 3

b How many matches make up design 17?
c One of the designs in this pattern uses 245 matches.
By solving an equation find which design this is.

5 In each case, change the subject of the given formula to the indicated letter.

a $M = \dfrac{k}{w}$, to k **b** $L = PQ$, to Q **c** $T = \dfrac{l}{m}$, to m

d $5(a - b) = c$, to a **e** $Q^2 - n = A$, to Q **f** $P = 3 - \dfrac{a}{b}$, to b

6 The formula for this 'quarter-ellipse' is $y = \dfrac{\sqrt{4 - x^2}}{2}$.

a Show that this formula changes to $x = 2\sqrt{1 - y^2}$.
b i Calculate y when $x = 1\cdot6$.
  ii Calculate x when $y = 0\cdot6$.

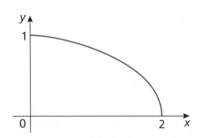

7 a $m = \dfrac{x}{y}$. What is the effect on m of

  i doubling y **ii** trebling x **iii** halving x?

b $k = n^2\sqrt{a}$. What is the effect on k of

  i doubling n **ii** multiplying a by 4   **iiii** dividing a by 9?

Revising Chapter 5   Equations and inequations

1 Solve these equations:

a $7 - y = 13y$ **b** $5(x - 2) = 15$ **c** $k + 2(k + 1) = 8$

d $12 - (x + 2) = x$ **e** $x^2 - 2x + 3 = x + x^2$ **f** $3x + 2x^2 = 2x^2 - x + 8$

2 The perimeter of this rectangular lawn is 48 metres.
  a Make an equation to represent this situation.
  b Solve the equation.
  c Find the dimensions of the lawn.

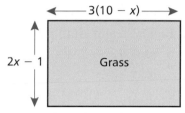

Lengths are in metres.

3 Simplify:

a $\dfrac{y}{3} \times 9$ **b** $\dfrac{2m}{3} \times \dfrac{9}{14}$ **c** $4 \times \dfrac{y}{2x}$ **d** $\dfrac{6}{5} \times \dfrac{5k}{3}$

4 Solve these equations:

a $\dfrac{k}{2} = 1$ **b** $\dfrac{y}{3} = \dfrac{1}{6}$ **c** $\dfrac{n}{8} = \dfrac{3}{4}$ **d** $\dfrac{2x}{5} = 3$

e $\dfrac{1}{7} = \dfrac{y}{14}$ **f** $\dfrac{1}{3} = \dfrac{2y}{9}$ **g** $\dfrac{7x}{8} = \dfrac{15}{16}$ **h** $\dfrac{1}{6} = \dfrac{5x}{8}$

i $\frac{1}{2}(x + 2) = 7$ **j** $\dfrac{x - 3}{4} = \dfrac{1}{2}$ **k** $\frac{2}{5}(m + 3) = m$ **l** $\dfrac{3x - 2}{4} = \dfrac{x + 1}{2}$

m $y - 3 = \frac{2}{5}y$ **n** $\dfrac{k}{5} = 3k - 14$ **o** $\frac{3}{4}x = x - 6$ **p** $\frac{1}{5}y = 2(y - 9)$

5 Scientists are testing new growth hormones on sheep. The results are shown below. For each sheep:

  a set up an equation that describes the situation (let the starting weight be x kg)

  b solve the equation to find the sheep's starting weight.

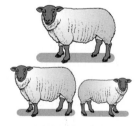

Sheep 1	Sheep 2	Sheep 3
In the 1st week there was a gain of 6 kg with the resulting weight being $1\frac{1}{5}$ times the starting weight.	Illness set in and there was a loss of 8 kg during the 1st week. The final weight was $\frac{5}{7}$ of its starting weight.	An increase of 4 kg during the first week. The starting weight was $\frac{5}{6}$ of this new weight.

Revising Chapter 6   Algebraic fractions

1 Which of the following are equivalent to $\dfrac{a}{2b}$?

   a $\dfrac{ab}{2b^2}$ **b** $\dfrac{5a^2}{10ab}$ **c** $\dfrac{3a}{6ab}$

   d $\dfrac{a+1}{2b+1}$ **e** $\dfrac{3az}{6bz}$ **f** $\dfrac{2b}{a}$

2 Factorise and simplify:

   a $\dfrac{3x+12}{5x+20}$ **b** $\dfrac{x^2-1}{3x+3}$ **c** $\dfrac{x^2+x-6}{x^2+3x}$ **d** $\dfrac{x^2+x-2}{x^2+5x+6}$

3 Multiply:

   a $\dfrac{a}{2}\times\dfrac{3}{b}$ **b** $3g\times\dfrac{4}{2g}$ **c** $\dfrac{x-1}{x^2}\times\dfrac{1}{x-2}$

4 Divide:

   a $\dfrac{2a}{3b}\div\dfrac{4a}{b}$ **b** $\dfrac{3x^2}{yz}\div\dfrac{2x}{3yz^2}$ **c** $2a\div\dfrac{4}{a}$ **d** $\dfrac{2x+3}{x+1}\div\dfrac{4x^2-9}{3}$

5 Add:

   a $\dfrac{a}{2}+\dfrac{a}{3}$ **b** $\dfrac{t}{5}+\dfrac{4}{t}$ **c** $\dfrac{1}{p}+\dfrac{1}{q}$ **d** $\dfrac{x}{x+1}+\dfrac{1}{3}$ **e** $2h+\dfrac{3}{h}$

6 Subtract:

   a $\dfrac{x}{4}-\dfrac{x}{5}$ **b** $\dfrac{2}{a}-\dfrac{3}{b}$ **c** $\dfrac{1}{d}-d$

7 In good conditions Peter travels 8 km per hour.
Going against the wind for 10 km he averages $(8-x)$ km/h.
For the journey back with the wind assisting him he averages $(8+x)$ km/h.
   a Form an expression in x for:
     i the time taken to cover the 10 km against the wind
     ii the return journey (wind assisted)
     iii the total time for the round trip.
   b If this total time is 4 hours calculate x to 1 d.p.

Revising Chapter 7   The straight line

1 Give the equation of the two gridlines which pass through the point $(3, -2)$.

2 The profile of the track on a corner is as shown.
What is the gradient of the slope?

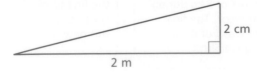

2 cm

2 m

3 Calculate the gradient of the line passing through:

 a (1, 5) and (5, 8)   **b** (−3, 2) and (4, 6)

 c (8, −2) and (8, 1)   **d** (6, −1) and (12, −1)

4 a Copy and complete the table.

 b Draw the line with equation $y = 3x − 1$.

 c What is its gradient?

 d Where does it cut the y axis?

x	0	1	2	3	4
$y = 3x − 1$	−1				

5 Write down the equation of the line with:

 a a gradient of −2, cutting the y axis at 5

 b a gradient of $\frac{1}{5}$, passing through (0, −2).

6 Draw a sketch of the following lines:

 a $y = 2x + 3$ **b** $y = −x − 4$ **c** $y = \frac{1}{3}x + 6$ **d** $y = −\frac{1}{4}x$

7 The more weight you put on a spring, the closer it gets to the ground. The graph shows how the height varies with the weight.

 a What is the height when there is no weight on the spring?

 b Work out the equation of the line, expressing the total weight in grams (W g) in terms of the height, h cm.

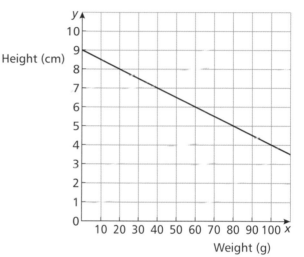

8 A teacher recorded the maths and physics marks of ten students.

Student no.	1	2	3	4	5	6	7	8	9	10
Maths mark	3	4	5	4	6	5	4	3	5	6
Physics mark	0	30	70	10	90	40	60	20	50	80

 a Make a scatter diagram to illustrate the data.

 b Add a line of best fit to the diagram.

 c Work out the equation of your line.

 d Use the equation to estimate the physics mark of a student who scored 4·5 for maths.

Revising Chapter 8 Quadratic equations

1 Solve the following quadratic equations.

 a $3x^2 + 13x + 4 = 0$ **b** $5x^2 + 11x + 2 = 0$

 c $3x^2 − 75 = 0$ **d** $64 − x^2 = 0$

 e $125 − 5x^2 = 0$ **f** $16x^2 − 49 = 0$

 g $98a − 72ax^2 = 0$ **h** $4x^2 + 40x + 84 = 0$

REVISE

2 Solve these quadratic equations, giving your answers correct to 2 decimal places.

  a $x^2 - 11x + 1 = 0$ **b** $3x^2 + 10x + 2 = 0$

  c $-2x^2 + 8x - 3 = 0$ **d** $8 - 7x - x^2 = 0$

3 This graph has the equation $y = x^3 - 3x - 1$.

  a It cuts the x axis where $y = 0$.
    Show that the equation $x^3 - 3x - 1 = 0$ has
    a solution between $x = 1$ and $x = 2$.
  b Find this solution correct to 2 decimal places.
  c Find the other two solutions of the
    equation $x^3 - 3x - 1 = 0$.

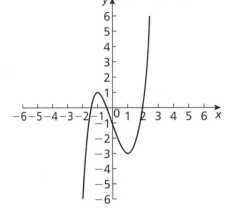

4 Solve these quadratic equations, giving your answers, where necessary, correct to
1 decimal place.

  a $\dfrac{x}{x + 1} = \dfrac{10}{x}$ **b** $\dfrac{x - 1}{x} = \dfrac{x - 1}{5}$

  c $x + \dfrac{3}{x} = 8$ **d** $\dfrac{1}{x} - \dfrac{1}{x + 1} = \dfrac{1}{30}$

5 The perimeter of a rectangle is 32 cm. Let x be the length of the rectangle.
  a Write down an expression for the breadth of the rectangle.
  b Given that the area is 28 cm², find the length and breadth of the rectangle.

Revising Chapter 9   Surds and indices

1 Simplify these expressions:

  a $x^{10} \times x^4$ **b** $3x^7 \times 4x^9$ **c** $27x^3 \div (9x)$ **d** $18x^7 \div (24x^6)$

  e $(3x^2)^4$ **f** $(9x^4)^2$ **g** $\dfrac{6x^3 \times 3x^7}{2x^6}$ **h** $8x^3 \times 2x^0$

2 Simplify, expressing your answer using positive indices:

  a x^{-11} **b** x^{-20} **c** $3x^{-10}$ **d** $\dfrac{x^{-7}}{2}$

  e $\dfrac{7x^{-3}}{6}$ **f** $x^5 \times x^{-6}$ **g** $3x^{-2} \times 4x^{-3}$ **h** $\dfrac{x^6}{x^4 \times x^{10}}$

  i $(x^8)^{-2}$ **j** $3(x^2)^{-4}$ **k** $\dfrac{(x^4)^{-5}}{6}$ **l** $\dfrac{5}{x^{-6}}$

  m $\dfrac{1}{6x^{-7}}$ **n** $\dfrac{10}{3x^{-10}}$ **o** $\dfrac{x^{-6}}{x^{-4}}$ **p** $(2x)^{-4}$

3 Simplify each expression:

a $x^4(x^3 - x^{11})$ **b** $x^{\frac{1}{3}}(x^{\frac{5}{3}} - x^{-\frac{10}{3}})$

4 Evaluate:

a $(15x^0)^2$ **b** $4^{\frac{7}{2}}$ **c** $9^{-\frac{5}{2}}$ **d** $\dfrac{1}{32^{\frac{3}{5}}}$

e $\dfrac{1}{64^{\frac{3}{2}}}$ **f** $\dfrac{2}{8^{-\frac{5}{3}}}$ **g** $\dfrac{8}{100^{-\frac{3}{2}}}$ **h** $\dfrac{2^{-6}}{3^{-2} \times 4^{-2}}$

5 Simplify each surd:

a $\sqrt{500}$ **b** $\sqrt{343}$ **c** $\sqrt{384}$ **d** $\sqrt{405}$

6 Simplify each expression, leaving your answer in surd form:

a $8\sqrt{2} - \sqrt{8}$ **b** $3\sqrt{2} + 2\sqrt{3} - 5\sqrt{2}$ **c** $\sqrt{96} - \sqrt{54}$

d $\sqrt{200} + \sqrt{800}$ **e** $\dfrac{\sqrt{243}}{\sqrt{192}}$ **f** $\sqrt{\dfrac{900}{1600}}$

7 Rationalise the denominators:

a $\dfrac{1}{\sqrt{22}}$ **b** $\dfrac{3}{\sqrt{7}}$ **c** $\dfrac{4}{6\sqrt{8}}$ **d** $\dfrac{\sqrt{5}}{3\sqrt{10}}$

e $\dfrac{\sqrt{3}+1}{\sqrt{2}}$ **f** $\dfrac{1}{\sqrt{6}-1}$ **g** $\dfrac{4}{3-\sqrt{8}}$ **h** $\dfrac{10}{\sqrt{6}-\sqrt{3}}$

Revising Chapter 10　Functions

1 a Given that $f(x) = x + 3x^2$, evaluate $f(-2)$.

　b $g(x) = \dfrac{2}{x^2}$. Find the value of $g(\frac{1}{2})$.

2 a $f(x) = \dfrac{a}{x}$. Find a given that $f(-2) = 4$.

　b $h(x) = 5^x$.

　　i What is the value of $h(4)$?　　**ii** Given that $h(x) = \sqrt{125}$, find x.

3 $f(x) = 2 - 5x$. Find $f(1 - a) + f(1 + a)$ in its simplest form.

4 $p(x) = ax + b$; $p(1) = 1$ and $p(-1) = -5$.
　a Write down two equations in a and b.
　b Solve the equations to find the values of a and b.
　c Write down the formula for p.

5 a Sketch the graph of the quadratic function $f(x) = 5 + 4x - x^2$.
　b Write down
　　i the maximum value of $f(x)$
　　ii the coordinates of the turning point
　　iii the equation of the axis of symmetry of the parabola
　　iv the values of x for which $f(x) = 0$.

REVISE

6 Use the given data to find the actual equations of these graphs.

They are of the form $y = \dfrac{a}{x}$, $y = ax^2$ or $y = a^x$.

a
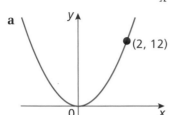

b
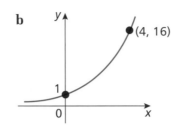

c
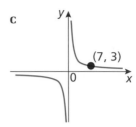

7 Every hour after its heating system is switched off, the temperature of a room drops 2 °C. The formula for the room temperature, T °C, after x hours is $T(x) = 25 - 2x$.
  a Calculate   **i** $T(2)$ **ii** $T(6)$
  b For how many hours is the heating switched off for the temperature to fall to 9 °C?

8 A company quadruples its profits (P) every year (t).
  Its profits in its first year were £1 million.
  a What were its profits in the fifth year?
  b Express the profits as an exponential equation.

Revising Chapter 11   Trigonometric graphs and equations

1 Sketch the graph of:
  a $y = \sin x°$ **b** $y = \cos x°$ **c** $y = \tan x°$ for $0 \leqslant x \leqslant 360$.

2 The line $y = -0.6$ is drawn and cuts the graph of $y = \cos x°$ at A and B. Find the coordinates of A and B.

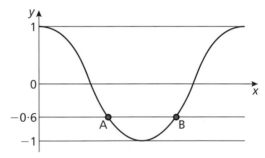

3 **a** Write down the maximum and minimum values of $3 \cos 2x°$.
  b What is the period of the graph of $y = 3 \cos 2x°$?
  c Sketch the graph of $y = 3 \cos 2x°$ for $0 \leqslant x \leqslant 360$.
  d Describe how the graph of $y = 2 + 3 \cos 2x°$ differs from the graph of $y = 3 \cos 2x°$.

4 Find the equations of these graphs. They are of the form $y = b \sin ax°$ or $y = b \cos ax°$.

a
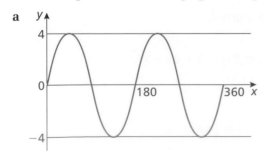

b
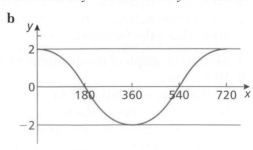

5 Find the equation of this graph. It is of the form $y = c + b \sin ax°$.

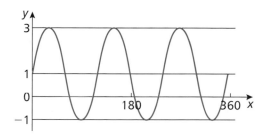

6 The height, h cm, of the end of the hour hand of a wall clock above the floor is given by the formula $h(t) = 160 + 8 \sin 30t°$, where t is measured in hours.

  a What are the maximum and minimum heights of the end of the hour hand above the floor?

  b After how many hours do the first maximum and minimum heights occur?

  c What is the length of the hour hand?

  d What is the period of h?

  e Calculate h after   **i** 1 hour   **ii** 4 hours.

  f Sketch the graph of $h(t) = 160 + 8 \sin 30t°$.

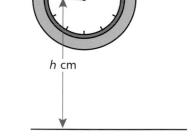

7 Solve algebraically the equations:

  a $\sin x° = 0.45$, for $0 \leqslant x \leqslant 360$

  b $3 + 4 \tan x° = 0$, for $0 \leqslant x \leqslant 360$

  c $6 \sin x° = 5 \cos x°$, for $0 \leqslant x \leqslant 360$

8 **a** Calculate $\cos x°$ for $0 \leqslant x \leqslant 180$, when $\sin x° = 0.9$.

  b What is the value of $\tan x°$ when $2 \sin x° - \cos x° = 0$ and x is an acute angle?

Revising Chapter 12   Trigonometry and triangle calculations

1 Calculate the value of x in each triangle.

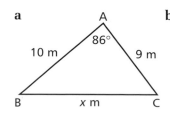

2 Two entrances, P and Q, to a mine are 180 m apart. The line joining P and Q is horizontal. The angle of depression from P to R, where the two shafts meet, is 16°. Section QR is 56 m long.

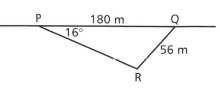

  a Calculate obtuse angle PRQ and hence find the angle of depression from Q to R.

  b How far is R below the surface of the mine?

REVISE

3 In △ABC, AB = 8 units, BC = 7 units
   and AC = 9 units.
   Show that cos A = $\frac{2}{3}$.

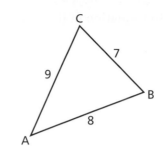

4 Two boats, P and Q, are 110 m apart.
   A helicopter at H is directly above the line
   joining P and Q. The angle of elevation
   from P to H is 67° and from Q to H is 58°.
   Calculate the height of the helicopter.

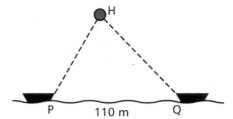

5 The diagrams represent two
   triangular plots of land.
   Which has the greater area?
   By how much?

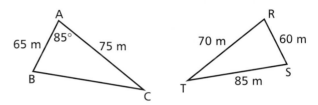

6 A plane flies from A on a bearing of 078°
   for a distance of 850 km to B where it
   changes course to 164° and flies a further
   1000 km to C.
   Calculate:
   a the distance the plane now is from A
   b its bearing from A
   c the course needed to fly the plane back
      to A.

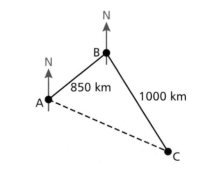

7 ABCD is a trapezium with AB parallel to
   DC. AB = 50 cm, AD = 30 cm, BC = 40 cm
   and diagonal BD = 65 cm.
   a Calculate the size of obtuse ∠BAD.
   b Calculate the height AE of the
      trapezium.
   c Calculate the area of the trapezium.

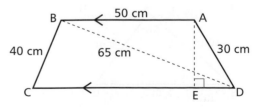

REVISE

14 Preparation for assessment

 Test A   Part 1

1. A box weighed $6 \cdot 31 \times 10^4$ grams. What would four such boxes weigh? Express your answer in scientific notation.

2. Split £96 in the ratio $5 : 3$.

3. Rationalise the denominator: $\dfrac{\sqrt{2}}{\sqrt{5}}$

4. The two triangles are similar.
   Find the value of x.
   All lengths are in centimetres.

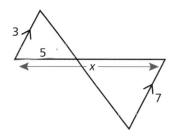

5. A selection of people at the cinema were asked their age.
   These were the responses.

25	18	46	24	17
48	21	29	17	16
44	34	17	32	41
15	38	50	40	43
25	17	15	47	24

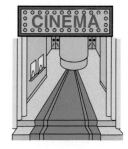

   Sort the data into ascending order using a stem-and-leaf diagram.

6. Simplify $\dfrac{x}{4} + \dfrac{3}{x}$.

7. This is a sketch of a straight line.
   What is its equation?

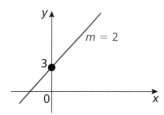

8. Solve these equations simultaneously:
   $2x - 3y = 14$
   $3x - y = 14$

9 Solve $7 - 3x = 25$.

10 $a = 3$, $b = -2$ and $c = -1$. Calculate the value of:
   a $a + b + c$ **b** $a - b - c$

11 Copy and complete this table where $y = 2x^2 - 6x - 8$.

x	-2	-1	0	1	2
y					

12 Sketch the graph of $y = 2 + 3 \cos \frac{1}{2}x°$.

13 Solve algebraically the inequality $6 - 3x \leqslant x + 2$.

Test A   Part 2

1 Vanessa deposits £14 000 in a savings account which pays an interest rate of 4·2% p.a. Calculate the total interest paid after 3 years if she:
   a withdraws the interest each year (simple interest)
   b leaves the interest in her account (compound interest).

2 $f(x) = px + q$; $f(2) = 7$ and $f(-3) = -13$.
   a Write down two equations in p and q.
   b Solve the equations to find the values of p and q.
   c Write down the formula for f.

3 Solve $\tan x° + 3 = 0$ for $0 \leqslant x \leqslant 360$.

4 The height, $h\,km$, of a satellite above the Earth after t hours is given by the formula $h(t) = 1200 + 100 \sin 15t°$.
   a What are the maximum and minimum heights of the satellite?
   b After how many hours is it first at its maximum height?
   c What is its height after
      i 10 hours   **ii** 24 hours?
   d What is the period of $h(t)$?
   e Sketch the graph of $h(t) = 1200 + 100 \sin 15t$.

5 A telephone help-line is manned by volunteers 24 hours a day. The number of hours each volunteer works per week depends on the number of volunteers.

Number of volunteers (n)	4	5	6	7	8
Time (t hours)	42	33·6	28	24	21

   a Make a new table of t against $\dfrac{1}{n}$.
   b Draw a graph to illustrate the new table.
      How does the graph show that t varies inversely as n?
   c Find a formula for t in terms of n.
   d If there were 12 volunteers how much time per week would each one need to volunteer?

REVISE

6 A railway tunnel is based on a circle as shown.
   a Calculate the height of the railway tunnel.
   b Calculate the size of angle AOB.
   c Calculate the area of triangle AOB.
   d Calculate the area of the cross-section of
      the tunnel.
   e The tunnel is 500 metres long. Calculate its volume.

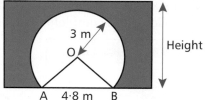

7 Calculate the mean and standard deviation of this set of marks.
   They are a random sample of the marks scored in a maths test.

62	10	20	69	50
18	100	47	29	43

8 An extension is being built onto a golf clubhouse
   so that people in the clubhouse can get a good
   view of the 18th green.
   The floor plan is shown here.
   a Calculate the area of the extension.
   b Calculate the cost of carpeting this extension
      with carpet that costs £11·99 per square metre.

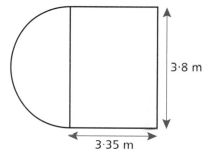

9 Joanne saved 50p, 20p and 10p coins in a savings jar.
   After a year, she had 280 coins.
   The ratio of 50p, 20p and 10p coins was 2 : 3 : 5 respectively.
   a How many 20p coins did she have?
   b How much money had she saved altogether?

10 a Show that the equation $x^3 - 2x + 1$ has a solution between $x = -2$ and $x = -1$.
   b Find this solution correct to 2 decimal places.

11 The diagram shows sales of a magazine at various outlets around a town before and
   after an advertising campaign.

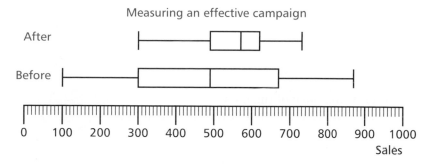

Measuring an effective campaign

Write a few sentences about the effectiveness of the advertising campaign.

12 A six-sided and a four-sided dice are cast at the same time.

  a Copy and complete the table to show all the possible outcomes.

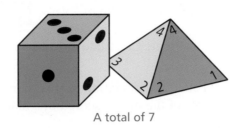

	1	2	3	4
1				
2				
3				
4				
5				
6				

A total of 7

  b What is the probability of scoring a total of 7?

13 For two numbers x and y there are three different **means** that can be calculated. Here are the formulae:

Arithmetic mean: $A = \dfrac{x + y}{2}$ Harmonic mean: $H = \dfrac{2xy}{x + y}$ Geometric mean: $G = \sqrt{xy}$

For $x = 2{\cdot}5$ and $y = 5$, calculate each of the three means to 2 decimal places and sort them into increasing order.

Test B Part 1

1 A sheet of paper is 8×10^{-5} m thick. How thick is a ream of 500 such sheets? Express your answer in standard form.

2 Calculate $8\frac{1}{3} - 2\frac{3}{5}$.

3 Simplify, expressing your answer using positive indices:
  a $x^4 \times x^{-7}$ **b** $(x^4)^{-7}$

4 Rationalise the denominator: $\dfrac{5}{2\sqrt{10}}$.

5 The outside temperature at several places round a school was measured in degrees Celsius.

4·5	2·3	4·0	1·8	2·3
3·4	4·3	4·5	1·6	3·3
3·4	4·2	4·9	4·5	2·8
4·0	1·4	3·0	4·5	4·0
3·2	1·6	2·3	2·3	2·2

Make a five-figure summary of the data.
(Hint: lowest score, highest score, quartiles and median.)

6 Simplify $\dfrac{3x}{4y} - \dfrac{x}{y}$.

7 Make a sketch of the line with equation $y = 3x + 4$.

REVISE

8 Solve the system of equations:

$y = 2x$

$x + y = 18$

9 Solve $5(x - 4) = 3 - 2(x + 1)$.

10 $f(x) = 9^x$. Evaluate:

   a $f(\frac{1}{2})$ **b** $f(-\frac{1}{2})$

11 Sketch $y = \cos 2x°$ for $0 \leqslant x \leqslant 360$.

12 A large loaf needs 500 g of flour; a medium loaf needs 400 g.
A baker can make 60 large loaves of bread from one sack of flour.
How many medium size loaves could be made from the sack?

13 Solve algebraically: $\dfrac{x}{2} = \dfrac{2 - 3x}{3}$

 Test B   Part 2

1 Copy and complete this table:

Item	Cost price	Selling price	% profit/loss
Bathroom suite	£480	£537·60	
Power shower	£125		32% profit
Accessories		£77·55	6% loss

2 Vic is paid a basic rate of £8·36 per hour.
Overtime is paid at time and three-quarters.
Each week he works 36 hours at the basic rate plus 3·5 hours of overtime.
  a Calculate his total earnings for
    i one week   **ii** a year.
  b He receives a pay rise of 3·5%. Calculate his new
    i weekly   **ii** annual wage.

3 △ABC is right-angled at B.
  a Calculate the perimeter of ABC.
  b By how much is ∠ACB smaller than ∠BAC?

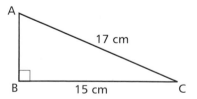

4 Solve $3 \cos x° = 2$ for $0 \leqslant x \leqslant 360$.

5 An artist wants to have his rectangular painting
displayed in a circular frame.
The painting measures 69 cm by 50 cm.

What is the diameter of the circular frame?

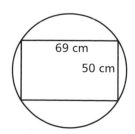

REVISE

6 Calculate the value of AB in the diagram.

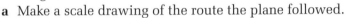

A

16 km

110°

C

23 km

B

7 The diagram shows the end view of a cylindrical tank which is partly filled with a liquid. The width of the surface is 90 cm and the surface is 15 cm below the centre of the circle.

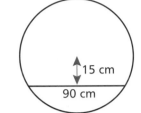

15 cm

90 cm

   a Calculate the circumference of the tank.

   b Calculate the area of the cross-section.

8 To try and get fitter, Jamie has been running and cycling each week in the ratio $2 : 5$. He ran 14 km in one week.

How many kilometres did he cover in total that week?

9 A plane flew for 850 km on a bearing of 050°. Turning, it followed a bearing of 145° for a further 600 km to its destination.

On the first leg of the return journey, it flew on a bearing of 220° for 250 km.

   a Make a scale drawing of the route the plane followed.

   b What bearing must the plane now take to get back to where it started?

   c Flying at a speed of 300 km/h, how long, to the nearest minute, will the plane take to complete the last part of its journey?

10 Evaluate $\dfrac{3 \cdot 24 \times 10^{6}}{1 \cdot 88 \times 10^{8}}$ correct to 3 significant figures.

11 The ages of the cars in a car park are sampled.

8	9	9	2	2
1	2	4	2	10

Calculate the mean and standard deviation of the ages of the cars in the car park.

12 Tiles with letters on them are shaken up in a bag. Letters are then picked out, noted and replaced.

W	P	S	D	G
A	V	U	U	I
L	O	U	O	F
F	J	O	I	N
L	O	B	O	F

Assuming this sample is typical of the distribution of letters in the bag, what is the probability that the next tile will have a vowel on it?

13 Examine this pattern …

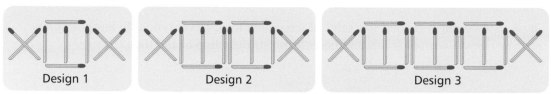

Design 1   Design 2   Design 3

a Copy and complete this table:

Design number n	1	2	3	4	5
Number of matches M					

b Find a formula for calculating the number of matches, M, in design n.

c One of the designs uses 219 matches.
By solving an equation, find which design this is.

 Test C   Part 1

1 A pin weighs 2×10^{-3} g. What is the weight of 64 such pins?
Express your answer in standard form.

2 Evaluate $3\frac{1}{4} \times 2\frac{1}{2}$.

3 Simplify, leaving your answers with positive indices.

    a $x^5 \div x^{20}$ **b** $x^2(x^{\frac{1}{2}} + x^{-3})$

4 Rationalise the denominator: $\dfrac{4}{2 + \sqrt{3}}$.

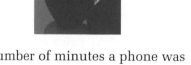

5 The cost, C pence, of Julie's phone calls varies directly as the time, t minutes, that she is connected.
For 16 minutes the charge is 12p.
a Find a formula for C in terms of t.
b How much is she charged for 24 minutes?
c How many minutes does she get for 90p?

6 The figures show the results of a survey into the number of minutes a phone was used per day over 25 days *before* an advertising campaign:

      67   79   67   66   78   71   78   72   74

      63   74   78   76   62   61   70   79   68

      65   63   68   75   60   68   63

After the campaign the figures were:

      77   100   96   100   87   95   95   74   76

      77   72   87   96   92   82   72   73   85

      71   74   82   89   70   78   70

Make a back-to-back stem-and-leaf diagram that could be used to help compare the data sets.

REVISE

7 Simplify by first factorising: $\dfrac{9x^2 - 4}{3x^2 + 2x}$

8 A café owner recorded her sales of ice-cream for different outside temperatures.
She drew a scatter graph and best-fitting straight line of the data.

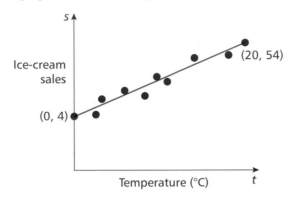

a Find the equation of the best-fitting straight line.
b Use it to estimate the number of sales when the outside temperature is 10 °C.

9 Solve the system of equations:
$y = x + 5$
$x + y = -1$

10 $f(x) = x^2 + 4x - 3$ and $g(x) = 3x + 3$.
Find for which values of x, $f(x) = g(x)$.

11 Sketch the graph of $y = 2 \sin \frac{1}{2}x°$ for $0 \leqslant x \leqslant 360$.

12 Solve $x^2 - 5x + 6 = 0$.

13 Solve algebraically: $8 = \frac{2}{3}(y - 6)$.

 Test C   Part 2

1 a The pre-VAT price of an iron is £24·75.
   What is the total cost, including VAT at 17·5%?
   b The price of an ironing-board, including VAT at 17·5%, is £13·99.
   Calculate the pre-VAT price.

2 Ms Morgan sells sports cars. Her basic monthly salary is £1360.
She is paid 1·5% commission on her sales.
   a Calculate her monthly salary when her sales total £85 000.
   b In one month her total salary is £2950.
   Calculate the total value of her sales for that month.

3 Calculate:
   a the area of the sail
   b the height, h m, of the sail, in two ways.

4 Solve $2 \sin x° + 0.5 = 0$ for $0 \leqslant x \leqslant 360$.

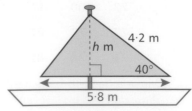

5 Two speedboats start from the same point and travel at the *same* speed.
One travels on a bearing of 010° and the other travels on a bearing of 100°.
After 20 minutes the direct distance between them is 24 miles.
  a How far has each speedboat travelled?
  b At what speed were the speedboats travelling?

6 The diagram shows a marker that is used to mark the leading throw in the hammer throwing competition at a Highland Games.
AC and BC are tangents to the circle at A and B respectively.
The radius of the circle is 16 cm and AC = BC = 30 cm.
Calculate the overall height of the marker.

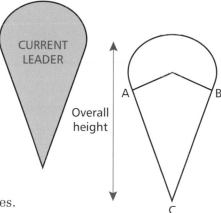

CURRENT LEADER

Overall height

7 Shona travelled 18 km in 2 hours and 15 minutes.
  a Calculate her average speed in kilometres per hour.
  b She needs to average 1 km/h faster than this.
    What time should she cover the distance in?

8 Evaluate, leaving your answer in standard form: $\dfrac{(6{\cdot}25 \times 10^{-4})^2}{1{\cdot}13 \times 10^6}$.

9 A tent manufacturer has produced two similar new tents for use in the mountains.
The larger tent is 3 metres in length and requires 16 m² of material to produce.
The smaller tent is 2 metres in length.
How much material is required to make this tent?

10 The shaded triangles in the trapezium are similar.
Calculate the exact value of: **a** x **b** y.

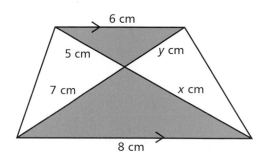

6 cm

5 cm

y cm

7 cm

x cm

8 cm

11 On the outside of a box of matches it states 'average contents 48 matches'.
Several samples were bought and their contents counted.

| 52 | 52 | 52 | 51 | 48 | 51 | 52 | 47 |
| 47 | 50 | 46 | 48 | 50 | 46 | 46 | |

  a Calculate the mean and standard deviation of the sample.
  b Comment on the claim.

REVISE

12 In a traffic survey the number of vehicles with different numbers of wheels is noted.

No. of wheels	Frequency
2	10
3	4
4	22
6	10
8 plus	14

If this distribution is typical of the traffic using this stretch of road, estimate the probability that the next vehicle along will have more than four wheels.

13 There are two local taxi firms in an area.

Fair Fares

80 pence per km
No other charges!

Speed Cabs

Initial charge 50 pence
Thereafter 75 pence per km

a Write down a formula to calculate the fare, F pence, to travel x km using a taxi from:
  i Fair Fares
  ii Speed Cabs
b On Monday Caitlin hires a Fair Fares taxi to travel home from work.
  The next day she does the same but uses a Speed Cabs taxi.
  She is surprised to find that they charged exactly the same fare for the journey.
  Calculate, by solving an equation, the distance from Caitlin's work to her home.

Test D Part 1

1 The radius of the Earth's orbit round the sun is 9.3×10^7 miles.
Express the diameter of this orbit in standard form.

2 Calculate: $8\frac{2}{3} \div 2\frac{1}{6}$.

3 Simplify: $2\sqrt{72} - \sqrt{50}$.

4 y varies inversely as the square root of x.
When $x = 36$, $y = 6.5$.
  a Find a formula for y in terms of x.
  b Calculate y when $x = 64$.
  c Calculate x when $y = 78$.

5 A teacher made this five-figure summary of the exam results:

$L = 17$, $Q_1 = 23$, $Q_2 = 34$, $Q_3 = 45$, $H = 55$.

Draw a box plot to illustrate the distribution of marks.

REVISE

6 Simplify by first factorising: $\dfrac{1 - 4x^2}{2x^2 + 3x + 1}$.

7 Find the equation of the line which passes through the points $(2, -3)$ and $(4, -7)$.

8 Solve this system of equations by substitution:
$3x - y = 13$
$y = 2x - 9$

9 A gardener creates a flower bed, ABC, in the shape of an isosceles triangle.
The perimeter of the bed is 12 m.
$AB = AC = x$ m. $\angle BAC = 120°$.
   a i Make a sketch of the bed.
   ii Express BC in terms of x.
   b Show that the *exact* value of the area of the flower bed can be expressed as $3x - \dfrac{x^2}{2}$.

10 Solve $x^2 - 5x + 6 = 0$.

11 If $a = -4$, $b = 3$ and $c = -6$, calculate the value of:
   a $\dfrac{c - a}{b}$ **b** $b^2 - c^2 - a^2$

12 Mr and Mrs Jones together earn £60 000. Mrs Jones earns more than her husband.
A mortgage company will lend them three times the greater salary plus twice the lower one.
The mortgage company is in fact willing to lend them £154 000.
Calculate their individual salaries.

13 Solve algebraically: $\dfrac{x + 1}{2} = \dfrac{2x + 3}{3x}$

 Test D   Part 2

   1 a Calculate the cost of buying the computer on HP.
   b Express the extra cost of buying on HP as a percentage, correct to 1 decimal place, of the cash price.

> **COMPUTER**
> **Cash price £899 or 12%**
> **deposit + 18 instalments**
> **of £49·99**

   2 a Alice's weekly wage has increased to £374·74. She has just received a rise of 2·5%. How much did she earn before the rise?
   b After a pay rise of 4·2%, Abdul's annual salary is £27 508·80. Calculate his monthly salary before the pay rise.

3 Two hikers are lost on Rannoch Moor.
One walks 6 km from P on a bearing of 320° to X.
The other walks 8 km from P on a bearing of 210° to Y.
  a How far apart are the hikers at X and Y?
  b What is the bearing of Y from X?
Do not use a scale drawing.

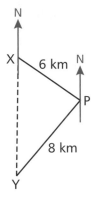

4 The line $y = 1 \cdot 4$ cuts the graph of
$y = 2 \sin x°$ at P and Q.
Solve $2 \sin x° = 1 \cdot 4$ to find the
coordinates of P and Q.

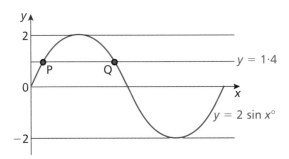

5 Mark is a birdwatcher. His portable
hide measures 1·2 metres by
1·2 metres by 2·05 metres. The hide
consists of aluminium rods covered
by canvas. To make it more sturdy, a
rod is fitted in the space diagonal.
What is the length of this rod?

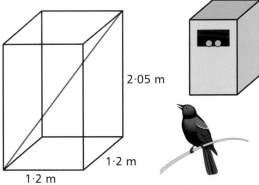

6 Two identical cylindrical vases
contain water. The radius of the
vases is 5 cm and their height is
33 cm. The vases are filled to a
depth of 25 cm and 20 cm. The
contents of the two cylindrical
vases are then poured into a
third vase which is a triangular
prism. The base of this vase is an equilateral triangle of side 15 cm.
Calculate the depth of the water once it has been transferred to the triangular prism.

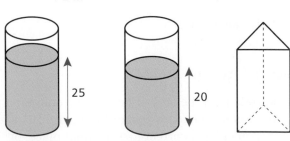

7 Rebecca drove for 1 hour 50 minutes at an average speed of 60 miles per hour.
She then drove at an average speed of 70 miles per hour for 1 hour.
What was her average speed for the whole journey?

REVISE

8 Solve $4x^2 + 21x + 5 = 0$.

9 A toy maker has two *similar* soft toys in stock.

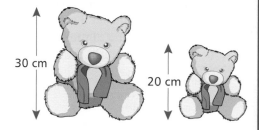

30 cm

20 cm

   a Calculate the reduction factor from the large to the small toy.

   b The larger toy requires 6 litres of stuffing. What volume of stuffing is needed for the smaller toy?

   c The smaller toy needs an area of 1000 cm² of material to make it. What area of material is required to make the larger toy?

10 Find the mean and standard deviation of:

    7·1   7·6   8·2   7·2   7·2   9·3   7·5

11 Two numbers add to make 8. Their difference is one twentieth of their sum. Calculate the two numbers.

12 There are four tills at a checkout.
When someone arrives at the checkout they randomly go to one of the tills.
This person is equally likely to be male or female.

   a Copy and complete the table.

	1	2	3	4
male	(male, 1)	(male, 2)		
female	(female, 1)			

   b A till is picked at random for a lucky prize.
      What is the probability that it is till 4 and a male?

13 **a** A train travels at 80 km/h on a journey of x km.
      Write an expression in terms of x for the time taken (in hours).

   b A bus travels at 40 km/h on a journey that is 5 km longer than the train's journey.
      Write an expression, in terms of x, for the time taken (in hours).

   c When journey times were compared, it was noted that the bus took 5 hours longer than the train.
      Set up an equation to represent this situation and solve it to find the length of the train's journey.

 Test E   Part 1

1 The speed of light is 3×10^8 metres per second.
How far does light travel in an hour? Express your answer in scientific notation.

2 Rationalise the denominator: $\dfrac{5}{\sqrt{5} + \sqrt{2}}$

REVISE

3 The cost of having a shower, C pence, varies directly as the wattage, w, and the length of time of the shower, t minutes.

Sean works out that 3 minutes with an 8 kilowatt shower costs him 15p.

a Find a formula connecting C, w and t.

b How much would it cost Sean if he used a 7 kilowatt shower for 4 minutes?

4 The heights of ten young plants are measured in millimetres.

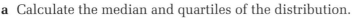

81	44	52	8	84
33	82	74	61	77

a Calculate the median and quartiles of the distribution.

b Calculate the semi-interquartile range.

5 Simplify: $\dfrac{(x+4)(x-3)}{2x^2-18}$

6 A car dealer keeps a record of the value of a particular model of car.
The scatter graph shows his data. A best-fitting straight line has been drawn.

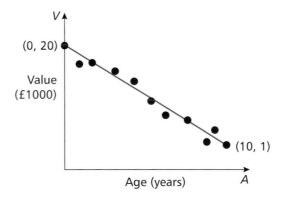

a Calculate the equation of the line, expressing V in terms of A.

b Estimate the value of a 6-year-old car.

7 Copy and complete this bank statement.

STATEMENT			
Date	Paid out	Paid in	Balance
3 Jan			7·56 DR
9 Jan		24·63	
15 Jan	142·08		
23 Jan		238·53	
27 Jan	196·43		

8 In the right-angled triangle $x = 3y$.

a Write down another equation involving x and y.

b Solve the equations.

9 Simplify $y + (-3y) - (-4y) - 8y$.

10 This graph is of the form $y = c + b \cos ax°$ or $y = c + b \sin ax°$.
Identify the values of a, b and c.

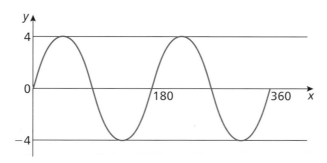

11 The diagram shows two similar triangles. Find the value of x.
All measurements are in centimetres.

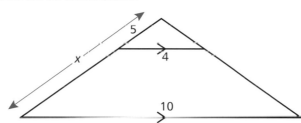

12 ABCDEFGH is a cuboid. AB = 3 cm, CG = 4 cm and AD = 12 cm.
Calculate the length of the space diagonal AG.

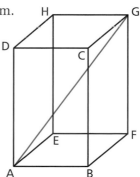

13 Make Q the subject of the formula $A = Q^2 - 2a$.

 Test E   Part 2

1 Before he goes on holiday Frank changes £280 into euro. He receives 414·40 euro.
   a Calculate the exchange rate.
   b On his return he changes his remaining 38 euro into pounds at the same rate.
   How much does he receive, correct to the nearest 10p?

2 **a** Ronnie sells his old car for £129. It has devalued by 85% since he bought it.
   How much did he pay for it?
   b However his flat which he paid £45 000 is now worth £52 000.
   Express the appreciation as a percentage of the amount he paid.

REVISE

3 The diagram represents the back section of
   a greenhouse. It is circular with a horizontal
   floor PQ 1·8 m wide. The radius OP of the
   cross-section is 1·4 m.
   a Calculate the height of the greenhouse, h m.
   b Calculate the angle of elevation from P to the
     top of the greenhouse, A.

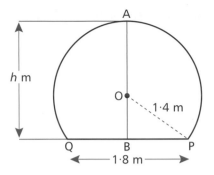

4 a Sketch the graph of $y = 3\cos x°$ for $0 \leqslant x \leqslant 360$.
   b On the same diagram draw the graph of $y = -2$ intersecting with $y = 3\cos x°$ at
     points A and B.
   c Calculate the coordinates of A and B.

5 Archie is marking out a piece of ground for his sheep pens.
   It is meant to be a rectangle, 8·4 metres by 11·2 metres.
   After marking the corners he measures a diagonal and finds
   that it is 14 metres.
   Are the corners right angles?

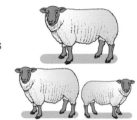

6 Dierdre's saucepan has a diameter of 17 cm.
   She is making a Shrewsbury sauce.
   When the sauce starts to simmer and reduce it is 4 cm deep.
   After simmering for 15 minutes the volume of the sauce has
   reduced by 45%.
   What is the volume of the sauce that remains?

7 Andrew is travelling at a constant speed of 24 km/h.
   a How long, in hours and minutes, will it take him to cover 38 kilometres?
   b If he increased his speed by 10%, what effect would it have on his time for this
     distance?

8 Solve $3x^2 + 14x + 8 = 0$.

9 Meg checked the wattage of the bulbs in her house.

       40 W   40 W   60 W   60 W   60 W   100 W   40 W   60 W   100 W

   Calculate the mean and standard deviation of the wattage of the bulbs.

10 The diagram shows a triangular plot of land.
   Calculate its area.

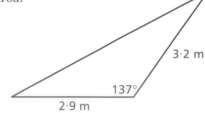

11 One glass of cola and five glasses of orange cost a total of £8·60 at a café.
Two colas and three oranges cost £7·40.
Let £x be the cost of a cola and £y be the cost of an orange.
a Form two equations in x and y.
b Find the total cost of two colas and an orange.

12 The number of absences per class is noted for one week.

6	11	6	9	9
6	12	9	7	4
8	10	7	11	5
9	11	5	6	2
5	12	10	8	8

a Make a frequency table of the distribution.
b If a class is selected at random, what is the probability that it will have had more than nine absences?

13 The area, A cm², of a trapezium is given by:

$$A = \frac{h(a + b)}{2}$$

a Change the subject of this formula to h.
b Calculate h for a trapezium with area 70 cm², a = 13 cm and b = 22 cm.

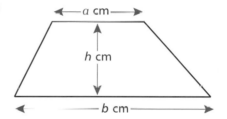

 Test F   Part 1

1 Simplify $\sqrt{48} + 2\sqrt{75}$.

2 Simplify $\dfrac{4x^2 - 4x - 3}{1 - x - 6x^2}$.

3 A gardener measured the distance round his onions in centimetres.

14·3	16·2	16·3	16·0	17·0
11·4	12·7	14·1	20·0	12·5
12·3	15·4	16·2	16·7	15·1
19·9	12·8	11·3	18·3	14·8

Give a five-figure summary of the data (highest, lowest, quartiles, median).

4 A line passes through (5, −2) and is parallel to y = 3x.
Find the equation of the line.

5 The sum of two numbers is −5. The difference between the numbers is 1.
a Let the larger number be x and the smaller be y.
Form two equations in x and y.
b Solve the system to discover the two numbers.

6 Solve $(x + 2)^2 - (1 - x)^2 = -5$.

7 A is an acute angle and $\sin A = \frac{3}{5}$.
Calculate the value of tan A.

8 This graph is of the form
$y = c + b \cos ax°$ or $y = c + b \sin ax°$.
Identify the values of a, b and c.

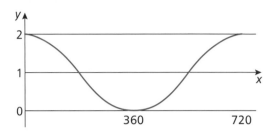

9 Solve $5x^2 - 245 = 0$.

10 Using the approximation that $\pi = \frac{22}{7}$,
calculate the perimeter of this shape
(a semicircle and a rectangle combined).

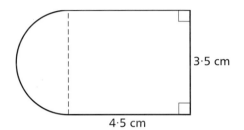

11 This figure is made from similar triangles.
Calculate the value of x.

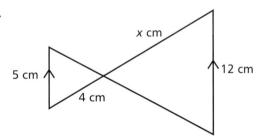

12 Prove this triangle is right-angled.

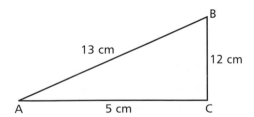

13 Solve algebraically to 2 decimal places: $\dfrac{3(x - 2)}{x} = \dfrac{x}{3}$.

 Test F Part 2

1 Calculate $(7·23 \times 10^4) \times (8·15 \times 10^{-2})$ correct to 3 s.f.

2 a George's car insurance premium is £965, less 40% no-claims bonus.
How much does he pay?
b Georgina pays £604·80. Her no-claims discount is 30%.
Calculate her premium before the discount.

3 Premier Parks own a plot of land valued at £100 000.
It has appreciated by 10% per year for 4 years.
Calculate the price, correct to the nearest £100, they paid for the land 4 years ago.

4 The shape of the front of this house is a triangle on
top of a rectangle. The eaves slope at angles
of 24° and 35° to the horizontal.
  a Calculate the height, h m, of the house.
  b Calculate the area of the triangular section.

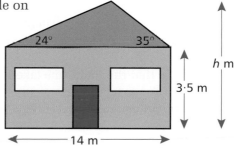

5 a What are the maximum and minimum values of $y = 4 + 2 \sin 3x°$?
  b What is the period of the graph of y?
  c What is the value of y when
     i $x = 30°$ **ii** $x = 21°$?
  d Sketch the graph of $y = 4 + 2 \sin 3x°$.

6 This is a sign used by a garden centre.
It is a semicircle of radius 55 cm and the
two chords are of equal length.
Calculate the area of the two darker segments.

7 A rhombus pendant has a perimeter of 22 cm and one of its diagonals is 6 cm.
Calculate its area.

8 This pig trough is a triangular prism.
Its volume is 151·2 litres.
Calculate the depth of the trough.

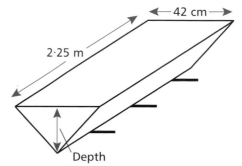

9 Calculate the solutions of $5x^2 - 9x - 1 = 0$, correct to 2 decimal places.

10 Golf balls are released from a height of 1 m onto a hard surface. The height, in
centimetres, to which they rebound is taken as a measure of their performance.

68	75	84	86	62	76	72	62	73	81	75	72	89
90	78	76	79	62	65	68	60	64	73	77	77	

  a Sort the data into ascending order using a stem-and-leaf diagram.
  b Find the median.
  c Compare it with the mean height.

11 Lauren can maintain a constant pace on flat land.
Going uphill takes 4 km/h off this pace.
Going downhill adds 4 km/h to the pace.
Over a long walk she does a total of 3 km uphill and
the same downhill.
It takes her 1 hour. (No walking is on the flat.)
Let x km/h represent the constant pace on level ground.

a Show that the total time taken $= \dfrac{3}{x+4} + \dfrac{3}{x-4}$.

b Form an equation and solve it to find Lauren's pace on level ground.

12 In the diagram shown, prove that $\sin x° = \frac{1}{4}$.

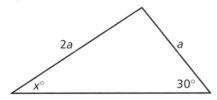

13 The equation of this 'quarter ellipse' is:

$$\frac{x^2}{36} + \frac{y^2}{9} = 1$$

a Show the steps that lead from this equation
to the formula:

$$x = 2\sqrt{9 - y^2}$$

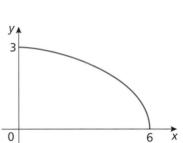

You must show each step of your working.

b When $y = 2.4$ calculate the value of x for a point (x, y) that lies on the ellipse.

Answers

1 Area and volume

Page 1 Exercise 1.1

1 a iii, iv, vi, ix, x   **b** i, ii, v, vii, viii

2 9 cm^2, 49 cm^2, 2025 mm^2

3 12 cm^2, 1632 mm^2, 22·4 m^2

4 a i 5 cm   **ii** 0·4 cm

     iii 300 cm   **iv** 1274 cm

     b i 480 mm   **ii** 629 mm

     iii 2000 mm   **iv** 28 880 mm

     c i 9 m   **ii** 3·85 m

     iii 6 m   **iv** 0·055 m

     d i 7000 m   **ii** 4500 m

     iii 2632 m   **iv** 82 470 m

     e i 5 km   **ii** 8·5 km

     iii 8·427 km   **iv** 0·063 km

5 a 500 ml   **b** 700 ml   **c** 600 ml

6 a 7 litres   **b** 3·5 litres

     c 4·25 litres   **d** 0·75 litre

7 a 2000 ml   **b** 4750 ml

     c 250 ml   **d** 9500 ml

8 a i 72 cm^3 **ii** 132 cm^2

     b i 72 cm^3 **ii** 114 cm^2

     c i 72 cm^3 **ii** 120 cm^2

9 a 6 kg   **b** 8·425 kg   **c** 7·25 kg

10 a 7000 g   **b** 3576 g   **c** 63 000 g

11 a 12·5 cm^2 **b** 15·54 cm^2 **c** 586·5 mm^2

12 a triangular prism   **b** cube

     c square-based pyramid   **d** cuboid

Page 4 Exercise 2.1

1 a 12 cm^2 **b** 7·955 cm^2 **c** 28·5 cm^2

     d 816·5 mm^2 **e** 126 cm^2 **f** 58·5 m^2

2 714 cm^2

3 1752 cm^2

4 200 cm^2

5 1560 cm^2

6 15 cm^2

Page 5 Exercise 2.2

1 a 10 cm^2 **b** 11·4 cm^2

     c 8·295 m^2 **d** 84 cm^2

2 a i 17·3 cm   **ii** 86·5 cm^2

     b i 7·5 cm   **ii** 37·4 cm^2

3 7·15 m^2

4 a 64 cm   **b** 0·64 m^2 or 6400 cm^2

5 a 17·8 mm   **b** 312·1 mm^2

Page 6 Exercise 3.1

1 a 21 cm^2 **b** 616 mm^2 **c** 19·32 cm^2

2 4 m^2

3 7·74 m^2

Page 8 Exercise 4.1

1 a 81 cm^2 **b** 49·5 cm^2

     c 225 mm^2 **d** 13·69 cm^2

2 areas are 2112 cm^2, 2193 cm^2 and 2137·5 cm^2

     – the middle one is biggest

3 4805 cm^2

4 a 28 125 m^2 **b** 2·8125 hectares

5 1·76 m^2

6 a $12·5 - 7·5 = 5$ cm^2 **b** $\frac{1}{2} \times 5 \times 2 = 5$ cm^2

Page 9 Exercise 5.1

1 a 56 cm^2 **b** 163·35 cm^2 **c** 940 mm^2

2 3 m^2

3 345 cm^2

4 4556 cm^2

Page 10 Exercise 5.2

1 a 520 mm^2 **b** 56 cm^2

     c 1596 cm^2 **d** 130·68 cm^2

2 a 48 mm   **b** 480 mm^2

3 a 3·5 cm   **b** 138 cm^2

4 a 4·5 cm   **b** 14·4 cm   **c** 113·4 cm^2

5 a 9·7 cm   **b** 10·6 cm   **c** 487·3 cm^2

Page 11 Exercise 6.1

1 1·48 m^2 **2** 936 cm^2 **3** 270 cm^2

4 a 8·4 cm^2 **b** 50·4 cm^2 **c** 13·2 cm^2

5 a 2798 cm^2 **b** 1800 cm^2

Page 12 Exercise 6.2

1 5555 cm^2

2 a 51 500 m^2 (3 s.f.)   **b** 5·15 hectares

3 4185 cm^2

4 112 cm

Page 14 Exercise 7.1

1 a 31·5 cm^3 **b** 2300 cm^3

     c 180 cm^3 **d** 306 cm^3

2 a 960 cm^3 **b** 960 cm^3

3 a 480 cm^3, 408 cm^2 **b** 4320 cm^3, 2136 cm^2

     c 960 cm^3, 720 cm^2 **d** 18 900 mm^3, 5712 cm^2

4 180 cm^3, 240 cm^2

5 a 6895·5 cm^2 **b** 27 540 cm^3 **c** 3 bags

7 168 cm^3, 248 cm^2

Page 15 Exercise 7.2   All answers given to 3 s.f.
1 a 5·63 cm   **b** 426·6 cm^2 **c** 366 cm^3
2 a 11·3 cm   **b** 891 cm^2 **c** 1130 cm^3
3 a 3·16 m^3 **b** 3·825 m^2
   c 1·86 m   **d** 11·2 m^2
4 a 10·4 cm   **b** 62·4 cm^2 **c** 20·8 cm
   d 1110 cm^2 **e** 1880 g

Page 17 Exercise 8.1
1 a 88 cm^2 **b** 1175 cm^2
   c 10 908 cm^2 **d** 220 cm^2
   e 10 m^2 or 103 673 cm^2
2 271 cm^2
3 74·6 m^2
4 a 352 cm^2 **b** 755 cm^2
   c 42 474 mm^2 **d** 199 cm^2
   e 4·5 m^2
5 523 cm^2
6 302 cm^2
7 a 8 cm
   b 324 cm^2

Page 18 Exercise 9.1
1 a 3420 cm^3 **b** 198 cm^3
   c 2070 cm^3 **d** 3·55 m^3
2 44 m^3
3 690 cm^3
4 285 000 cm^3
5 volumes are 1248 cm^3, 1208 cm^3 and 825 cm^3
   – the one on the left will burn longest

Page 19 Exercise 9.2
1 0·508 m^3
2 almost 3 (2·99)
3 44·2 litres
4 d = 9 cm, h = 9 cm

Page 19 Exercise 10.1
1 a 4·31 m^2
   b 8·61 m^3
   c i 7·56 m^3 **ii** 1·05 m^3 **iii** 8·61 m^3
2 a 2·16 m   **b** 25·3 m^2
3 a 29·0 m^3 **b** 31·1 m^2 **c** 15·3 m^2
4 a 8880 cm^3
   b 2600 cm^2
5 8·36 m^3, 20·4 m^2
6 a 41·5 litres   **b** 4050 cm^3 **c** 16 600 cm^2
7 2200 cm^3

Page 22 Revise
1 a 80 cm^2 **b** 20 cm^2
   c 58·1 cm^2 **d** 630 mm^2
2 a 136 cm^2 **b** 209 cm^2 **c** 551 mm^2
3 248 cm^2
4 a 225 cm   **b** 213 000 cm^3 **c** 30 700 cm^2
5 a 961 cm^3 **b** 603 cm^2
6 a 4810 cm^3 **b** 1720 cm^2
7 47 025 cm^3

2 Money

Page 24 Exercise 1.1
1 a i 90p   **ii** £2·80   **iii** £34·10   **iv** £64·00
   b i 27p   **ii** £4·33   **iii** £25·98   **iv** £60·00
2 $38\frac{3}{4}$ hours
3 a £4·25   **b** £3·92   **c** £1·56   **d** £5·97
4 a £30   **b** £28   **c** £3·90   **d** £13·70
   e £4·25   **f** £619·50   **g** £1275   **h** £13·12
5 a £24   **b** £2100   **c** £278   **d** £278·20
6 a i 0·25   **ii** 0·175   **iii** 0·07   **iv** 2·5
   b i 3%   **ii** 62·5%   **iii** 40%   **iv** 12·5%
7 a £1123·20   **b** £1·62   **c** £41·69
8 a 1659   **b** 840

Page 25 Exercise 2.1
1 a £800   **b** £400   **c** £4800
   d £50   **e** £2800   **f** £800
2 a £43·98   **b** £74·99   **c** £1921·38
   d £23 500   **e** £22·72   **f** £6326
   g £9·41   **h** £84·65
3 a £1605·44   **b** £645·46
4 a £179·82   **b** £215·88
5 a £8·40   **b** £1025·50
   c £1·33   **d** £0·14
6 a £10 246·80 **b** £426·95
7 a $\frac{4}{5}, \frac{24}{25}, \frac{9}{20}, \frac{2}{3}, \frac{5}{8}, \frac{1}{40}$, $2\frac{3}{4}$, $3\frac{1}{6}$, $4\frac{31}{50}$; 0·8, 0·96, 0·45,
   0·66…, 0·625, 0·025, 2·75, 3·166…, 4·62; 80%,
   96%, 45%, 66·66…%, 62·5%, 2·5%, 275%,
   316·6…%, 462%
8 a £194·40   **b** 33·3% profit
   c £280   **d** 5% loss
9 82·5p
10 a 240 bags
   b compare cost of 720 bags: 2 × 5·99 = £11·98
      with 3 × £3·99 = £11·97

Page 27 Exercise 3.1
1 a £1470·60   **b** £7353
2 £123·75   **3** £38 280
4 a £2260   **b** £2725
5 a £249·38   **b** £12 967·76
6 £580
7 39 (38·5) hours
8 a Mrs Smith by £2674·50   **b** £35 629·50

Page 28 Exercise 3.2
1 a £294   **b** £308·25
2 a £2383·33   **b i** £31 080   **ii** £2590
3 a £35 700   **b i** £3123·75   **ii** £37 485
4 a £19 251   **b i** £370·21   **ii** £10·28
5 a £9·36   **b i** £243·36   **ii** £12 654·72
   c 4%
6 a £1350, £2354·40   **b** £74·40 per month
7 a £17 314·50 − £20 826·36
   b i £1442·88   **ii** £1735·53
8 £26 400

Page 30 Exercise 4.1
1 a 7·5 h   b 37·5 h
2 a 8·75 h   b 35 h
3 a 28·5 h   b £234·84
4 a 8, 8, 8·25, 7·75, 8·25
   b 40·25 h   c £338·91
5 a 18 00   b 14 00   c 13 15
   d 08 45   e 16 15

Page 31 Exercise 4.2
1 £9·60, £19·50, £10·47, £10·30, £19·11
2 a £12·90   b £58·05
3 a £28·80   b £28·80   c £57·60
4 a £322   b £57·50   c £370 50
5 8, 7·75, 8·25, 8·25, 7·5, total 39·75 h; £324·87

Page 32 Exercise 5.1
1 a £79·92   b £1998
2 a £175   b £1125
3 a £3·40   b £27·50
4 a £900   b £26 160
5 £125
6 a £187·50   b £60 000
7 a £302·50   b 6
8 a £2150; £168 000

Page 34 Exercise 6.1
1 a £587·40   b £176·69   c £410·71
2 a £2833·39   b £935·05   c £1898·34
3 Commission £75; gross pay £2028·70; superannuation £121·72; deductions £621·60; net pay £1407·10
4 Gross pay £559·47; basic pay £487·42; income tax £93·47

Page 35 Exercise 7.1
1 a No   b yes, 11p
2 a £260   b £28·60
3 a i £519   ii £190   b £58·99
4 £1507
5 a £3100·05   b £258·34
6 £386
7 £3·91
8 £750 000

Page 37 Exercise 8.1
1 Amy pays tax.
2 a £7255   b £0   c £29 538   d £12 108
3 a 10%   b £150
4 a i £2020   ii £500
   b i £202   ii £110   c £312
5 a £1000   b £100
6 a £16 800   b i £202   ii £3251·60
   c £3453·60
7 £255, £25·50, £0, £25·50; £5255, £202; £711·70, £913·70; £15 255, £202, £2911·70, £3113·70; £25 255, £202, £5111·70, £5313·70
8 a £18 720   b £2832·10   c £15 887·90

Page 38 Exercise 8.2
1 £2680
2 a £6765   b £36 145
3 £46 250; 10% of £2020 = £202; 22% of (£31 400 − £2020) = £6463·60; 40% of (£46 250 − £31 400) = £5940; total = £12 605·60
4 £112 550; 10% of £2020 = £202; 22% of (£31 400 − £2020) = £6463·60; 40% of (£112 550 − £31 400) = £32 460; total = £39 125·60
5 a £94 705·60   b £165 294·40
6 a £21 711·60   b £1809·30   c £4470·70
7 a £23 705·60   b 29·6%
8 a i £4047·60   ii £2173·05   iii £1470
   b £16 809·35   c 68·6%

Page 39 Brainstormer
£49 321

Page 40 Exercise 9.1
1 a Highland Building Society
   b £1502·75
   c 13579246
2 a £144, £219, £398
   b £35·80, £28·50 DR, £118
   c £75·35 DR, £82·65, £13·34 DR
   d £01·03, £33·46 DR, £590·62
3 a £20·62 DR
   b £559·38
   c £91·39
   d £108·61 DR
4 a £1722·21
   b £178·34
   c £454·62 DR
   d £45·38
   e £40·21 DR

Page 41 Exercise 10.1
1 a £8   b £18·80
2 a £525   b £2625
3 a £150   b £12·50   c £125
4 £3, £9·90, £112, £2812·50
5 3%
6 2·1%
7 £7400

Page 43 Exercise 10.2
1 £240, £4240; £254·40, £4494·40
2 a £621   b £642·74   c £665·23
3 £1373·88
4 a £936   b £981·65
5 a £595·51   b £131·41
   c £440·90   d £175·51
6 £15 007·30
7 8 years
8 £3600

Page 44 Brainstormer

£10 957 969·46

Page 45 Exercise 11.1

1 £535

2 £560

3 **a** £92 000   **b** £99 360

4 **a** £4500   **b** £3600   **c** £3060

5 65%

6 40%

7 **a** 7·5%,8·5%   **b** the sculpture

8 **a** Appreciated by 24%

   b appreciated by 25%

   c depreciated by 5%

   d appreciated by 47·25%

Page 46 Exercise 11.2

1 £5800

2 **a** £156 800   **b** depreciated by 10·4%

3 **a** £105 000   **b** £115 500   **c** 20%

4 £500

5 £50 000

6 £300 000

7 £8 000 000

8 **a** £19 200   **b** 1·56%

Page 48 Revise

1 **a** £267·75   **b** £7·65

2 **a** £32 976   **b i** £2954·10   **ii** £35 449·20

3 £445·10

4 **a** £2675   **b i** £340   **ii** £25 000

5 **a** £599·65   **b** £157·56   **c** £442·09

6 **a i** £50·05   **ii** £45·50

   b i £2613·05   **ii** £4927·60

   c i £3110·05   **ii** £10 327·60

7 £41·02 DR, £44·36, £23·03 DR

8 £14 090·90

9 £145 000

10 £19 268·48

3 Similarity

Page 50 Exercise 1.1

1 12 m, 250 m, 3250 m, 71·5 km

2 **a** pupil's own drawing

   b 600 m

3 **a** 2 m × 1·5 m; 0·5 m × 1 m

   b 1·5 cm × 3 cm

4 pupil's own drawings

Page 51 Exercise 2.1

1 **a** 1 : 4000   **b** 1 : 15 000

   c 1 cm rep. 60 m   **d** 1 cm rep. 2 km

2 **a** 1 : 1000, 4 m   **b** 1 : 20 000, 1 km

   c 1 : 50 000, 1·25 km   **d** 1 : 80 000, 2·4 km

3 **a** 1 : 500, 16 cm   **b** 1 : 4000, 6 cm

   c 1 : 200 000, 2·5 cm   **d** 1 : 150 000, 6 cm

4 **a** 30 km, 40 km   **b** 50 km

5 450 m, 1050 m, 1200 m, 825 m

6 **a i** 120 m   **ii** 90 m   **iii** 100 m

   b 310 m (nearest 10 m)   **c** 0·3 cm

7 4·375 m

Page 53 Exercise 3.1

1 **b** 2845 km

2 **b** 4860 km   **c** 236°

3 **b** 1543 miles   **c** 319°

4 **b** 8·72 km   **c** 244°   **d** 2 h 11 min

5 **b** 287 miles   **c** 041°

6 **b** 21·5 miles, 17·5 miles   **c** 12·4 miles

7 **b** 048°   **c** 0·8 km   **d** 24 km/h

8 **b** 148°, 69°   **c** Paul, 2 min

Page 55 Exercise 4.1

1 **a** 1 cm rep. 6 m **b** 1 : 600

2 1 cm rep. 4·5 m **b** 1 : 450

3 1 : 800 000

4 1 : 5050

5 **a** 1 : 80   **b** 10

6 1 : 55 200 000   **b** 2·25 cm

Page 58 Exercise 5.1

1 **a** 3, 18 cm   **b** $\frac{8}{5}$, $\frac{64}{5}$ cm   **c** $\frac{5}{3}$, $\frac{20}{3}$ cm

   d $\frac{5}{3}$, 10 cm   **e** $\frac{2}{3}$, 6 cm   **f** $\frac{1}{5}$, 3 cm

2 **a** 2, ∠BAC = ∠EDF, ∠ACB = ∠DFE, ∠CBA = ∠FED

   b 3, ∠KLJ = ∠HGI, ∠KJL = ∠GHI, ∠LKJ = ∠HIG

   c $\frac{4}{3}$, ∠MNO = ∠QPR, ∠NMO = ∠PQR, ∠MON = ∠PRQ

   d $\frac{3}{2}$, ∠STU = ∠VWX, ∠TSU = ∠WVX, ∠TUS = ∠UXV

3 **a** $3\frac{3}{4}$ cm   **b** $13\frac{1}{3}$ cm   **c** 7·5 cm   **d** 2 cm

4 **a** 4 cm, 4·9 cm   **b** 2 cm, 4·5 cm

Page 60 Exercise 5.2

1 **b i** ED   **ii** CE   **iii** DC

2 **a** 2, 10   **b** 3, 15

   c 2·5, 15   **d** 1·25, 12·8

3 **a i** 0·9 m   **ii** 1·5 m   **iii** 1·05 m

   b i 0·8 m   **ii** 1·4 m   **iii** 1·2 m

4 **a** 2   **b** 5·4   **c** $v = 1·12$, $w = 2·75$

5 **a** 6·75   **b** $6\frac{2}{3}$, $3\frac{1}{3}$ **c** 8   **d** $\frac{18}{11}$

6 **a** 9   **b** $3\frac{1}{3}$ **c** 7·5

   d 5   **e** 7   **f** 5·7

Page 63 Exercise 6.1

1 **a** 6   **b** 6   **c** 12

2 **a** 40 cm   **b** 1·5 cm

3 6 m

4 **a** 1   **b** $3\frac{1}{3}$ **c** 2   **d** 2·25

5 2 m

6 **a** 1 m   **b** 1·2 m

7 **a** 1·56 m   **b i** 0·89 m   **ii** 0·67 m

Page 66 Exercise 7.1

1 a i 1·4   **iii** 5·6 cm
  b i $\frac{5}{3}$ **iii** 11 cm
2 a i 1·8   **ii** $\frac{5}{6}$
  b i 9 cm   **ii** 6 cm   **iii** 16·2 cm
3 a the angle is common to both rectangles
  b i 2   **ii** 7·6
4 1·6 m   **b** 1·375   **c** 2·475 m
5 a 36 cm   **b** 52·5 cm
6 a 1·4 m   **b** $\frac{4}{7}$ **c** 60 cm
7 a 7·65 m   **b** 4·59 m   **c** 84·72 m

Page 68 Exercise 8.1

1 a 4, 16   **b** 10 cm²   **c** 160 cm²
2 73·5 cm²
3 a 1·30 m²   **b** 0·325 m²
4 800
5 a 35·4 cm   **b** 7·1 cm
6 0·6 m
7 £404·60
8 4

Page 70 Exercise 9.1

1 a 94 cm²   **b** 846 cm²
2 a 9·42 m²   **b** 21·2 m²
3 a 314 cm²   **b** 707 m²
4 a 240 cm²   **b** 96 000 m²
5 a 22·5 cm   **b** 5300 cm²
6 a i $\frac{1}{4}$ **ii** $\frac{1}{2}$
  b i 25·0 cm   **ii** 12·5 cm
7 a 15·59 cm²   **b** 103·2 cm²   **c** 45·9 cm²

Page 72 Exercise 10.1

1 81 cm³
2 a 162 m³
  b 12 m²
3 no, different ratios
4 40 cm
5 a 0·54 m³
  b 24 cm
6 13·5 cm
7 a 47 cm³
  b yes, it is cheaper per cm³

Page 73 Brainstormer

  a 3
  b i 2 : 1   **ii** 3 : 1   **iii** 1 : 2
    iv 1 : 3   **iv** n : 1   **vi** 1 : n

Page 75 Revise

1 a 1·5 km   **b** 8 km
  c 1·7 km   **d** 10·4 km
2 a 12 cm, 5 cm   **b** 7·5 cm   **c** 32 cm
3 $x = 4$ cm, $y = 9$ cm, $w = 30$ cm
4 9450 cm³
5 241 cm²
6 a i 2 m   **ii** $\frac{1}{4}$ **iii** 0·393 m²
  b i 0·785 m²   **ii** 0·049 m²

4 Formulae

Page 78 Exercise 1.1

1 a 11   **b** 2   **c** 14   **d** 11
  e 13   **f** 8   **g** 4   **h** 0
  i 2   **j** 3   **k** 1   **l** 1
  m 25   **n** 14   **o** 18   **p** 13
  q 7   **r** 9   **s** 1   **t** 5
  u 10   **v** 84   **w** $\frac{1}{2}$ **x** 3
2 a 18   **b** −6   **c** 6   **d** 12
  e −27   **f** −3   **g** 81   **h** 9
  i 3   **j** 3   **k** 27   **l** 0
  m 36   **n** 162   **o** 36   **p** 36
  q 3   **r** 2   **s** 2   **t** 1
  u −1   **v** $-\frac{1}{2}$ **w** 54   **x** 4
3 a $5x - 10$ **b** $12a + 12b$
  c $km + 2kn$ **d** $x^2 - xy$
  e $-w + z$ **f** $-6c + 9d$
  g $15m - 10m^2$ **h** $-4nm + 12n^2$
4 a $11 - 3x$ **b** $-k - 6$
  c $-8 + 5n$ **d** $k + 5$
  e $-17 + 7x$ **f** $-4a + 2b$
5 a $x = 3$ **b** $y = -11$ **c** $a = 6$
  d $w = 15$ **e** $k = -6$ **f** $x = 1$
  g $y = 7$ **h** $x = 2$ **i** $a = -2$
6 a $A = \pi r^2$ **b** $C = \pi D$ **c** $C = 2\pi r$
  d $P = 4x$ **e** $A = x^2$
  f $P = 2m + 2n$ or $2(m + n)$ **g** $A = mn$
7 a $V = k^3$
  b $S = 6k^2$
  c $E = 12k$
8 a $V = lbh$
  b $S = 2lb + 2lh + 2bh$ or $S = 2(lb + lh + bh)$
  c $E = 4l + 4b + 4h$ or $4(l + b + h)$
9 a i 18
    ii 5
    iii 14th term
  b i nth term $= 7n - 4$
    ii 346
    iii $7n - 4 = 157$; 23rd term

Page 79 Challenge

$A = 1, A = m - n, m = 6$ and $n = 5$;
$A = 2, A = m - n^2, m = 6$ and $n = 2$;
$A = 3, A = m^2 - n^2, m = 2$ and $n = 1$;
$A = 4, A = \dfrac{8}{m + n}, m = 1$ and $n = 1$;
$A = 5, A = \dfrac{m + n}{2}, m = 6$ and $n = 4$;
$A = 6, A = m^2 - n, m = 3$ and $n = 3$;
$A = 7, A = m + n^2, m = 3$ and $n = 2$;
$A = 8, A = m^2 + n, m = 2$ and $n = 4$;
$A = 9, A = m + n, m = 7$ and $n = 2$;
$A = 10, A = mn, m = 5$ and $n = 2$

Page 80 Exercise 2.1

1 a i $1 \cdot 90 \, \text{m}^2$ **ii** $2 \cdot 85$ litres
   b i $1 \cdot 68 \, \text{m}^2$ **ii** $2 \cdot 52$ litres
   c i $1 \cdot 18 \, \text{m}^2$ **ii** $1 \cdot 77$ litres

2 a i $A = \pi r^2$ **ii** $211 \cdot 2 \, \text{cm}^2$
   b i $D = ST$ **ii** $88 \, \text{km}$
   c i $r = \sqrt{\dfrac{A}{\pi}}$ **ii** $8 \cdot 0 \, \text{cm}$
   d i $V = lbh$ **ii** $152 \, \text{cm}^3$
   e i $S = \dfrac{D}{T}$ **ii** $64 \, \text{mph}$
   f i $C = \pi D$ **ii** $74 \, \text{cm}$
   g i $h = \dfrac{V}{lb}$ **ii** $5 \, \text{cm}$
   h i $D = \dfrac{C}{\pi}$ **ii** $1 \cdot 8 \, \text{cm}$

3 a $15 \cdot 2 \, \text{cm}^2$ **b** $39 \cdot 3 \, \text{cm}^2$
   c $62 \, \text{cm}^2$ **d** $173 \, \text{cm}^2$

4 a $1340 \, \text{cm}^3$ **b** $235 \, \text{cm}^3$
   c $1710 \, \text{cm}^3$ (all to 3 s.f.)

Page 81 Exercise 2.2

1 a/b Mr Henderson: $25 \cdot 9$ ideal weight;
     Mr Lewis: $31 \cdot 5$ obese;
     Ms Lawson: $19 \cdot 0$ underweight;
     Mrs Smith: $26 \cdot 0$ slightly overweight

2 a i lens B   **ii** 6 units
   b i lens A for eyepiece; $2\frac{1}{2}$ times
     ii telescope A: $\times 62\frac{1}{2}$; telescope B: $\times 63\frac{1}{2}$;
     telescope B

3 a New York (by $2 \, °F$)   **b** $180 \, °F$
   c $-459 \cdot 67 \, °F$

4 a Formula 1: 151; formula 2: 180
     (formula 2 considerably higher)
   b formula 1: 300; formula 2: 302
     (the two formulae are roughly equal)

5 Fork 1: A; Fork 2: G; Fork 3: F$^{\#}$; Fork 4: C$^{\#}$

Page 84 Challenge

a $-4 \cdot 5 \, °C$
b $-35 \cdot 9 \, °C$; $-41 \cdot 5 \, °C$;
     situation 2 feels colder by $5 \cdot 6 \, °C$

Page 84 Exercise 3.1

1 a $P = n - m$ **b i** 2 **ii** $4 \cdot 5$

2 a $A = \dfrac{x + y}{2}$ **b i** $50 \cdot 5$ **ii** $59 \cdot 3$

3 a $S = 23\,000 + 1500y$ **b i** $26\,000$ **ii** $32\,000$
4 a $M = 250 - 30h$ **b i** 160 **ii** 70
5 a $C = 25h + 22A$ **b i** 455 **ii** 904
6 a $P = D + am$ **b i** $12\,000$ **ii** $12\,320$
7 a $M = 5n - 1$ **b i** 84 **ii** 279
8 a $T_n = 8n + 3$; $T_{17} = 139$
   b $T_n = 9n - 3$; $T_{23} = 204$
   c $T_n = 12n - 10$; $T_{47} = 554$
   d $T_n = 23n - 6$; $T_{100} = 2294$

Page 86 Challenge

a i $360°$ **ii** $720°$ **iii** $900°$
b $S = 180(n - 2)$
c i $90°$ **ii** $120°$ **iii** $128\frac{4}{7}°$
d $A = \dfrac{180}{n}(n - 2)$
e i $135°$ **ii** $150°$

Page 87 Exercise 3.2

1 a i $P = 2\pi x + 4x$ **ii** $A = 4x^2 - \pi x^2$
   b i $P = 8x - 8$ or $P = 8(x - 1)$
     ii $A = 4x - 4$ or $A = 4(x - 1)$
   c i $P = 4\pi x + 2\pi$ or $P = 2\pi(2x + 1)$
     ii $A = 2\pi x + \pi$ or $A = \pi(2x + 1)$
   d i $P = 2\pi x + 4\sqrt{2}x$ or $P = 2x(\pi + 2\sqrt{2})$
     ii $A = (\pi - 2)x^2$

2 a i $5 \times 2x$
     ii $(2x)^2 - \frac{1}{2} \times 2x \times \sqrt{3}x$ (and simplify)
   b i $\pi x + 2 \times \pi \times \dfrac{x}{2}$
     ii $\pi \times \left(\dfrac{x}{2}\right)^2 - 2 \times \pi \times \left(\dfrac{x}{4}\right)^2$ (and simplify)
   c i $2x + 2 \times \pi \times \dfrac{x}{2}$
     ii $x^2 - 2 \times \pi \times \left(\dfrac{x}{4}\right)^2$

Page 88 Exercise 4.1

1 a $x = mk$ **b** $x = q - r$ **c** $x = b - a$
   d $x = \sqrt{a}$ **e** $x = \dfrac{A}{b}$ **f** $x = k - m$
   g $x = \dfrac{d}{c}$ **h** $x = 2m - n$ **i** $x = \sqrt{\dfrac{M}{2}}$
   j $x = \dfrac{R}{T}$ **k** $x = \sqrt{n - 3}$ **l** $x = \dfrac{k - 4}{3}$
   m $x = k^2$ **n** $w^2 - 2$ **o** $x = \dfrac{m^2}{9}$

2 a $P = n - m$ **b** $P = 5k$ **c** $P = a - b$
   d $P = \sqrt{w}$ **e** $P = \sqrt{2w}$ **f** $P = \dfrac{2}{C}$
   g $P = \sqrt{\dfrac{Q}{3}}$ **h** $P = 2(a - 1)$ **i** $P = \dfrac{x}{e}$
   j $P = \dfrac{y + 3}{2}$ **k** $P = \dfrac{r}{Q}$ **l** $P = \dfrac{x - 5}{y}$

3 a i $b = \dfrac{A}{l}$ **ii** $2 \cdot 15$ **b i** $h = \dfrac{V}{lb}$ **ii** 4
   c i $m = \dfrac{F}{C}$ **ii** $1 \cdot 2$ **d i** $Q = \dfrac{P}{K}$ **ii** $1 \cdot 5$
   e i $x = \dfrac{P - 2y}{2}$ **ii** $1 \cdot 6$ **f i** $b = \dfrac{2A}{h}$ **ii** $3 \cdot 2$

4 a $R = \dfrac{E}{I}$ **b** $I = \dfrac{W}{E}$ **c** $E = \sqrt{RW}$
   d $W = I^2 R$ **e** $R = \dfrac{W}{I^2}$

5 a $l = \dfrac{A}{b}$; $b = \dfrac{A}{l}$ **b** $R = \dfrac{E}{I}$; $E = IR$

c $f = \dfrac{F}{M}$; $F = Mf$ **d** $A = \dfrac{V}{l}$; $l = \dfrac{V}{A}$

e $m = \dfrac{E}{c^2}$; $c^2 = \dfrac{E}{m}$

Page 90 Challenge

a It should hold for *all* connected graphs

b i $V = E - R + 1$ **ii** $R = E - V + 1$
iii $E = V + R - 1$

c i 1   **ii** 9   **iii** 8

d pupil's own drawings

Page 91 Exercise 4.2

1 a i $r = \sqrt{\dfrac{A}{4\pi}}$ **ii** 3·09 cm

b i $r = \sqrt[3]{\dfrac{3V}{4\pi}}$ **ii** 6·20 cm

2 a $x = \dfrac{q - p}{3}$ **b** $x = 2(y - 3)$

c $x = \dfrac{2y - 5}{3}$ **d** $x = \dfrac{A}{2} + 3$ or $x = \dfrac{A + 6}{2}$

e $x = 3B - 5$ **f** $x = 5 - 6k$

g $x = \dfrac{2}{w} - 1$ or $\dfrac{2 - w}{w}$

h $x = 1 - \dfrac{a}{6}$ or $x = \dfrac{6 - a}{6}$

i $x = b - \dfrac{c}{a}$ or $\dfrac{ab - c}{a}$

j $x = \dfrac{a}{c} - b$ or $\dfrac{a - bc}{c}$

k $x = c - \dfrac{b}{a}$ or $\dfrac{ac - b}{a}$

l $x = \dfrac{a}{cd} - \dfrac{b}{c}$ or $\dfrac{a - bd}{cd}$

3 a $r = \sqrt{\dfrac{3V}{\pi h}}$ **b** 9·77 cm

4 a $h = \dfrac{(3600A)^2}{w}$ **b** 182 cm

5 a 3-fin: $h = \dfrac{A}{0.17(d + 0.5)}$; 4-fin: $h = \dfrac{A}{0.13(d + 0.5)}$
b 3-fin: 27·1; 4-fin: 35·5; so the 4-fin by 8·4 inches

6 a Parabaloid: $r = \sqrt{\dfrac{2V}{\pi h}}$; neiloid: $r = \sqrt{\dfrac{4V}{\pi h}}$
b parabaloid: $r = 5·64$; neiloid: $r = 7·98$; the neiloid by 2·34 cm

7 a $d = \sqrt{\dfrac{GMm}{F}}$ **b** $l = \dfrac{T^2 g}{4\pi^2}$ **c** $l = \sqrt{\dfrac{85\,000w}{f}}$

d $D = 4\sqrt{\dfrac{V}{L}} + 4$ (other versions possible)

e $U = \dfrac{4(L + 27)}{5}$ **f** $h = \dfrac{A}{2\pi r} - r$

Page 93 Challenge

a $\frac{1}{4}$

b $h = R$

c $h = 6370$ km

d approx. 2640 km $((\sqrt{2} - 1)R)$

Page 94 Exercise 5.1

1 a doubled   **b** halved
   c halved   **d** doubled
   e multiplied by 4   **f** multiplied by 3

2 The higher ball has coefficient of restitution half that of the other

3 a $\frac{2}{3}$ of note A   **b** it is a D fork

4 a add $\frac{1}{2}$ hour (32 min)   **b** $7\frac{1}{2}$ min (divided by 8)

Page 97 Revise

1 a 62·8 cm   **b** 40 cm²   **c** 67·5 cm²
   d 26 cm   **e** 314 cm²

2 a $R = 400$ **b** both the same (200 ohms)

3 47 m²

4 a $S = 27\,000 + 1200n$
   b 33 000

5 a $M = 6n + 2$ **b** 224

   c $n = \dfrac{M - 2}{6}$ **d** 49th design

6 a $V = \frac{1}{4}\pi D^2 h$ **b** $v = \frac{1}{24}\pi D^2 h$

7 a $y = kb$ **b** $y = \dfrac{c}{r}$

   c $y = \dfrac{3}{2} - m$ or $y = \dfrac{3 - 2m}{2}$

   d $y = \sqrt{B + 3}$ **e** $y = \dfrac{w^2}{25}$ **f** $y = 3(Q - 2)$

5 Equations and inequations

Page 98 Exercise 1.1

1 a $x = 3$ **b** $y = 10$ **c** $a = -3$ **d** $w = -4$
   e $x = -3$ **f** $k = -7$ **g** $m = -14$ **h** $n = -16$
   i $y = 5$ **j** $w = 4$ **k** $c = 11$ **l** $f = 5$

2 a $x = -\frac{2}{3}$ **b** $x = -\frac{3}{2}$ **c** $x = -\frac{5}{4}$ **d** $a = \frac{3}{7}$
   e $a = \frac{3}{2}$ **f** $k = \frac{4}{3}$ **g** $m = -\frac{5}{2}$ **h** $n = -\frac{7}{2}$
   i $k = -\frac{2}{5}$ **j** $n = \frac{1}{3}$ **k** $m = -\frac{3}{7}$ **l** $x = -\frac{2}{3}$
   m $y = \frac{3}{4}$ **n** $x = \frac{1}{2}$ **o** $w = -\frac{1}{2}$ **p** $x = -\frac{1}{3}$

3 a $4x + 12$ **b** $-3y + 6$ **c** $-2x - 3$
   d $-14 + 7y$ **e** $8 + 3w$ **f** $2 - 3x$
   g $3 + 2k$ **h** $3 - 3m$ **i** $8x - 2$
   j $2n + 3$ **k** $2m - 2$ **l** $2w + 9$

4 a 6   **b** 12   **c** 9   **d** 20
   e 42   **f** 30   **g** 48   **h** 252
   i $2x$ **j** $2x^2$ **k** $6xy$ **l** $4x^2 y$

5 a $x^2 + x - 6$ **b** $y^2 - 3y - 2$
   c $2w^2 - 7w + 3$ **d** $m^2 - 10m + 25$
   e $3m^2 - 6m + 3$ **f** $-x^2 + 2x + 4$
   g $-x^2 + 12x - 25$ **h** $4n$
   i $5m^2 + 24m + 7$

Page 99 Exercise 2.1

1 a $x = 1$ b $x = -3$ c $x = -3$
d $x = 1$ e $x = 1$ f $x = 4$
g $x = 2$ h $x = 4$ i $x = -1$
j $x = -1$ k $x = -2$ l $x = -2$

2 a $x = 1$ b $y = 0$ c $n = 4$
d $m = -4$ e $k = 6$ f $w = -6$
g $y = -4$ h $w = 1$ i $x = -1$
j $x = 2$ k $w = 3$ l $k = 5$
m $m - 5$ n $n = 5$ o $r = 4$
p $c = 4$ q $q = 5$ r $x = 2$

3 a $f = 1$ b $x = 3$ c $y = 0$
d $x = 2$ e $m = -\frac{1}{2}$ f $x = 8$
g $k = -1$ h $w = 1$ i $y = -\frac{1}{2}$
j $n = \frac{1}{2}$ k $w = \frac{3}{2}$ l $a = -\frac{2}{3}$

4 a i $4x + 8 = 48$
  ii $x = 10$
  iii $16\,\text{cm} \times 8\,\text{cm}$
 b i $6x - 8 = 34$
  ii $x = 7$
  iii $5\,\text{cm} \times 12\,\text{cm}$
 c i $10x + 4 = 34$
  ii $x = 3$
  iii $9\,\text{cm} \times 8\,\text{cm}$
 d i $4x + 8 = 5x + 3$
  ii $x = 5$
  iii $2\,\text{cm} \times 12\,\text{cm}$
 e i $6x + 26 = 12x + 2$
  ii $x = 4$
  iii $17\,\text{cm} \times 8\,\text{cm}$
 f i $-2x + 38 = 50 + 10x$
  ii $x = -1$
  iii $9\,\text{cm} \times 11\,\text{cm}$

Page 100 Challenges

A a $(8x - 4) + 1 = (-2x + 12)$; $x = 1.5$;
   triangular: 2·5 cm, 2·5 cm and 3 cm;
   rectangular: 2·5 cm × 2 cm
 b triangular: 3 cm² (height 2 cm using Pythagoras' theorem);
   rectangular: 5 cm²; the rectangular by 2 cm²
B Size I: 2 m × $\frac{1}{2}$ m; Size II: 1·5 m × 1 m

Page 101 Exercise 2.2

1 a $x = 3$ b $x = 4$
 c $x = 7$ d $x = -3$
 e $x = -4$ f $x = -2$
 g $x = -\frac{1}{3}$ h $x = -2$
 i $x = 3$ j $x = 5$
 k $x = -1$ l $x = -2$

2 a $x = 5$; 3 cm, 4 cm and 5 cm
 b $x = 8$; 5 cm, 12 cm and 13 cm
 c $x = 10$; 9 cm, 12 cm and 15 cm
 d $x = \frac{7}{2}$; $1\frac{1}{2}$ cm, 2 cm and $2\frac{1}{2}$ cm
 e $x = 2$; 3 cm, 4 cm and 5 cm
 f $x = \frac{5}{2}$; 5 cm, 12 cm and 13 cm

Page 103 Exercise 3.1

1 a 2 b 2 c $\frac{3}{2}$ d $\frac{3}{4}$
 e 3 f $\frac{5}{3}$ g $\frac{4}{3}$ h $\frac{3}{4}$
 i $\frac{4}{5}$ j $\frac{15}{7}$ k $\frac{28}{5}$ l $\frac{3}{2}$

2 a $2x$ b $\frac{3y}{2}$ c $\frac{x}{4}$ d $\frac{a}{4}$
 e 3 f 2 g 2 h $\frac{2}{3}$
 i $\frac{2y}{5}$ j $9x$ k $20w$ l 17

3 a $y = 6$ b $x = 4$ c $k = 2$ d $m = 2$
 e $x = 6$ f $n = 8$ g $w = 6$ h $x = 9$
 i $y = 6$ j $r = 6$ k $n = \frac{4}{3}$ l $e = \frac{2}{3}$
 m $k = 4$ n $r = 28$ o $a = \frac{3}{5}$ p $n = 10$

4 a $y = 4$ b $x = 17$ c $m = 7$ d $n = 11$
 e $y = 15$ f $k = 2$ g $w = 3$ h $n = 3$
 i $x = 1$ j $y = 2$ k $a = 1$ l $k = 2$
 m $y = 9$ n $m = 30$ o $w = 15$

5 a i £$(x - 300)$
   ii $\frac{1}{2}(x - 300) = 200$
   iii £700
 b i £$(x - 250)$
   ii $\frac{1}{3}(x - 250) = 180$
   iii £790

6 a $x = 15$ b $k = 12$ c $m = 2$
 d $n = 3$ e $y = 5$ f $w = 10$
 g $m = 14$ h $x = 6$ i $a = 9$

7 a i $x + 24 = \frac{5}{3}x$ ii 36 cm
 b i $x + 18 = \frac{3}{2}x$ ii 36 cm
 c i $x + 30 = \frac{7}{3}x$ ii 22·5 cm

8 a i $x - 36 = \frac{2}{3}x$ ii 108 litres
 b i $x - 60 = \frac{1}{2}x$ ii 120 litres
 c i $x - 96 = \frac{2}{5}x$ ii 160 litres

Page 105 Challenges

A $\frac{1}{3}x - 20\,000 = \frac{1}{4}x$;
  leaves £240 000 each receives £80 000
B $x + 1200 + \frac{3}{5}(x + 1200) = 4x$;
  there are now 3200 ($x = 800$)

Page 106 Exercise 4.1

1 a $y = 42$ b $m = 12$ c $w = \frac{24}{5}$ d $k = 20$
 e $a = 12$ f $k = -7$ g $x = \frac{7}{3}$ h $f = 2$
 i $y = 3$ j $c = 4$ k $h = -2$ l $x = -3$

2 a $x = 68$ b $a = 10$ c $x = 4$
 d $y = 15$ e $a = 20$ f $x = 50$
 g $n = 24$ h $x = 7$ i $x = 21$

3 a i $\frac{1}{2}(x + 3)$ ii $\frac{1}{3}(x - 4)$
 b $\frac{1}{2}(x + 3) - \frac{1}{3}(x - 4) = 8$; 31 litres

4 a 100 b 90

5 a $\dfrac{d}{15}$ hours b $\dfrac{d}{20}$ hours
 c 12 minutes is $\frac{1}{5}$ of 1 hour so the difference is $\frac{1}{5}$
 d 12 km

6 $\dfrac{d}{3} - \dfrac{d}{5} = 15$; $d = 112\frac{1}{2}$; 225 miles in total

Page 108 Exercise 5.1

1 **a** $v \leqslant 20$ **b** $W < 5$ **c** $M \geqslant 6 \cdot 5$
  d $L \geqslant 4$ **e** $16 \leqslant t \leqslant 25$ **f** $3 \cdot 5 \leqslant C \leqslant 15$

2 **a** $v \leqslant 20$ **b** $v \leqslant 60$ **c** $20 < v \leqslant 30$
  d $30 < v \leqslant 60$ **e** $v > 30$ **f** $20 < v \leqslant 30$

3 **a i** $a < 0, b < 0$
    ii $a > 0, b > 0$
    iii $a > 0, b < 0$
  b i $a > 2, b > 1$
    ii $a < 2, b > 1$
    iii $a < 2, b < 1$
    iv $a > 2, b < 1$

4 **a** $a < c, b > d$ **b** $e < g, f < h$ **c** $k < n, m > p$

5 **a** $x < 8$ **b** $y > 3$ **c** $m + 2 < 6$

Page 109 Challenges

A Add a to both sides of the second balance.
  This gives $ac > aa$ but $aa > bc$ so $ac > bc$.
  Remove c from both sides so $a > b$.

B Balance 1 and 3 give $ad > dc$ and $dc > bc$ so
  $ad > bc$ but $bd > ad$ (middle balance) so $bd > bc$.
  Now remove b from both sides giving $d > c$.

Page 110 Exercise 6.1

1 **a** $x \geqslant -1$ **b** $y < 7$ **c** $k \leqslant -4$ **d** $x < 3$
  e $y \geqslant 3$ **f** $w \leqslant 5$ **g** $n > 2$ **h** $c \geqslant 2$
  i $k < -3$ **j** $t \leqslant -12$ **k** $x \leqslant 2$ **l** $m > -10$
  m $a \leqslant 0$ **n** $z \leqslant -6$ **o** $k < 7$ **p** $x \leqslant 8$

2 **a** $x \leqslant 4$ **b** $x < 3$ **c** $k > 4$ **d** $m \geqslant -2$
  e $m \leqslant -2$ **f** $x > -4$ **g** $y \geqslant -3$ **h** $k < 2$
  i $n \leqslant 1$ **j** $t > \frac{1}{2}$ **k** $w \leqslant \frac{1}{5}$ **l** $a \geqslant \frac{1}{2}$
  m $c > -\frac{1}{2}$ **n** $e < -\frac{1}{2}$ **o** $x \leqslant \frac{1}{2}$ **p** $w \geqslant -\frac{3}{2}$

3 **a** $y > 2$ **b** $k < -3$ **c** $w \geqslant -4$ **d** $m \leqslant 7$
  e $y \geqslant -11$ **f** $n \leqslant 12$ **g** $c > \frac{1}{2}$ **h** $y < -\frac{2}{3}$

4 **a** $x < -4$ **b** $y > 2$ **c** $k \leqslant 3$ **d** $n \geqslant 3$
  e $x \leqslant 3$ **f** $n \geqslant -2$ **g** $w \geqslant -2$ **h** $k > -4$
  i $m \leqslant 1$ **j** $x < 4$ **k** $k \leqslant -5$ **l** $n \geqslant 5$

5 **a i** $x \geqslant -1$ **ii** $-1, 0, 1, 2, 3$
  b i $x < -1$ **ii** $-3, -2$
  c i $x > 2$ **ii** 3
  d i $x \leqslant 0$ **ii** $-3, -2, -1, 0$
  e i $x > 1$ **ii** $2, 3$
  f i $x < 2$ **ii** $-3, -2, -1, 0, 1$
  g i $x \leqslant -2$ **ii** $-3, -2$
  h i $x \geqslant 3$ **ii** 3
  i i $x \geqslant 0$ **ii** $0, 1, 2, 3$
  j i $x \leqslant -2$ **ii** $-3, -2$
  k i $x > -2$ **ii** $-1, 0, 1, 2, 3$
  l i $x \leqslant -3$ **ii** -3

6 **a** $x > -2$ **b** $k \leqslant 1$ **c** $m \geqslant 2$
  d $k \geqslant -1$ **e** $x > -2$ **f** $y \leqslant 2$
  g $x \geqslant 2$ **h** $w < -\frac{1}{2}$ **i** $a \leqslant 4$
  j $x > 2$ **k** $p \geqslant 10$ **l** $x < \frac{1}{2}$

Page 111 Exercise 6.2

1 **a** $x > 3$ **b** $w < 6$ **c** $y > -3$
  d $m \geqslant 1$ **e** $k \leqslant 2$ **f** $n \geqslant \frac{1}{2}$

g $a > 2$ **h** $k < -1$ **i** $x \geqslant 5$
j $x \geqslant 2$ **k** $x \geqslant 3$ **l** $x \geqslant 2$
m $x \leqslant -\frac{1}{2}$ **n** $x \geqslant -\frac{5}{3}$ **o** $x \geqslant -3$
p $x \leqslant 4$ **q** $x \geqslant \frac{7}{2}$ **r** $x \leqslant -3$

Page 112 Challenge

$x = 10, y = 6$ and $z = 9$

Page 113 Revise

1 **a** $x = 2$ **b** $y = 6$ **c** $k = -1$
  d $m = 1$ **e** $y = 1$

2 **a** $10x + 4 = 34$
  b $x = 3$ **c** $10 \text{ cm} \times 7 \text{ cm}$

3 **a** $(2x + 3)^2 = (2x + 2)^2 + 5^2$ **b** $x = 5$
  c perimeter $= 30$ cm, area $= 30$ cm^2

4 **a** $x = 5$ **b** $x = 5$

5 **a** $2x$ **b** $16x$ **c** $6x - 4$

6 **a** $x = 10$ **b** $y = 6$ **c** $x = 6$
  d $k = 8$ **e** $y = 28$

7 **a** $x - 250$ **b** $\frac{5}{6}(x - 250) = 350$
  c 670

8 Range A: $P < 10$;
  Range B: $10 \leqslant P \leqslant 25$;
  Range C: $P > 25$

9 **a** $m \geqslant 5$ **b** $k > -5$ **c** $n \leqslant -3$ **d** $x < 2$
  e $x > 2$ **f** $y > 5$ **g** $x \leqslant 7$

10 $n = 2, 3$ or 4

6 Algebraic fractions

Page 116 Exercise 1.1

1 **a** $1, 2, 3, 4, 6, 8, 12, 24$
  b $1, 2, 3, 4, 6, 12, x, 2x, 3x, 4x, 6x, 12x$
  c $1, a, b, ab, a^2, a^2b$
  d $1, 2, 4, x, 2x, 4x, y, 2y, 4y, xy, 2xy, 4xy$

2 **a** $2(x + 6y)$ **b** $4x(3 - y)$
  c $3x(2y - x)$ **d** $(x - 2)(x + 2)$
  e $(2x - 1)(2x + 1)$ **f** $(3x - 7y)(3x + 7y)$
  g $(x + 2)(x + 1)$ **h** $(x - 3)(x + 2)$
  i $(3x + 2)(2x + 1)$

3 **a** $\frac{18}{24}$ **b** $\frac{36}{45}$ **c** $\frac{28}{35}, \frac{10}{35}$
  d between $\frac{16}{24}$ and $\frac{18}{24}$ lies $\frac{17}{24}$

4 **a i** $5x$ **ii** $\frac{5}{5}$ **iii** $\frac{5}{7}$ **iv** $\frac{5}{9}$
  b i $2x$ **ii** $\frac{2}{5}$ **iii** $\frac{2}{7}$ **iv** $\frac{2}{9}$
  c i $6x^2$ **ii** $\frac{6}{25}$ **iii** $\frac{6}{9}$ **iv** $\frac{6}{49}$
  d i $\frac{3}{2}$ **ii** $\frac{3}{2}$ **iii** $\frac{3}{2}$ **iv** $\frac{3}{2}$

5 **a** $\frac{9}{20}$ **b** $\frac{23}{20}$ **c** $\frac{20}{21}$ **d** $\frac{49}{40}$
  e $\frac{1}{8}$ **f** $\frac{13}{24}$ **g** $\frac{8}{21}$ **h** $\frac{1}{18}$

6 **a** $\frac{1}{20}$ **b** $\frac{6}{20}$ **c** $\frac{4}{21}$ **d** $\frac{15}{40}$

7 **a** $\frac{4}{5}$ **b** $\frac{8}{15}$ **c** $\frac{7}{3}$ **d** $\frac{24}{25}$

8 $\frac{3}{10}$

9 $\frac{1}{3}$

10 $\frac{9}{10}$ km/m

Page 117 Exercise 2.1

1 a $\dfrac{2}{2x}$ **b** $\dfrac{x+1}{x(x+1)}$ **c** $\dfrac{x}{x^2}$

2 a, b, e, f, g

3 a $\dfrac{2}{2x}, \dfrac{x}{2x}$ **b** $\dfrac{6}{3x}, \dfrac{1}{3x}$

c $\dfrac{y}{xy}, \dfrac{x}{xy}$ **d** $\dfrac{2y}{3xy}, \dfrac{x}{3xy}$

e $\dfrac{1}{x^2}, \dfrac{x}{x^2}$ **f** $\dfrac{2x}{x(x+1)}, \dfrac{x+1}{x(x+1)}$

g $\dfrac{x}{x(x+1)}, \dfrac{2(x+1)}{x(x+1)}$ **h** $\dfrac{8}{4(x+1)}, \dfrac{3(x+1)}{4(x+1)}$

i $\dfrac{x+1}{(x+1)(x+2)}, \dfrac{x+2}{(x+1)(x+2)}$

j $\dfrac{2x}{x(1-x)}, \dfrac{3(1-x)}{x(1-x)}$ **k** $\dfrac{x^2}{7x}, \dfrac{7}{7x}$

l $\dfrac{10x}{15}, \dfrac{3(4x+1)}{15}$

4 a $\dfrac{1}{4}$ **b** $\dfrac{2}{3}$ **c** $\dfrac{1}{2}$ **d** $\dfrac{1}{5x}$

e $\dfrac{1}{3}$ **f** $\dfrac{a}{b}$ **g** $\dfrac{x}{4}$ **h** $\dfrac{x}{2}$

i $\dfrac{p}{3}$ **j** $\dfrac{8}{b}$ **k** $\dfrac{5}{3xy}$ **l** $\dfrac{b}{2a}$

m $\dfrac{x}{5y}$ **n** $\dfrac{2r^2}{3p}$ **o** $\dfrac{6e}{f}$

5 a $\dfrac{a}{3}$ **b** $\dfrac{4}{5}$ **c** $\dfrac{x}{4}$ **d** $\dfrac{x+1}{3}$

e $\dfrac{x-1}{x-3}$ **f** $\dfrac{1}{a+1}$ **g** $\dfrac{1}{p+q}$ **h** $\dfrac{1}{4}$

6 a 4 **b** 5 **c** $\dfrac{1}{7}$ **d** $\dfrac{1}{4}$

e $\dfrac{3}{5}$ **f** $\dfrac{2}{5}$ **g** $\dfrac{3}{2}$ **h** $\dfrac{x}{2}$

Page 119 Exercise 2.2

1 a $\dfrac{4}{x-1}$ **b** $\dfrac{1}{1+x}$ **c** $\dfrac{x}{x-2}$

d $\dfrac{1}{4a-1}$ **e** $\dfrac{4}{2x-3}$ **f** $4(x+3)$

g $\dfrac{3(1+2x)}{4}$ **h** x^2+1

2 a $\dfrac{3}{x+1}$ **b** $\dfrac{1}{x+1}$ **c** $\dfrac{3}{x+4}$

d $\dfrac{1}{x-1}$ **e** $\dfrac{1}{x+1}$ **f** $\dfrac{3}{3+x}$

g $\dfrac{7}{1-x}$ **h** $\dfrac{x}{x+5}$ **i** $\dfrac{x}{5-x}$

j $\dfrac{x-2}{x}$ **k** $\dfrac{x-2}{y}$ **l** $\dfrac{x-1}{x}$

3 a $\dfrac{x+4}{x+1}$ **b** $\dfrac{2x+1}{x-1}$ **c** $\dfrac{4+3x}{1-x}$

d $\dfrac{3x+2}{x+1}$ **e** $\dfrac{x-1}{3x+1}$ **f** $\dfrac{x-1}{x+1}$

g $\dfrac{1-2x}{x}$ **h** $\dfrac{5+x}{x}$

i $\dfrac{(1+x)(1+x^2)}{x+2}$ **j** $\dfrac{(x+2)(x^2+4)}{x+8}$

k $-1-x$ **l** $\dfrac{(x-3)(x+3)}{(x+2)(x-1)}$

Page 120 Exercise 3.1

1 a $\dfrac{ab}{4}$ **b** $\dfrac{2x}{3y}$ **c** $\dfrac{3k}{4t}$ **d** $\dfrac{1}{xy}$

e $\dfrac{6}{kl}$ **f** $\dfrac{mn}{16}$ **g** $\dfrac{2mn}{5}$ **h** $\dfrac{8}{3mn}$

i $\dfrac{1}{2}$ **j** 9 **k** $\dfrac{6}{t^2}$ **l** $\dfrac{15}{x}$

m 1 **n** $\dfrac{ab}{xy}$ **o** $\dfrac{4}{x^3}$ **p** 1

q 3 **r** 12 **s** 10 **t** $\dfrac{z^2}{2}$

u $12k$ **v** 3 **w** $\dfrac{15n}{m}$ **x** 1

2 a $\dfrac{x^2}{6}$ **b** 2 **c** $\dfrac{x^2}{4}$ **d** 10 **e** $\dfrac{y}{2}$

3 a $4x$ **b** $\dfrac{y+1}{2}$ **c** z **d** $\dfrac{x(y+2)}{4-x}$

e $\dfrac{1}{y+4}$ **f** $\dfrac{y^2}{(2y+5)(2-x)}$

4 $\dfrac{2x+2}{4x+6} \times \dfrac{2x+3}{x+1} = \dfrac{2(x+1)(2x+3)}{2(2x+3)(x+1)} = 1$

5 a i $p = \dfrac{x}{k}$ **ii** $q = \dfrac{y}{k}$

b i $A = \dfrac{xy}{2}$ **ii** $A = \dfrac{xy}{2k^2}$ **c** $\dfrac{1}{k^2}$

6 a

row 1:	$\dfrac{1}{a^2}$	$\dfrac{1}{ab}$	$\dfrac{1}{ac}$	$\dfrac{c}{a}$	$\dfrac{b}{a}$	1
row 2:	$\dfrac{1}{ab}$	$\dfrac{1}{b^2}$	$\dfrac{1}{bc}$	$\dfrac{c}{b}$	1	$\dfrac{a}{b}$
row 3:	$\dfrac{1}{ac}$	$\dfrac{1}{bc}$	$\dfrac{1}{c^2}$	1	$\dfrac{b}{c}$	$\dfrac{a}{c}$
row 4:	$\dfrac{c}{a}$	$\dfrac{c}{b}$	1	c^2	bc	ac
row 5:	$\dfrac{b}{a}$	1	$\dfrac{b}{c}$	bc	b^2	ab
row 6:	1	$\dfrac{a}{b}$	$\dfrac{a}{c}$	ac	ab	a^2

b ii $\dfrac{1}{a} < \dfrac{1}{b}$ **c** $\dfrac{1}{a} < \dfrac{1}{b} < \dfrac{1}{c}$

d i perfect squares **ii** each entry 1
iii symmetry **iv** reciprocals
7 a 1 **b** 1 **c** 1 **d** 1 **e** 1 **f** 1 **g** 1 **h** 1
8 product is always 1

Page 122 Exercise 4.1

1 a $\dfrac{1}{8}$ **b** $\dfrac{1}{3x}$ **c** $\dfrac{x}{8}$ **d** $\dfrac{3}{4x}$

e $\dfrac{1}{32}$ **f** $\dfrac{1}{5y}$ **g** $\dfrac{y}{15}$ **h** $\dfrac{1}{7t}$

2 a $\dfrac{3}{2}$ **b** $\dfrac{6}{5}$ **c** $\dfrac{2}{y^2}$ **d** $\dfrac{9}{10}$

e $\dfrac{3}{2}$ **f** $3xy$ **g** $\dfrac{2a}{3b}$ **h** $\dfrac{3}{x}$

i $\dfrac{c}{b}$ **j** $\dfrac{3x}{4b}$ **k** $\dfrac{1}{a^2}$ **l** 1

3 a $\dfrac{4(x+2)}{3(x+3)}$ **b** $\dfrac{36}{35}$ **c** $\dfrac{5}{2k}$ **d** $\dfrac{5(x-1)}{2}$

e $(y+7)(x-2)$ **f** $\dfrac{t-3}{2(1+a)}$

g $\dfrac{x-1}{x+1}$ **h** $\dfrac{1}{2}$

4 a $\dfrac{14\,000\,000}{x}$

b $\dfrac{14\,000\,000}{x+1}$

c $\dfrac{14\,000\,000}{x+1} \div \dfrac{14\,000\,000}{x} = \dfrac{x}{x+1}$

5 a i $a = \dfrac{6}{x}$

b i $b = \dfrac{20}{a}$

ii $b = 20 \div a = 20 \div \dfrac{6}{x} = \dfrac{20x}{6} = \dfrac{10x}{3}$

6 a row 1:   1 $\dfrac{b}{a}$ $\dfrac{c}{a}$ $\dfrac{1}{ac}$ $\dfrac{1}{ab}$ $\dfrac{1}{a^2}$

row 2: $\dfrac{a}{b}$ 1 $\dfrac{c}{b}$ $\dfrac{1}{bc}$ $\dfrac{1}{b^2}$ $\dfrac{1}{ab}$

row 3: $\dfrac{a}{c}$ $\dfrac{b}{c}$ 1 $\dfrac{1}{c^2}$ $\dfrac{1}{bc}$ $\dfrac{1}{ac}$

row 4: ac bc c^2 1 $\dfrac{c}{b}$ $\dfrac{c}{a}$

row 5: ab b^2 bc $\dfrac{b}{c}$ 1 $\dfrac{b}{a}$

row 6: a^2 ab ac $\dfrac{a}{c}$ $\dfrac{a}{b}$ 1

b down from left to right: all entries 1;
up fom left to right: all entries squares
c left-to-right 'mirror' image

Page 124 Exercise 5.1

1 a $\dfrac{7x}{12}$ **b** $\dfrac{7a}{10}$ **c** $\dfrac{11t}{28}$

d $\dfrac{13k}{6}$ **e** $\dfrac{31x}{30}$ **f** $\dfrac{4x+3}{12}$

g $\dfrac{4p+15}{20}$ **h** $\dfrac{7q+15}{21}$ **i** $\dfrac{16g+35}{56}$

j $\dfrac{3n+4}{9}$ **k** $\dfrac{xy+15}{5y}$ **l** $\dfrac{wv+4}{4v}$

m $\dfrac{2rk+10}{5k}$ **n** $\dfrac{6hg+12}{8g}$ **o** $\dfrac{27mn+56}{63n}$

p $\dfrac{x^2+10}{2x}$ **q** $\dfrac{t^2+5}{5t}$ **r** $\dfrac{2p^2+63}{9p}$

s $\dfrac{15g^2+8}{24g}$ **t** $\dfrac{30h^2+22}{55h}$

2 a $\dfrac{2x+y}{xy}$ **b** $\dfrac{3b+2a}{ab}$ **c** $\dfrac{5w+t}{tw}$

d $\dfrac{4z+2y}{yz}$ **e** $\dfrac{m+n}{mn}$ **f** $\dfrac{3y+2x}{2xy}$

g $\dfrac{2f+3e}{6ef}$ **h** $\dfrac{5t+8r}{20rt}$ **i** $\dfrac{21x+8y}{3xy}$

j $\dfrac{27v+2t}{3tv}$ **k** $\dfrac{11}{3x}$ **l** $\dfrac{11}{10g}$

m $\dfrac{9}{4k}$ **n** $\dfrac{67}{14b}$ **o** $\dfrac{19}{15x}$

p $\dfrac{2x+5}{2y}$ **q** $\dfrac{x^2+y^2}{xy}$ **r** $\dfrac{9a^2+2b^2}{6ab}$

s $\dfrac{30k^2+3h}{5kh}$ **t** $\dfrac{23v}{5t}$

3 a $\dfrac{R_1+R_2}{R_1R_2}$ **b** $\dfrac{R_1R_2}{R_1+R_2}$ **c** $\dfrac{35}{12}$ ohms

4 a $\dfrac{u+v}{uv}$ **b** $f = \dfrac{uv}{u+v}$ **c** $\dfrac{30}{11}$

Page 126 Exercise 5.2

1 a $\dfrac{7x+11}{10}$ **b** $\dfrac{11w-12}{12}$ **c** $\dfrac{3g}{4}$

d $\dfrac{11k+25}{15}$ **e** $\dfrac{13r-42}{21}$ **f** $\dfrac{11h-5}{6}$

g $\dfrac{p^2+p+5}{5p+5}$ **h** $\dfrac{t^2-t+6}{2t-2}$ **i** $\dfrac{6f^2+9f+5}{10f+15}$

j $\dfrac{4x+3}{x^2+x}$ **k** $\dfrac{2x}{x^2-1}$ **l** $\dfrac{2x+1}{x^2+x-6}$

m $\dfrac{4x-1}{3x-1-2x^2}$ **n** $\dfrac{13x-26}{2x^2-2x-15}$

o $\dfrac{11x+13}{6x^2+11x+3}$ **p** 0

q $\dfrac{-3x-1}{x^2-6x+5}$ **r** $\dfrac{2x^2+x+1}{x^2-1}$

2 a $\dfrac{x+30}{7}$ **b** $\dfrac{6k+5}{k+1}$ **c** $\dfrac{2t}{t-1}$

d $\dfrac{4m-1}{m}$ **e** $\dfrac{14d-4}{2d-1}$ **f** $\dfrac{x+2}{x+1}$

3 a $\dfrac{2a+a_1+a_2}{(a+a_1)(a+a_2)} = \dfrac{2a+a_1+a_2}{a^2+aa_1+aa_2+a_1a_2}$

b/c
$$= \dfrac{2a+a_1+a_2}{a^2+aa_1+aa_2+a^2} = \dfrac{2a+a_1+a_2}{a(2a+a_1+a_2)} = \dfrac{1}{a}$$

4 a $T_{up} = \dfrac{5}{v-3}$; $T_{down} = \dfrac{5}{v+3}$; $T_{total} = T_{up}+T_{down}$

b $\dfrac{10v}{v^2-9}$

c $\dfrac{25}{36}$ hours = 41 min 40 s

5 a row 1: $\dfrac{2}{a}$; $\dfrac{a+b}{ab}$; $\dfrac{a+c}{ac}$; $\dfrac{1+ac}{a}$; $\dfrac{1+ab}{a}$; $\dfrac{1+a^2}{a}$

     row 2: $\dfrac{a+b}{ab}$; $\dfrac{2}{b}$; $\dfrac{b+c}{bc}$; $\dfrac{1+bc}{b}$; $\dfrac{1+b^2}{b}$; $\dfrac{1+ab}{b}$

     row 3: $\dfrac{a+c}{ac}$; $\dfrac{b+c}{bc}$; $\dfrac{2}{c}$; $\dfrac{1+c^2}{c}$; $\dfrac{1+bc}{c}$; $\dfrac{1+ac}{c}$

     row 4: $\dfrac{1+ac}{a}$; $\dfrac{1+bc}{b}$; $\dfrac{1+c^2}{c}$; $2c$; $b+c$; $a+c$

     row 5: $\dfrac{1+ab}{a}$; $\dfrac{1+b^2}{b}$; $\dfrac{1+bc}{c}$; $b+c$; $2b$; $a+b$

     row 6: $\dfrac{1+a^2}{a}$; $\dfrac{1+ab}{b}$; $\dfrac{1+ac}{c}$; $a+c$; $a+b$; $2a$

b down fom left to right: all entries $2 \times$ first number;
up from left to right:
all entries of form $\dfrac{1+x^2}{x}$

c 'mirror' image with diagonal going down from left to right as axis

Page 127 Exercise 6.1

1 a $\dfrac{a}{6}$ **b** $\dfrac{4b}{21}$ **c** $\dfrac{c}{24}$

d $\dfrac{18a}{35}$ **e** $\dfrac{-7g}{40}$ **f** $\dfrac{7t-2}{14}$

g $\dfrac{2u-5}{6}$ **h** $\dfrac{v-2}{4}$ **i** $\dfrac{16h-9}{12}$

j $\dfrac{7k-10}{35}$ **k** $\dfrac{ab-6}{3b}$ **l** $\dfrac{cd-15}{5d}$

m $\dfrac{3ef-20}{4f}$ **n** $\dfrac{35xy-3}{15x}$ **o** $\dfrac{2bc-1}{4c}$

p $\dfrac{a^2-21}{3a}$ **q** $\dfrac{b^2-8}{4b}$ **r** $\dfrac{3c^2-1}{2c}$

s $\dfrac{45k^2-8}{20k}$ **t** $\dfrac{h^2-75}{5h}$

2 a $\dfrac{b-a}{ab}$ **b** $\dfrac{2y-5x}{xy}$ **c** $\dfrac{3t-7k}{kt}$

d $\dfrac{5k-6p}{pk}$ **e** $\dfrac{3ac-2b}{bc}$ **f** $\dfrac{1}{6a}$

g $\dfrac{17}{4a}$ **h** $\dfrac{2b^2-a^2}{ab}$ **i** $\dfrac{5g-3}{g^2}$

j $\dfrac{2k-1}{k^2}$ **k** $\dfrac{5a-1}{a}$ **l** $\dfrac{3k-2}{3k}$

m $\dfrac{3-8g}{2g}$ **n** $\dfrac{3-35r}{5r}$ **o** $\dfrac{k^2-1}{k}$

p $\dfrac{3t^2-3}{t}$ **q** $\dfrac{x-xy}{y}$ **r** $\dfrac{15p}{4}$

s $\dfrac{6fg-3f}{2g}$ **t** $\dfrac{4a^3-b^3}{a^2b^2}$

3 a $\dfrac{v}{u}$ **b** 3   **c i** $\dfrac{u-v}{uv}$ **ii** $f = \dfrac{uv}{u-v}$

4 a $x-2$ **b i** $\dfrac{x}{10}$ **ii** $\dfrac{x-2}{12}$

c $\dfrac{x+10}{60}$ **d** 110 m

5 a $x+1$ **b i** $\dfrac{20}{x}$ **ii** $\dfrac{20}{x+1}$

c $\dfrac{20}{x^2+x}$ **d** £4

Page 129 Exercise 6.2

1 a $\dfrac{x-10}{12}$ **b** $\dfrac{k+6}{6}$ **c** $\dfrac{h+3}{4}$

d $\dfrac{2f-3}{15}$ **e** $\dfrac{1}{3}$ **f** $\dfrac{13m-15}{10}$

g $\dfrac{d^2+d-3}{3d+3}$ **h** $\dfrac{a^2-a-18}{9a-9}$ **i** $\dfrac{12v^2+3v-4}{6v}$

j $\dfrac{1}{a^2+a}$ **k** $\dfrac{1}{p^2+p}$ **l** $\dfrac{4}{h^2-2h-3}$

m $\dfrac{x}{x^2-3x+2}$ **n** $\dfrac{1}{x^2+3x+2}$ **o** $\dfrac{g^2-g}{6g^2+5g+1}$

2 a $\dfrac{18-v}{4}$ **b** $\dfrac{1}{1-a}$ **c** $\dfrac{3}{b-1}$

d $\dfrac{-k-2}{k+1}$ **e** $\dfrac{7+2x}{2+x}$ **f** $\dfrac{5d-3d^2}{3d-4}$

3 a $\dfrac{x^2-2x+30}{6x}$ **b** $\dfrac{x^2+x+2}{2x^2+2x}$ **c** $\dfrac{6-x}{6x}$

d $\dfrac{x^2+y^2-2xy}{xy}$ **e** $\dfrac{x^2+6x+6}{2x+x^2}$ **f** $\dfrac{1+2k-6k^2}{6k+3}$

4 a i $\dfrac{120}{x-2}$ **ii** $\dfrac{120}{x+2}$

b $\dfrac{480}{x^2-4}$ **c** 10   **d** 15

5 a $\dfrac{2}{x}, \dfrac{x}{2}$ **b** $\dfrac{4-x^2}{2x}$ **c** 1

6 a row 1: $0; \dfrac{b-a}{ab}; \dfrac{c-a}{ac}; \dfrac{1-ac}{a}; \dfrac{1-ab}{a}; \dfrac{1-a^2}{a}$

row 2: $\dfrac{a-b}{ab}; 0; \dfrac{c-b}{bc}; \dfrac{1-bc}{b}; \dfrac{1-b^2}{b}; \dfrac{1-ab}{b}$

row 3: $\dfrac{a-c}{ac}; \dfrac{b-c}{bc}; 0; \dfrac{1-c^2}{c}; \dfrac{1-bc}{c}; \dfrac{1-ac}{c}$

row 4: $\dfrac{ac-1}{a}; \dfrac{bc-1}{b}; \dfrac{c^2-1}{c}; 0; c-b; c-a$

row 5: $\dfrac{ab-1}{a}; \dfrac{b^2-1}{b}; \dfrac{bc-1}{c}; b-c; 0; b-a$

row 6: $\dfrac{a^2-1}{a}; \dfrac{ab-1}{b}; \dfrac{ac-1}{c}; a-c; a-b; 0$

b down from left to right: all entries 0;
up from left to right: all entries of form

$\dfrac{1-x^2}{x}$ or $\dfrac{x^2-1}{x}$

c with diagonal going down fom left to right as
axis, image of x is $-x$

Page 132 Revise

1 a $\dfrac{15}{5y}$ **b** $\dfrac{3y}{y^2}$ **c** $\dfrac{3x}{xy}$

2 a, d and **f**

3 a $\dfrac{(x+5)(x-1)}{(x+2)(x-1)} = \dfrac{x+5}{x+2}$

b $\dfrac{(3x-1)(3x+1)}{(3x-1)(x+4)} = \dfrac{3x+1}{x+4}$

c $\dfrac{(2x+1)(x-1)}{x(2x+1)} = \dfrac{x-1}{x}$

4 a $\dfrac{xy}{15}$ **b** $\dfrac{3}{4}$ **c** $\dfrac{10}{3}$ **d** $\dfrac{3}{x}$

5 a $\dfrac{15}{14}$ **b** $\dfrac{2c}{ab}$ **c** k^2 **d** $\dfrac{x}{(x+1)^2}$

6 a $\dfrac{12x}{35}$ **b** $\dfrac{g^2+15}{3g}$ **c** $\dfrac{3n+4m}{mn}$

d $\dfrac{5x+1}{x^2-1}$ **e** $\dfrac{x^2+1}{x}$

7 a $\dfrac{4a}{21}$ **b** $\dfrac{y-x}{xy}$ **c** $\dfrac{3-4x}{x}$

d $\dfrac{2}{x^2-1}$ **e** $\dfrac{g^2-2}{g}$

8 a i $\dfrac{9}{10-x}$ **ii** $\dfrac{9}{10+x}$ **iii** $\dfrac{180}{100-x^2}$

b 2 km/h

9 a $\dfrac{10\,000\,000}{x}$ **b** $\dfrac{10\,000\,000}{x-4}$

c $\dfrac{40\,000\,000}{x^2-4x}$ **d** 5

7 The straight line

Page 134 Exercise 1.1

1 a $(1, 4)$
  b i 4rd **ii** 2nd **iii** 3rd
  c i 1st and 4th
   ii 3rd and 4th
   iii 2nd and 3rd
  d i 3 units
   ii 5 units

2 a

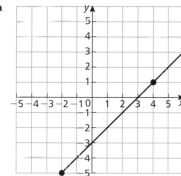

b i on **ii** below **iii** above
c iii
d P

3 a

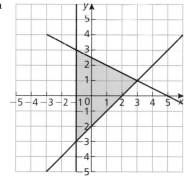

b $(-1, 3), (-1, -3), (3, 1)$
c i outside **ii** inside **iii** on

4 a

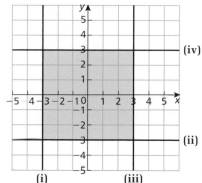

b $(-3, 3), (3, 3), (3, -3), (-3, -3)$

5 a/b

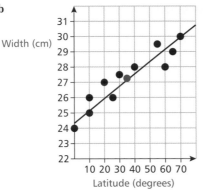

Width (cm) — Latitude (degrees)

c strong positive
d orange point on grid
e on line

Page 136 Exercise 2.1

1 a $x = -4$ b $x = 0$ c $x = 2$
  d $y = 4$ e $y = 0$ f $y = -2$
2 a It's the x axis b it's the y axis
  c i x axis ii parallel
   iii y axis iv parallel
3 a i 2 units ii 5 units
   iii $(b - a)$ units iv $4 - (-4) = 8$
  b i 2 units ii $(b - a)$ units
   iii $5 - (-1) = 6$ units
4 a i $(4, 6)$ ii $(3, -2)$ iii $(0, 0)$
  b (a, b)
5 a 5 units b 9 units c 45 units^2
  d i 32 units^2 ii 3 units^2 iii 4 units^2
  e 56 units^2
6 a $y = 3$ b $x = 5$ c $x = 6$ d $y = 0$
  e $y = 7$ f $x = -2$ g $y = -4$ h $x = 2$
  i $x = 1$ j $y = -7$
7 a $x = -4, x = 3, y = 4, y = -1$
  b a line passing through $(-4, 4)$ and $(3, -1)$;
   a line passing through $(3, 4)$ and $(-4, -1)$

Page 138 Brainstormer

$(b - a)(d - c)$ units^2

Page 139 Exercise 3.1

1 a $\frac{1}{5}$ b $\frac{3}{2}$ c $\frac{24}{31}$ d $\frac{3}{4}$ e 0 f $\frac{8}{5}$
2 a i 9·2 cm ii 0·6
  b i 85·1 cm ii 0·2
  c i 7·7 cm, 6·4 cm ii 1·2
3 only b
4 a 6 m b 125 m
  c 600 m d undefined (vertical descent)
5 a 1 b $\frac{2}{7}$ c undefined d $\frac{4}{1} = 4$
  e $\frac{1}{4}$ f $\frac{3}{5}$ g $\frac{2}{5}$ h $\frac{1}{8}$

Page 141 Exercise 4.1

1 a $m_{PQ} = 1; m_{RS} = \frac{1}{3}; m_{EF} = -\frac{1}{4}; m_{UV} = -\frac{2}{3}; m_{JK} = 0;$
  m_{GH} is undefined

2 a parallel
  b both $-\frac{1}{5}$ in the first diagram and $\frac{1}{2}$ in the second
  c true
3 i $\frac{1}{3}$ ii $\frac{1}{4}$ iii $\frac{2}{9}$ iv $\frac{1}{5}$ v $\frac{1}{3}$
  i and v parallel

4 a

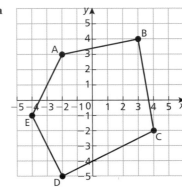

  b i $\frac{1}{5}$ ii -6 iii $\frac{1}{2}$ iv -2 v 2
  c $\frac{5}{7}$
5 a i $\frac{1}{4}$ ii $\frac{1}{6}$ iii $\frac{1}{5}$ iv $\frac{1}{5}$ b UVW
6 a

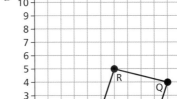

  b i 3 ii 3 iii $\frac{1}{4}$ iv $\frac{1}{4}$
  c parallelogram

Page 143 Exercise 4.2

1 a 3 b $\frac{4}{3}$ c 3
  d $-\frac{3}{2}$ e $-\frac{4}{7}$ f -1
  g undefined h 0 i -1
2 a $m_{MN} = -\frac{5}{2}; m_{NP} = -\frac{1}{6}; m_{PQ} = -\frac{5}{2}; m_{QM} = -\frac{1}{6}$
  b parallelogram c rhombus
3

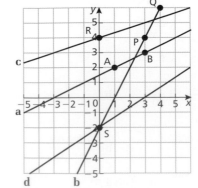

4 a i $m_{AB} = \frac{1}{6}$; $m_{BC} = -6$ **ii** 90°
    iii $m_{AC} = -\frac{5}{7}$; $m_{BD} = \frac{7}{5}$ **iv** 90°
  b i $m_{AB} = -3$; $m_{BC} = \frac{1}{3}$ **ii** 90°
    iii $m_{AC} = 2$; $m_{BD} = -\frac{1}{2}$ **iv** 90°
  c For perpendicular lines AB and CD,

$$\text{if } m_{AC} = a \text{ then } m_{BC} = -\frac{1}{a}$$

Page 144 Exercise 5.1

1 a 0, 3, 6, 9, 12; $m = 3$ **b** 1, 2, 3, 4; $m = 1$
  c 0, 1, 2, 3, 4; $m = \frac{1}{2}$ **d** 0, 1, 2, 3, 4 ; $m = \frac{1}{3}$

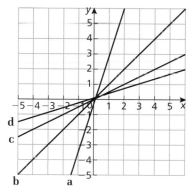

2 a/c

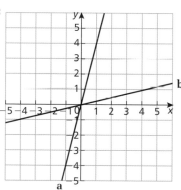

  b $m = 4$ **d** $m = \frac{1}{4}$
  e by inspection, gradient $= m$ **f** origin
3 a $y = 3x$ **b** $y = 5x$ **c** $y = -4x$
  d $y = \frac{1}{5}x$ **e** $y = -\frac{2}{3}x$
4 a 1, 2, 3, 4, 5
  b

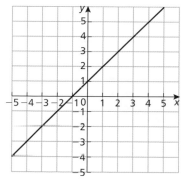

  c 1   **d** (0, 1)

5

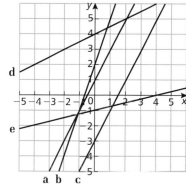

  a ii 2   **iii** (0,1)
  b ii 3   **iii** (0, 2)
  c ii 2   **iii** (0, −3)
  d ii $\frac{1}{2}$ **iii** (0, 4)
  e ii $\frac{1}{4}$ **iii** (0, −1)
6 a c **b** (0, c)
  c i (0, 2)   **ii** (0, 1)   **iii** (0, −2)
    iv (0, −4)   **v** (0, 0)   **vi** (0, −1)
    vii (0, 5)   **viii** (0, 4)
  d i and **iii**; **ii** and **v**

Page 146 Exercise 6.1

1 a $m = 2$; $c = 3$ **b** $m = -3$; $c = -6$
  c $m = \frac{2}{5}$; $c = 1$ **d** $m = -1$; $c = 0$

2 a (graph, $m = 1$, through 4)   **b** (graph, $m = 2$, through −3)
  c (graph, $m = -1$, through 5)   **d** (graph, $m = -3$, through −1)
  e (graph, $m = \frac{1}{3}$, through 1)   **f** (graph, $m = -\frac{2}{3}$, through −2)

3 a $y = 2x + 3$ **b** $y = -x + 1$ **c** $y = 3x - 2$
4 a $y = 2x + 2$ **b** $y = -3x + 6$ **c** $y = \frac{1}{5}x - 1$
5 a $y = 5x - 4$ **b** $y = -2x + 6$
  c $y - \frac{1}{3}x - 1$ **d** $y = -\frac{4}{5}x + 3$
6 a i $\frac{1}{2}$ **ii** 1   **iii** $y = \frac{1}{2}x + 1$
  b i $y = x + 2$ **ii** $y = 4x - 1$
    iii $y = -4x - 3$ **iv** $y = -\frac{1}{2}x + 5$
7 a $y = 2x + 1$ **b i** and **ii**   **c** above
8 a $y = 3x - 2$
  b i above   **ii** between
    iii below   **iv** between

Page 147 Exercise 6.2

1 a $y = 2x + 3$ b $y = -2x + 1$ c $y = \frac{1}{2}x - 4$
  d $y = \frac{1}{4}x + 2$ e $y = -\frac{1}{3}x + 1$ f $y = -\frac{2}{5}x - 3$

2 a $y = -x + 6$ b above
  c i above   ii on   iii below

3 a -5 b $y = 6x - 5$

4 a $y = 5x - 3$ b $y = -x + 4$
  c $y = 3x - 6$ d $y = \frac{1}{2}x + 3$

5 a $y = 2x + 3$ b $y = -3x + 5$ c $y = \frac{1}{2}x - 2$

Page 149 Exercise 7.1

1 a $(0, 2)$, $(10, 10)$ b $0{\cdot}8$
  c $C = 0{\cdot}8T + 2$ d £11·60

2 a $(0, 1000)$, $(50, 200)$ b -16
  c $V = -16T + 1000$ d 62·5 minutes

3 a

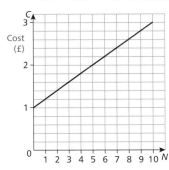

  b $C = 0{\cdot}2N + 1$

4 a

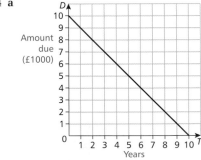

  b $D = -1000T + 10\,000$ c year 10

Page 151 Exercise 8.1

Students' lines of best fit may differ slightly from those given in these answers.

1 a i and ii

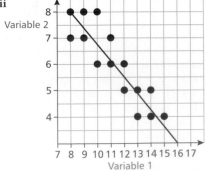

  iii $y = -\frac{5}{8}x + 13$ iv 6·4

b i and ii

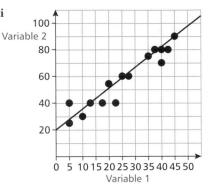

  iii $y = 1{\cdot}6x + 20$ iv 58·4

2 a\b

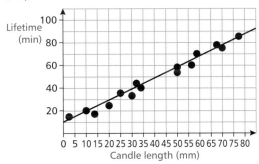

  c $T = C + 10$ d 60 min

3 a/b

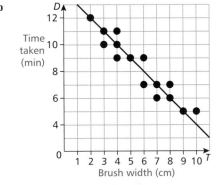

  c $T = -W + 14$ d 8 min

4 a strong positive

b

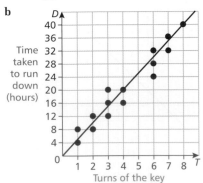

  c $y = 5x$

Page 154 Revise

1 a $x = -4$ **b** $x = 1$ **c** $y = -3$

2 a $\frac{1}{2}$ **b** $-\frac{3}{2}$

c 0 **d** undefined

3 a 4 **b** -3

c $\frac{1}{4}$ **d** $-\frac{1}{2}$

e undefined **f** 0

4 a 1, 3, 5, 7, 9

b

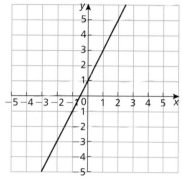

c 2

d (0, 1)

5 a $y = x - 1$

b $y = -3x + 2$

c $y = \frac{1}{2}x - 3$

d $y = -\frac{2}{3}x$

6 a **b**

c **d**

7 a 300 g **b** $W = 50F + 300$

8 a/b Students' answers may differ slightly.

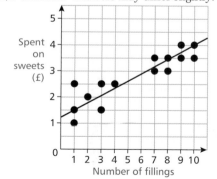

c $S = \frac{5}{18}F + \frac{11}{9}$

d 2·6 fillings, i.e. between 2 and 3 fillings

8 Quadratic equations

Page 156 Exercise 1.1

1 a 3 **b** -4 **c** 1

d $\frac{1}{3}$ **e** $\frac{10}{3}$ **f** $-\frac{2}{3}$

2 a $3x - 12$ **b** $-8 + 6x$ **c** $6x - 3x^2$

3 a $5x - 4$ **b** $6x - 10$ **c** $2x^2 + 5x$

4 a $2(3x + 4)$ **b** $4(3y - 2)$ **c** $a(8a + 1)$

d $2t(t + 3)$ **e** $3x(x - 3)$ **f** $2h(3 - h)$

g $5k(3k - 1)$ **h** $6ab(b + a)$

5 a $(a - b)(a + b)$ **b** $(k - 1)(k + 1)$

c $(5 - x)(5 + x)$ **d** $(y - 6)(y + 6)$

e $(x - 7y)(x + 7y)$ **f** $(10a - 8b)(10a + 8b)$

g $(7s - 11t)(7s + 11t)$ **h** $(1 + 4t^2)(1 - 2t)(1 + 2t)$

6 a $(x + 3)(x + 2)$ **b** $(x + 4)(x + 3)$

c $(y + 6)(y + 2)$ **d** $(x + 3)(x + 7)$

e $(x - 6)(x + 3)$ **f** $(y - 5)(y + 2)$

g $(s - 6)(s - 4)$ **h** $(x - 6)(x - 2)$

i $(2x + 1)(x + 2)$ **j** $(3y - 4)(y + 2)$

k $(3x - 2)(2x - 3)$ **l** $(2s + 3)(2s - 9)$

Page 157 Exercise 2.1

1 a 0, 3 **b** 0, 3

2 a 0, 8 **b** 0, 8

3 a 0, 5 **b i** 8 **ii** 8

c 0, 5 **d** 2·5 seconds

4 a 42, 48, 50, 48, 42, 32, 18, 0 **b** 0, 10

c i 1, 9 **ii** 2, 8 **iii** 3, 7 **iv** 5

Page 159 Exercise 3.1

1 a 0, -1 **b** 0, -2 **c** 0, 3 **d** 0, 4

e 0, -5 **f** 0, 4 **g** 0, $\frac{5}{2}$ **h** 0, $-\frac{7}{2}$

i 0, $\frac{3}{4}$ **j** 0, $\frac{2}{3}$ **k** 0, $\frac{11}{4}$ **l** 0, $\frac{7}{5}$

2 a 0, -3 **b** 0, 5 **c** 0, 12 **d** 0, -2

e 0, 2 **f** 0, 3 **g** 0, -3 **h** 0, 4

i 0, $\frac{5}{2}$ **j** 0, -2 **k** 0, $\frac{4}{3}$ **l** 0, 4

m 0, 15 **n** 0, 3 **o** 0, $-\frac{5}{6}$

3 a ± 4 **b** ± 5 **c** ± 10

d $\pm \frac{3}{2}$ **e** $\pm \frac{5}{2}$ **f** $\pm \frac{7}{3}$

g $\pm \frac{2}{3}$ **h** $\pm \frac{5}{6}$ **i** $\pm \frac{4}{7}$

4 a ± 2 **b** ± 5 **c** ± 7 **d** ± 8

e ± 10 **f** ± 11 **g** ± 12 **h** ± 9

i $\pm \frac{1}{4}$ **j** $\pm \frac{2}{9}$ **k** $\pm \frac{4}{5}$ **l** $\pm \frac{1}{4}$

m $\pm \frac{1}{2}$ **n** $\pm \frac{5}{9}$ **o** ± 3 **p** 0

q ± 2 **r** $\pm \frac{7}{6}$

5 a 45 metres **b** 5 seconds

Page 160 Exercise 3.2

1 a ± 2 **b** ± 2 **c** ± 3 **d** ± 2

e ± 4 **f** ± 1 **g** ± 2 **h** ± 3

i ± 3 **j** ± 10 **k** ± 3 **l** ± 5

2 a $\pm \frac{3}{2}$ **b** $\pm \frac{3}{2}$ **c** $\pm \frac{5}{3}$ **d** $\pm \frac{3}{2}$

e $\pm \frac{5}{2}$ **f** $\pm \frac{3}{2}$ **g** $\pm \frac{3}{2}$ **h** $\pm \frac{5}{9}$

i $\pm \frac{2}{3}$ **j** $\pm \frac{7}{5}$ **k** $\pm \frac{5}{7}$ **l** $\pm \frac{5}{6}$

m $\pm \frac{3}{5}$ **n** $\pm \frac{2}{5}$ **o** $\pm \frac{4}{5}$

3 a 30 metres **b** 60 metres

Page 161 Exercise 4.1

1 a $-2, -1$ **b** $-\frac{1}{2}, 2$ **c** $3, -4$
d $2, -3$ **e** $\frac{2}{5}, -\frac{1}{2}$ **f** $-\frac{2}{5}, \frac{3}{4}$

2 a $-1, -4$ **b** $-4, -2$ **c** $-5, -2$ **d** $-4, -5$
e $-6, -2$ **f** $-4, -7$ **g** $-5, -6$ **h** $-7, -2$
i $-6, -4$ **j** $-5, -7$ **k** $-2, -2$ **l** $-6, -3$

3 a $-2, 1$ **b** $-3, 2$ **c** $4, -2$ **d** $5, -3$
e $4, -1$ **f** $-6, 3$ **g** $4, 1$ **h** $4, 2$
i $5, 3$ **j** $4, 3$ **k** $4, 5$ **l** $3, 6$
m $7, -4$ **n** $9, -5$ **o** $6, 5$

4 a $-\frac{1}{2}, -2$ **b** $-\frac{1}{3}, -4$ **c** $-\frac{1}{5}, -2$ **d** $-\frac{1}{4}, -3$
e $-\frac{1}{6}, -2$ **f** $-\frac{1}{2}, -3$ **g** $\frac{1}{2}, 2$ **h** $\frac{1}{3}, 2$
i $\frac{1}{5}, 3$ **j** $\frac{1}{2}, -\frac{1}{3}$ **k** $\frac{2}{3}, -\frac{3}{2}$ **l** $\frac{1}{4}, -\frac{3}{2}$
m $\frac{4}{3}, -\frac{1}{2}$ **n** $\frac{3}{2}, -\frac{1}{4}$ **o** $\frac{1}{4}, \frac{2}{3}$

5 a 2, 8 metres **b** 10 metres

Page 162 Exercise 4.2

1 a $-4, -2$ **b** $-1, -7$ **c** $6, -2$
d $\frac{9}{2}, -\frac{7}{2}$ **e** $\frac{8}{3}, -2$ **f** $\frac{5}{2}, -2$
g $\frac{11}{3}, -3$ **h** $\frac{12}{5}, -\frac{8}{5}$ **i** $\frac{11}{4}, -\frac{5}{4}$

2 a $-\frac{1}{2}, 3$ **b** $\frac{4}{3}, 2$ **c** $-3, \frac{8}{3}$ **d** $\frac{1}{2}, -\frac{5}{3}$
e $2, -\frac{11}{3}$ **f** $\frac{1}{3}, \frac{5}{2}$ **g** $-1, \frac{2}{3}$ **h** $4, -\frac{3}{2}$
i $2, \frac{4}{5}$ **j** $\frac{3}{2}, -4$ **k** $-2, 2\frac{2}{3}$ **l** $\frac{1}{6}, 4$
m $0, -2$ **n** $-3, -1$

3 a ± 4 **b** $\pm\frac{1}{2}$ **c** $1, -3$
d $\frac{4}{3}, -\frac{2}{3}$ **e** $5, 0$ **f** $8, -2$
g $2, -3$ **h** $7, -8$ **i** $3, 1$

Page 163 Challenge

a $(x-1)(x-3) = 0$ **b** $(x-4)(x-3) = 0$
c $(x-2)(x+3) = 0$ **d** $(x+2)(x-5) = 0$

Page 164 Exercise 5.1

1 b i 3 cm **ii** 4 cm
2 a $x - 4$ **b** $-10, -6$
3 a i $x + 3$ **ii** $x + 1$ **iii** $(x+3)(x+1)$
b i 6 m by 6 m **ii** 9 m by 7 m
4 a $\dfrac{110}{x}$ **b** 10, 11
5 a 1 second, 10 seconds
b no, maximum height is 484 metres

Page 164 Exercise 5.2

1 a 20 **b** 25
2 a 1 second, 4 seconds
b 5 seconds
3 a $s + 5$
b i $\dfrac{150}{s+5}$ **ii** $\dfrac{150}{s}$
c $\dfrac{150}{s} - \dfrac{150}{s+5} = 1$ **e** 30 mph, 25 mph
4 a $42 - x$ **b** 18 **c** 72 cm
5 a i $x - 4$ **ii** $x + 4$ **iii** $\dfrac{1}{x-4}$
iv $\dfrac{1}{x+4}$ **v** $\dfrac{1}{x-4} + \dfrac{1}{x+4}$
b 16 mph

Page 166 Exercise 6.1

1 a $1, 2, 2$ **b** $3, -5, 1$ **c** $4, 1, -3$
d $2, -1, -1$ **e** $1, 0, -1$ **f** $2, 1, 0$
g $1, 3, 2$ **h** $5, -3, 4$ **i** $-1, 1, 1$
j $-2, -1, 6$ **k** $-2, 0, 9$ **l** $-1, 1, 0$
m $-1, -2, 4$ **n** $5, 0, 0$ **o** $1, 0, 0$

2 a $x^2 - 2x + 5 = 0, 1, -2, 5$
b $2x^2 + 5x - 1 = 0, 2, 5, -1$
c $x^2 + 6x - 4 = 0, 1, 6, -4$
d $10x^2 - 2x - 2 = 0, 10, -2, -2$
e $x^2 + x - 4 = 0, 1, 1, -4$
f $2x^2 - 2x + 1 = 0, 2, -2, 1$

3 no term in x^2

4 a $1, -5, -12$ **b** $3, 4, 0$
c $4, 0, 6$ **d** $0, 6, -5$, not a quadratic
e $2, 0, 0$ **f** $2, 1, -3$

5 a $x^2 + x - 1 = 0$ **b** $x^2 + x - 1 = 0$

Page 168 Exercise 7.1

1 a $-2, -1$ **b** $2, 1$ **c** $-4, -3$
d $4, 3$ **e** $-2, 1$ **f** $-1, -\frac{1}{2}$
g $-\frac{1}{3}, 2$ **h** $3, \frac{1}{4}$

2 a $-4, -5$ **b** $\frac{2}{3}, 4$ **c** $2, -\frac{1}{5}$

3 a $-0.30, -6.70$ **b** $1.14, -2.64$
c $1.85, -0.18$ **d** $0.64, -0.39$
e $6.16, -0.16$ **f** $2.37, -3.37$
g $1.70, -1.37$ **h** $0.62, -9.62$
i $0.27, -2.77$ **j** $0.45, -4.45$
k $-0.05, -4.95$ **l** $-0.31, -3.19$

4 a $\sqrt{(-16)}$ no solution **b i** $\sqrt{0}$ **ii** 1 root

5 a $0.41, -2.41$ **b** $0.78, -1.28$ **c** $1.87, -1.87$
d $2.62, 0.38$ **e** $3, -4$ **f** $2, -3$

6 a $x = 1 + \dfrac{1}{x}$ **b** $x^2 - x - 1 = 0$ **c** 1.62

Page 168 Exercise 7.2

1 a 3·55 s, 0·45 s **b** 4 s **c** 20 metres
2 a 5·3 m **b** 7·3 m
3 a $4x^2 + 18x$ **b** $4x^2 + 18x = 150$
c 4·27 cm, 8·54 cm, 3 cm
4 a $3 + 2x, 5 + 2x$
b $(3 + 2x)(5 + 2x)$
c 4·84 m by 6·84 m
5 a i $x + 5$ **ii** $\dfrac{300}{x}$ **iii** $\dfrac{180}{x+5}$
b $\dfrac{300}{x} + \dfrac{180}{x+5} = 18$ **c** 25p, 30p

Page 172 Exercise 8.1

1 b 2·7
2 b 1·16 **c** -2.16
3 b 1·73
4 b -1.30
5 a 1·46
6 b -2.7
7 b 1·2
8 a 1·32 **b** -1.32

Page 174 Revise

1 a $0, \frac{3}{8}$ **b** $0, \frac{7}{10}$ **c** ± 3
 d $0, -\frac{a}{5}$ **e** $0, -\frac{5}{2}$ **f** $0, \frac{4}{9}$

2 a $-1, -9$ **b** $12, -2$ **c** $6, 5$
 d $-7, 4$ **e** $\frac{5}{2}, -\frac{3}{4}$ **f** $\frac{6}{5}, -\frac{3}{2}$
 g $\frac{7}{2}, -\frac{5}{3}$ **h** $\frac{3}{4}, \frac{5}{2}$ **i** $\frac{2}{5}, \frac{3}{7}$

3 a $-10.91, -0.09$ **b** $-0.27, 7.27$ **c** $-0.23, 2.90$
 d $0.63, 6.37$ **e** $0.42, -5.92$ **f** $-0.92, 0.78$

4 b 0.63

5 a $4, 1$ **b** $2.82, 0.18$

6 a $12 - 2x$ **b** $x(12 - 2x)$
 c 4.41 metres or 1.59 metres

9 Surds and indices

Page 175 Exercise 1.1

1 a $4x$ **b** x^3 **c** $x^4 + x^3$
 d $10x^2$ **e** $3x$ **f** 6

2 a 9 **b** 12 **c** 24 **d** 20
 e 36 **f** 6 **g** 2 **h** 12

3 a 64 **b** 16 **c** 125 **d** -8
 e 3 **f** 3 **g** 2

4 a 3.21×10^3 **b** 5.4367×10^5
 c 6.78×10^6 **d** 3.1×10^{-4}
 e 7.21×10^{-6}

5 a 7.13×10^8 **b** 1.3161×10^5
 c 1.695×10^{11} **d** 4.41×10^6

6 $1, 4, 9, 16, 25, 36, 49, 64, 91, 100, 121, 144$

7 a $x^2 - 4$ **b** $y^2 - 9$ **c** $4x^2 - 9$ **d** $a^2 - b^2$

Page 176 Exercise 2.1

1 a x^8 **b** x^5

2 a 10^6 **b** 10^8 **c** 10^5 **d** 10^7

3 a x^4 **b** x^7 **c** x^8 **d** x^{15}
 e x^9 **f** a^{16} **g** b^{12} **h** f^{22}
 i x^7 **j** x^{10} **k** x^{15} **l** x^{20}
 m d^{15} **n** y^{16} **o** a^{22} **p** f^{33}
 q 4^{13} **r** 7^{16} **s** 13^{21} **t** 60^{26}

4 a $8x^2$ **b** $12x^3$ **c** $30x^6$ **d** $6x^{10}$
 e x^6 **f** $8x^7$ **g** $24a^{10}$ **h** $60a^6$
 i $40x^7$ **j** $32a^9$ **k** $56b^9$ **l** $12c^{15}$
 m $20r^{13}$ **n** $30t^{12}$ **o** $-24s^{12}$ **p** $12c^{13}$

5 a 3.2×10^6 **b** 3×10^{11} **c** 8.8×10^{11}
 d 9.6×10^{11} **e** 2.64×10^{13} **f** 2.625×10^{10}

Page 177 Exercise 3.1

1 a a^3 **b** x^6

2 a x^5 **b** x^5 **c** x^7 **d** x^5
 e x^{11} **f** a^9 **g** b^4 **h** b^{10}
 i c^8 **j** c^{10} **k** s^8 **l** t^{17}
 m 10^2 **n** 10^8 **o** 2^3 **p** 3^4
 q $4t^2$ **r** $5x^3$ **s** $5x$ **t** $12x^3$
 u 0.6^6 **v** 1.3^3 **w** $\left(\frac{3}{4}\right)^3$ **x** $(3x)^3$

3 a x^6 **b** x^5 **c** x^4 **d** a^5
 e $6s^3$ **f** $10t^5$ **g** $\frac{4}{3s^2}$ **h** $4a$
 i y **j** a^2 **k** vw^2 **l** q^8

Page 178 Exercise 4.1

1 a $\dfrac{1}{1000}, \dfrac{1}{10\,000}, \dfrac{1}{100\,000}$
 b $10^{-3}, 10^{-4}, 10^{-5}$

2 $2^3, 2^2, 2^1, 2^0, 2^{-1}, 2^{-2}, 2^{-3}$

3 a $\dfrac{1}{x^3}$ **b** $\dfrac{1}{x^6}$ **c** $\dfrac{1}{x^7}$ **d** $\dfrac{1}{a^8}$
 e $\dfrac{1}{a^{10}}$ **f** $\dfrac{2}{s^3}$ **g** $\dfrac{7}{z^3}$ **h** $\dfrac{3}{d^4}$
 i $\dfrac{1}{2a^9}$ **j** $\dfrac{1}{6e^5}$ **k** $\dfrac{1}{3y^8}$ **l** $\dfrac{1}{10g^7}$
 m $\dfrac{2}{3x^4}$ **n** $\dfrac{4}{7f^5}$ **o** $\dfrac{5}{9x^8}$ **p** $\dfrac{3}{5d^7}$
 q x^3 **r** x^8 **s** g^0 **t** n^7
 u $2c^9$ **v** $5m^{10}$ **w** $5x^6$ **x** $\dfrac{3d^8}{5}$
 y $\dfrac{7t^{12}}{11}$ **z** $\dfrac{13x^{12}}{12}$

4 a x^{-4} **b** x^{-1} **c** a^{-10} **d** b^{-13}
 e $3x^{-13}$ **f** $5x^{-9}$ **g** $7y^{-10}$ **h** $3x^{-11}$
 i $\dfrac{x^{-4}}{2}$ **j** $\dfrac{x^{-7}}{3}$ **k** $\dfrac{a^{-5}}{8}$ **l** $\dfrac{b^{-6}}{7}$
 m $\dfrac{3x^{-5}}{4}$ **n** $\dfrac{2x^{-12}}{5}$ **o** $\dfrac{4x^{-9}}{7}$ **p** $\dfrac{3a^{-7}}{8}$
 q $\dfrac{1}{x^{-3}}$ **r** $\dfrac{1}{y^{-8}}$ **s** $\dfrac{5}{x^{-4}}$ **t** $\dfrac{8}{y^{-5}}$
 u $\dfrac{1}{5s^{-8}}$ **v** $\dfrac{1}{9r^{-13}}$ **w** $\dfrac{4}{9d^{-14}}$ **x** $\dfrac{6}{5y^{-15}}$

5 a $\dfrac{1}{x^5}$ **b** $\dfrac{1}{x^3}$ **c** $\dfrac{1}{x^3}$ **d** $\dfrac{1}{x^2}$
 e $\dfrac{1}{3x^{12}}$ **f** $\dfrac{5}{2}$ **g** $\dfrac{2}{3t^5}$ **h** $\dfrac{8}{3s^4}$

6 a $\dfrac{1}{x^{-12}}$ **b** $\dfrac{1}{x^{-15}}$ **c** $\dfrac{1}{x^{-7}}$ **d** $\dfrac{1}{x^{-7}}$

7 a $x^7 + x^{-2}$ **b** $x^1 + x^{-1}$
 c $x^{-4} + x^{-1} + x + x^4$ **d** $x + x^5$
 e $x^{-9} + x^4$ **f** $x^4 + x^{-5} + x + x^{-8}$

Page 179 Exercise 5.1

1 a x^8 **b** x^{12} **c** x^{20}
 d x^{15} **e** x^{56} **f** x^{-4}
 g x^{-12} **h** x^{-35} **i** x^{-12}
 j x^{-20} **k** x^{-6} **l** x^{-8}
 m x^{-18} **n** x^{-14} **o** x^{-32}

2 a $\dfrac{1}{x^8}$ **b** $\dfrac{1}{x^{15}}$ **c** $\dfrac{1}{x^{42}}$ **d** $\dfrac{1}{x^{24}}$
 e $\dfrac{1}{x^{18}}$ **f** $\dfrac{1}{x^{30}}$ **g** $\dfrac{1}{s^{15}}$ **h** $\dfrac{1}{16x^{12}}$
 i $\dfrac{9}{d^8}$ **j** $\dfrac{s^6}{16}$

3 a $\dfrac{1}{x^6}$ b $\dfrac{1}{x^{20}}$ c $\dfrac{1}{x^{42}}$ d $\dfrac{1}{x^{28}}$

e $\dfrac{1}{x^{32}}$ f $\dfrac{x^{20}}{243}$ g $\dfrac{x^{18}}{8}$ h x^{12}

i $\dfrac{1}{x^{15}}$ j $\dfrac{x^4}{16}$ k $\dfrac{1}{x^8}$ l $\dfrac{1}{x^8}$

m $\dfrac{125}{27x^{12}}$ n $\dfrac{625}{81x^{12}}$ o $\dfrac{16}{x^{16}}$

4 a 1 b 27 c 1 d x^6

e 1 f 6 g 7 h x^{-12}

5 a x^4 b $\dfrac{1}{x^{13}}$ c $\dfrac{1}{x^3}$ d $\dfrac{4x^{11}}{3}$

e $81x^{12}$ f $\dfrac{2x^4}{81}$ g x^3 h $\dfrac{1}{x^{18}}$

Page 181 Exercise 6.1

1 a $\dfrac{1}{x^{\frac{1}{2}}}$ b $x^{\frac{5}{12}}$ c $\dfrac{1}{x^{\frac{1}{12}}}$ d $\dfrac{2x^{\frac{3}{10}}}{3}$

e $\dfrac{1}{x^{\frac{1}{6}}}$ f $\dfrac{1}{a^{\frac{11}{12}}}$ g $\dfrac{1}{a^{\frac{13}{15}}}$ h $\dfrac{5}{2}x^{\frac{5}{4}}$

2 a $x^{\frac{1}{2}}$ b $x^{\frac{1}{3}}$ c x^2 d $3x^{\frac{1}{2}}$

e $16x^{\frac{4}{3}}$ f $16x^{\frac{3}{2}}$ g $81x^{\frac{4}{3}}$ h x^{-1}

i $8x^{\frac{3}{5}}$ j $\dfrac{1}{5x}$ k $\dfrac{1}{12x}$ l $\dfrac{3}{2}x^{\frac{5}{2}}$

m $\dfrac{3}{5x^{\frac{4}{3}}}$ n $\dfrac{11}{3x}$ o $\dfrac{2}{49x}$

3 a $\sqrt[3]{x^3}$ b $\sqrt[3]{x^5}$ c $\sqrt[3]{x^{-4}}$ d $\sqrt[2]{(3x)^2}$

e $4\sqrt[4]{x^5}$ f $10\sqrt{x^{-1}}$ g $12\sqrt[4]{x^{-3}}$ h $8\sqrt[7]{x^5}$

i $\sqrt[4]{x^5}$ j $\sqrt{x^7}$ k $\sqrt[5]{x^{11}}$ l $\sqrt[5]{(2x)^8}$

m $7\sqrt[7]{x^9}$ n $8\sqrt[6]{x^{-7}}$ o $10\sqrt[8]{x^9}$ p $\sqrt[5]{(4x)^{-6}}$

4 a 3 b 3 c 3 d 10

e 5 f 2 g 5 h 7

i $\frac{1}{2}$ j $\frac{1}{7}$ k $\frac{1}{8}$ l $\frac{1}{4}$

m $\frac{1}{32}$ n 16 o 243 p $\frac{1}{8}$

q $\frac{1}{8}$ r 189 s 64 t 48

5 a $x^{2\frac{1}{2}}+x^{2\frac{1}{3}}$ b $x^{4\frac{1}{2}}+x^{3\frac{1}{3}}$ c $x^{7\frac{3}{4}}+x^{5\frac{3}{4}}$

d $x^{\frac{7}{6}}+x^{\frac{11}{12}}$ e $\dfrac{1}{x^{4\frac{1}{2}}}-\dfrac{1}{x^2}$ f $\dfrac{1}{x^{2\frac{1}{3}}}-x^{3\frac{2}{3}}$

g $x^{1\frac{1}{2}}+x^{\frac{5}{3}}$ h $3x^{\frac{1}{4}}-2x^{\frac{3}{10}}$ i $\dfrac{1}{x^{\frac{1}{4}}}-\dfrac{3}{x^{\frac{1}{3}}}$

6 a $4x^4$ b $2x$ c $2a$ d $\dfrac{1}{8x^{\frac{3}{2}}}$

e $2x^{\frac{1}{2}}$ f $\dfrac{x^{\frac{2}{3}}}{3}$ g $5a^2$ h $\dfrac{8x^{\frac{1}{6}}}{9}$

Page 182 Investigation

a $3i,\ 4i,\ 5i$ b $-1i,\ 1,\ 1i,\ -1,\ -1i$

Page 184 Exercise 7.1

1 a, b, d, i

2 $\sqrt{1\cdot21}=1\cdot1$ which can be written as a fraction, i.e. $\frac{11}{10}$

3 a 5 b $5\sqrt3$ c $4\sqrt6$

4 a $3\sqrt2$ b $2\sqrt2$ c $2\sqrt6$ d $3\sqrt3$

e $10\sqrt2$ f $4\sqrt2$ g $6\sqrt2$ h $7\sqrt3$

i $8\sqrt2$ j $3\sqrt5$ k $2\sqrt7$ l $9\sqrt5$

m $6\sqrt5$ n $4\sqrt7$ o $7\sqrt5$

5 a $3\sqrt7$ b $5\sqrt6$ c $5\sqrt3$ d $9\sqrt3$

6 a i $6\sqrt3$ ii $4\sqrt3$

7 a $8\sqrt2$ b $\sqrt6\sqrt{11}$ c $\sqrt{13}$ d $3\sqrt2$

e $\sqrt5$ f $2\sqrt6$ g $2\sqrt2$ h $5\sqrt6$

i $6\sqrt3$ j $6\sqrt5$ k $5\sqrt{11}$ l $5\sqrt{10}$

m $9\sqrt7$ n $12\sqrt3 + 12\sqrt2$

o $4\sqrt3 + 17\sqrt2$

8 a 3 b 7 c 20 d 110

e 90 f 4 g 10 h $4\sqrt{10}$

9 a 30 b $48\sqrt3$ c $30\sqrt2$

d $24\sqrt5$ e 80 f $36\sqrt2$

10 a $16 - 8\sqrt3$ b $5\sqrt3 - 5$ c $6\sqrt2 - \sqrt6$

d $4\sqrt3 - 3$ e $2\sqrt5 - 15$ f $5 - 4\sqrt5$

g $\sqrt{10x} - 7\sqrt{(2x)}$

h $6\sqrt{(3x)} - 2\sqrt{6x^2}$

i $20 + 5\sqrt{2x} + 4\sqrt{3x} + \sqrt{6x^2}$

j $30 - 5\sqrt{7x} + 6\sqrt{5x} - \sqrt{35x^2}$

k 1

l 2

11 a 23 b 4 c -9 d 1

e -2 f 5 g 19 h 55

i -70, difference of two squares

12 $4\sqrt{13}$ cm

13 $\sqrt2$

14 a 50 cm b $10\sqrt{29}$ cm

Page 186 Exercise 8.1

1 a 2 b $2\sqrt2$ c 2 d 3

e 5 f $2\sqrt2$ g 4 h 5

i $2\sqrt3$ j $2\sqrt5$ k $3\sqrt2$ l $\sqrt{11}$

m $\frac{1}{5}$ n $\frac{1}{4}$ o $\dfrac{1}{2\sqrt3}$ p $\dfrac{1}{2\sqrt2}$

q $\frac{5}{3}$ r $\dfrac{1}{2\sqrt{10}}$ s $\dfrac{1}{2\sqrt5}$ t $\dfrac{1}{5\sqrt2}$

2 a $0\cdot4$ b $0\cdot6$ c $0\cdot8$ d $0\cdot2$

e $0\cdot05$ f $0\cdot09$ g $0\cdot007$ h $0\cdot001$

3 a 10 b $\frac{18}{5}$ c 3

d $\frac{5}{4}$ e $\frac{1}{4}$ f $\frac{1}{10}$

4 a $\frac{2}{3}$ b $\frac{2}{3}$ c $\frac{2}{3}$ d $\frac{3}{4}$

e $\frac{2}{3}$ f $\frac{6}{7}$ g $\frac{9}{7}$ h $\frac{8}{5}$

i $\frac{4}{3}$ j $\frac{2}{5}$ k $\frac{7}{5}$ l $\frac{1}{2}$

m $\frac{5}{7}$ n $\frac{2}{3}$ o $\frac{3}{2}$ p $\frac{2}{5}$

Page 188 Exercise 9.1

1 a $\dfrac{\sqrt{5}}{5}$ **b** $\dfrac{\sqrt{2}}{2}$ **c** $\dfrac{\sqrt{7}}{7}$ **d** $\dfrac{\sqrt{8}}{8}$

e $\dfrac{2\sqrt{3}}{3}$ **f** $\dfrac{3\sqrt{5}}{5}$ **g** $\sqrt{2}$ **h** $\sqrt{7}$

i $\dfrac{\sqrt{5}}{15}$ **j** $\dfrac{\sqrt{3}}{21}$ **k** $\dfrac{\sqrt{8}}{64}$ **l** $\dfrac{\sqrt{7}}{63}$

m $\dfrac{2\sqrt{5}}{15}$ **n** $\dfrac{5\sqrt{7}}{21}$ **o** $\dfrac{3\sqrt{2}}{10}$ **p** $\dfrac{2\sqrt{5}}{35}$

q $\dfrac{\sqrt{10}}{2}$ **r** $\dfrac{\sqrt{21}}{3}$ **s** $\dfrac{3\sqrt{10}}{5}$ **t** $\dfrac{\sqrt{15}}{3}$

2 a $\frac{1}{3}(\sqrt{3}+3)$ **b** $\frac{1}{2}(\sqrt{10}+2\sqrt{2})$ **c** $1-\dfrac{4\sqrt{3}}{3}$

d $1-\dfrac{2\sqrt{5}}{5}$ **e** $\frac{1}{5}(\sqrt{10}+3\sqrt{5})$ **f** $\dfrac{3\sqrt{6}}{2}+2\sqrt{2}$

g $\dfrac{3\sqrt{6}}{2}+1$ **h** $\dfrac{\sqrt{10}}{2}+\dfrac{\sqrt{6}}{4}$

Page 189 Exercise 9.2

1 a $1-\dfrac{\sqrt{2}}{2}$ **b** $\frac{1}{4}(\sqrt{5}-1)$ **c** $\frac{1}{2}(3-\sqrt{7})$

d $\dfrac{5-\sqrt{3}}{22}$ **e** $\frac{1}{2}(3+\sqrt{5})$ **f** $\frac{2}{7}(4+\sqrt{2})$

g $\frac{3}{11}(5+\sqrt{3})$ **h** $\frac{7}{34}(6+\sqrt{2})$ **i** $2(1+\sqrt{2})$

j $6(\sqrt{5}-2)$ **k** $2(\sqrt{3}+1)$ **l** $-\frac{7}{2}(\sqrt{2}+2)$

2 a $\sqrt{3}-\sqrt{2}$ **b** $\frac{1}{3}(\sqrt{3}+\sqrt{2})$ **c** $\dfrac{\sqrt{2}}{6}$

d $\frac{1}{4}(\sqrt{7}+\sqrt{3})$ **e** $\frac{3}{4}(\sqrt{6}+\sqrt{2})$ **f** $\frac{5}{4}(\sqrt{7}+\sqrt{3})$

g $\frac{2}{5}(\sqrt{8}-\sqrt{3})$ **h** $\frac{5}{2}(\sqrt{5}+\sqrt{3})$ **i** $\frac{2}{3}(\sqrt{5}+\sqrt{2})$

j $\dfrac{\sqrt{6}}{2}$ **k** $\frac{4}{19}(2\sqrt{5}+1)$ **l** $\frac{3}{41}(3\sqrt{5}-2)$

m $\frac{3}{37}(3\sqrt{5}-2\sqrt{2})$ **n** $\frac{2}{5}(2\sqrt{2}+\sqrt{3})$

o $\frac{4}{23}(3\sqrt{6}+2\sqrt{2})$ **p** $\frac{1}{10}(2\sqrt{7}+\sqrt{3})$

Page 189 Challenge

a $2\sqrt[3]{3}$ **b** $3\sqrt[3]{3}$ **c** $4\sqrt[3]{5}$ **d** $2\sqrt[4]{2}$ **e** $3\sqrt[4]{2}$ **f** $2\sqrt[5]{2}$

Page 191 Revise

1 a x^{15} **b** $12x^{13}$ **c** $3x^3$ **d** $\dfrac{1}{x}$

e $\dfrac{6}{x^8}$ **f** $\dfrac{5}{x^5}$ **g** $8x^6$ **h** $\dfrac{1}{x^6}$

i $\dfrac{3x^3}{2}$ **j** 81 **k** $\dfrac{1}{27}$ **l** $x^{\frac{7}{6}}+x^{\frac{11}{12}}$

m $6x+4x^{\frac{9}{2}}$

2 a $\dfrac{1}{x^{-11}}$ **b** $\dfrac{1}{x^{-3}}$ **c** x^{-15} **d** 7^{-3}

e $2x^{-4}$ **f** $\dfrac{5x^{-5}}{243}$ **g** $\dfrac{6}{x^{-20}}$ **h** $\dfrac{3}{2x^4}$

3 a 32 **b** 2 **c** $\frac{1}{27}$ **d** 75

4 a $0{\cdot}08$ **b** $0{\cdot}004$

5 a $5\sqrt{10}$ **b** $4\sqrt{2}$ **c** $7\sqrt{3}$ **d** $7\sqrt{6}$

e $-\sqrt{5}$ **f** $\dfrac{5\sqrt{47}}{47}$ **g** $2\sqrt{6}-6$ **h** $30-10\sqrt{10}$

6 a $\dfrac{\sqrt{7}}{7}$ **b** $\dfrac{3\sqrt{10}}{20}$ **c** $\dfrac{\sqrt{10}}{12}$

d $(2-\sqrt{3})$ **e** $\frac{5}{4}(\sqrt{5}+1)$ **f** $\dfrac{\sqrt{3}(\sqrt{7}-2)}{3}$

g $\dfrac{2\sqrt{14}-\sqrt{21}}{5}$ **h** $\dfrac{-2\sqrt{15}+3\sqrt{10}}{6}$

10 Functions

Page 192 Exercise 1.1

1 a 20 **b** -9 **c** $3a$ **d** a^2+2

2 a 4 **b** 1 **c** -2 **d** -5

3 a -1 **b** -11 **c** 8 **d** -125

4 Straight line: **a, d, i**; quadratic: **b, c, f, h**; neither: **e, g**

5 a 3 **b** $(0,-2)$ **c** $(\frac{2}{3},0)$

d
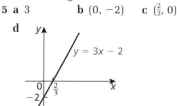

6
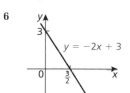

7 a $x=-3{\cdot}0,2{\cdot}0$ **b** $x=-0{\cdot}3,1{\cdot}5$ **c** $x=-2{\cdot}7,0{\cdot}2$

8 a 4 **b** -3 **c** -28

9 a 12 seconds   **b** 32 m

   c i $2\frac{1}{3}$ s and $9\frac{2}{3}$ s

     ii one answer as the ball is rising and the other as the ball is falling

   d 17 m

10 a i -30

     ii $(0,1500)$

     ii $H=-30t+1500$

   b 50 s

Page 195 Exercise 2.1

1 a $\{3,6,9,12\}$

   b $\{-5,-3,-1,1,3\}$

   c $\{2,3,6,11,18\}$

2 a $\{0,5,10\}$ **b** $\{-4,0,4\}$ **c** $\{5,15,25\}$

3 a Subtract 7   **b** divide by 8   **c** times (-2)

4
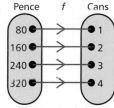

f means divide by 80

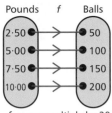

Pounds   f   Balls

2·50 → 50
5·00 → 100
7·50 → 150
10·00 → 200

f means multiply by 20

6 a

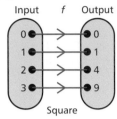

Input   f   Output

0 → 0
1 → 1
2 → 4
3 → 9

Square

b

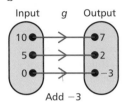

Input   g   Output

10 → 7
5 → 2
0 → −3

Add −3

c

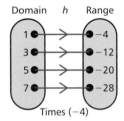

Domain   h   Range

1 → −4
3 → −12
5 → −20
7 → −28

Times (−4)

d

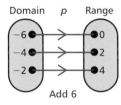

Domain   p   Range

−6 → 0
−4 → 2
−2 → 4

Add 6

e

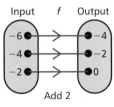

Input   f   Output

−6 → −4
−4 → −2
−2 → 0

Add 2

f

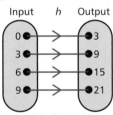

Input   h   Output

0 → 3
3 → 9
6 → 15
9 → 21

Double then add 3

7 a i {1, 2, 3}   **ii** {26, 27, 28}
  b add 25
  c

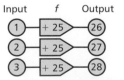

Input   f   Output

1 → +25 → 26
2 → +25 → 27
3 → +25 → 28

f: 1 ⟶ 26
f: 2 ⟶ 27
f: 3 ⟶ 28

Add 25

f

1 → 26
2 → 27
3 → 28

Add 25

8 a Square
  b half
  c subtract 7

Page 197 Exercise 3.1

1 a i 3   **ii** 0   **iii** −3
  b i 12   **ii** 0   **iii** −6
2 a i 7   **ii** 11   **iii** 3
  b i 2   **ii** 10   **iii** 26
3 a i 9   **ii** 1   **iii** 6
  b i 3   **ii** 0   **iii** 9
4 a i 1500   **ii** 2500   **iii** 3500
  b the distance travelled in 7 hours
5 a i 8
  ii 125
  b the volume of a cube of side 3 cm
6 a i 154
   ii $\frac{22}{7}$
  b the area of a circle of radius 7 units
7 $x = \frac{1}{2}$
8 $x = 5$
9 $x = 4$
10 $a = -6$
11 $t = -3$

Page 198 Exercise 3.2

1 **a i** $f(t) = t^2 - 4t$
     ii $f(a) = a^2 - 4a$
  b i 5
    ii −3
2 **a** 4   **b** 2   **c** 8   **d** 4
3 **a** $f(a) = 1 - a^2$
  b i 0   **ii** −3   **iii** −3
4 a i 18   **ii** 27
  b 7 hours
5 a i 12   **ii** 9   **iii** 6
  b i £72
   ii the number of CDs Eve can buy if each
     costs £8
6 **a** 8 ÷ 0 is not a real number
  b i 8   **ii** −8   **iii** 4
   iv 16   **v** $\frac{16}{3}$ **c** $x = -3$
7 **a** $\frac{1}{2}$ **b** 8   **c** 18
8 **a** $f(2a) = 6a$
  b $f(a - 1) = 3a - 3$
9 **a** $g(2a) = 4a + 1$
  b $g(a + 1) = 2a + 3$
10 a $h(2t) = 4t^2$
  b $h(t + 3) = t^2 + 6t + 9$
11 2
12 a $a + b = 8, 2a + b = 14$
  b $a = 6, b = 2$
  c $f(x) = 6x + 2$
13 a $4p + q = 10, 6p + q = 16$
  b $p = 3, q = -2$
  c $g(x) = 3x - 2$
14 a $-3a + b = 8, 4a + b = -6$
  b $a = -2, b = 2$
  c $t(x) = 2 - 2x$

Page 201 Exercise 4.1

1 $y = 7, 6, 5, 4, 3, 2, 1, 0, -1$

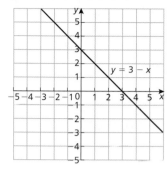

2 $y = 5, -3, -1, 1, 3, 5, 7$

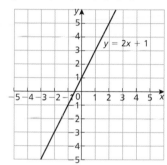

3 a $y = 3x - 1$ **b** $y = 8 - 2x$

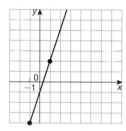

c $y = 4x - 3$

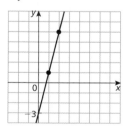

4 a $y = 9, 4, 1, 0, 1, 4, 9$ **b** $(0, 0)$
    c $x = 0$

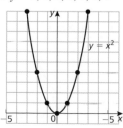

5 a $y = 19, 12, 7, 4, 3, 4, 7, 12, 19$

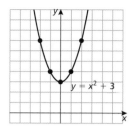

b i 3   **ii** $(0, 3)$ **iii** $x = 0$
c No

6 a $y = -11, -4, 1, 4, 5, 4, 1, -4, -11$

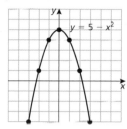

b i 5   **ii** $(0, 5)$ **iii** $x = 0$
c $-\sqrt{5}$ and $\sqrt{5}$

7 a $(\frac{3}{2}, -25)$ **b** -25 **c** $x = \frac{3}{2}$
    d i -1 and 4 **ii** 0 and 3

8 a $(1, 18)$ **b** 18   **c** $x = 1$
    d i -2 and 4 **ii** 0 and 2

9 a $y = -8, 0, 6, 10, 12, 12, 10, 6, 0, -8$
    b i the maximum value is $12\frac{1}{4}$; TP is $(-\frac{3}{2}, 12\frac{1}{4})$
        ii -5 and 2

10 a $y = x^2 - 3x$

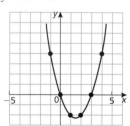

b i minimum value of $h(x)$ is $-2\frac{1}{4}$;
        TP is $(1\frac{1}{2}, -2\frac{1}{4})$
    ii 0 and 3

11 a $y = 4 - 4x^2$

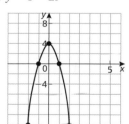

b i maximum value of $p(x)$ is 4; TP is $(0, 4)$
    ii -1 and 1

12 a $y = x^2 - 6x + 9$

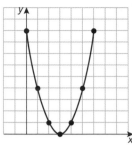

b i minimum value of $q(x)$ is 0; TP is (3, 0)
  ii 3

13 a

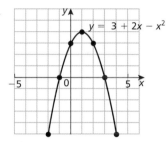

$y = 3 + 2x - x^2$

b i maximum value is 4; TP is (1, 4)
  ii −1 and 3

Page 203 Exercise 4.2

1 a 36 m after 3 seconds
  b 6 seconds
  c i 33 m
    ii 11 m
  d 1 s and 5 s

2 a

  b i $d = 31$
    ii $t = 2·4$ or $2·5$
  c i $d = 11·25$
    ii $t = 2·8$

3 a

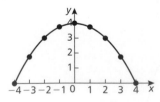

$y = 4 - \dfrac{x^2}{4}$

  b i 2·4   **ii** 0·9
  c 4 m
  d 8 m
  e 6·3 − 6·4 m

4 a

s	0	20	40	60	80	100
D	0	2	8	18	32	50

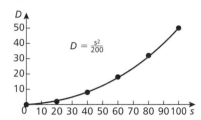

$D = \dfrac{s^2}{200}$

c i 12·5 m and 40·5 m
  ii 71 km/h and 89 km/h

Page 205 Exercise 5.1

1 a

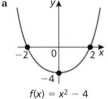

$f(x) = x^2 - 4$

b

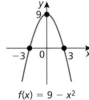

$f(x) = 9 - x^2$

c

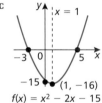

$f(x) = x^2 - 2x - 15$

d

$f(x) = x^2 - 2x - 8$

e

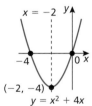

$y = x^2 + 4x$

f

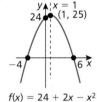

$f(x) = 24 + 2x - x^2$

g

$f(x) = x^2 - 7x + 6$

h

$f(x) = x^2 - 3x$

i

$f(x) = 5x - x^2$

j

$f(x) = x^2 - 6x + 9$

k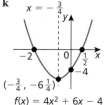
$$x = -\tfrac{3}{4}$$
$$\left(-\tfrac{3}{4}, -6\tfrac{1}{4}\right)$$
$$f(x) = 4x^2 + 6x - 4$$

l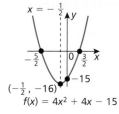
$$x = -\tfrac{1}{2}$$
$$\left(-\tfrac{1}{2}, -16\right)$$
$$f(x) = 4x^2 + 4x - 15$$

m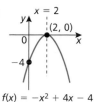
$$x = 2$$
$$f(x) = -x^2 + 4x - 4$$

n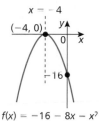
$$x = -4$$
$$f(x) = -16 - 8x - x^2$$

2 a/b i
$$y = x^2$$

ii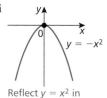
$$y = -x^2$$
Reflect $y = x^2$ in the x axis.

iii
$$y = x^2 - 1$$
Translate $y = x^2$ one unit down.

iv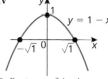
$$y = 1 - x^2$$
Reflect $y = x^2$ in the x axis then translate it one unit up.

v
$$y = 2x^2$$
Stretch $y = x^2 \times 2$ in the y direction.

vi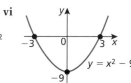
$$y = x^2 - 9$$
Translate $y = x^2$ nine units down.

c i
$$y = x^2 + 1$$

ii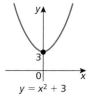
$$y = x^2 + 3$$

iii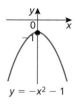
$$y = -x^2 - 1$$

iv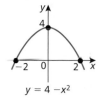
$$y = 4 - x^2$$

Page 205 Brainstormer

a i −2 and 4; 0 and 5

ii $x < -2$, $x > 4$; $0 < x < 5$

iii $-2 < x < 4$; $x < 0$, $x > 5$

iv $x > 1$; $x < 2\cdot5$ **v** $x < 1$; $x > 2\cdot5$

b e.g. $y = x^2 - 2x - 8$; $y = 5x - x^2$

Page 206 Exercise 6.1

1a/b y: $-1, -1\cdot5, -3, -4, -6, -9, -12, -24, -36,$
$36, 18, 12, 9, 6, 4, 3, 1\cdot5, 1$

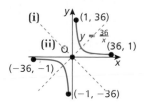
(i) (1, 36)
$y = \frac{36}{x}$
(ii) (36, 1)
(−36, −1)
(−1, −36)

c As x increases, y decreases and the graph gets closer and closer to the x axis.

d As x decreases, y increases, and the graph gets closer and closer to the y axis.

e $y - x$ and $y = -x$.

2 a

x	−16	−8	−4	−2	−1	1	2	4	8	16
y	−1	−2	−4	−8	−16	16	8	4	2	1

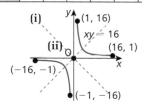
(i) (1, 16)
$xy = 16$
(ii) (16, 1)
(−16, −1)
(−1, −16)

b $y = x$ and $y = -x$ are axes of symmetry; the origin is the centre of symmetry

3

x	−16	−8	−4	−2	−1	1	2	4	8	16
y	1	2	4	8	16	−16	−8	−4	−2	−1

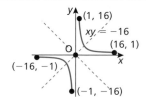
(1, 16)
$xy = -16$
(16, 1)
(−16, −1)
(−1, −16)

4 a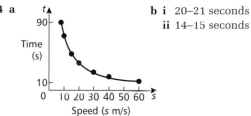
Speed (s m/s)

b i 20–21 seconds
ii 14–15 seconds

c 24 m/s **d** 720 m **e** $st = 720$

5 a

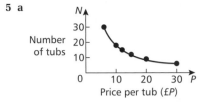

b £7   **c** £180   **d** $PN = 180$

6 a

Length (L m)	15	20	25	30	40	48	50	60	80
Breadth (B m)	80	60	48	40	30	25	24	20	15

b

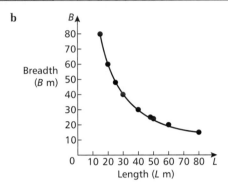

c $LB = 1200$

Page 208 Exercise 7.1

1 a

x	-1	0	1	2	3
y	$\frac{1}{3}$	1	3	9	27

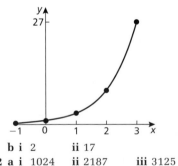

b i 2   **ii** 17
2 a i 1024   **ii** 2187   **iii** 3125   **iv** 6561
3 b i 3   **ii** 11   **iii** 14
4 a 256   **b** 4096   **c** 5·28   **d** 0·189
5 a 2·24   **b** 8·10   **c** 65·7   **d** 0·0894
6 a i 2   **ii** 4   **iii** 8   **iv** 16
    v 2^{10} or 1024   **b** $N = 2^t$
7 a £3^5 or £243   **b** $A = 3^y$ **c** £3^0 or £1

Page 209 Challenge

$A = 2^n$, where A is the amount in pence Alec's dad has to pay and n is the number of boulders removed. No, unless his dad is very rich!
a £10 485·76   **b** £10 737 000
    c £11 259 000 000 000 or £1·1259 × 10^{13}

Page 209 Exercise 8.1

2 a Reciprocal   **b** quadratic   **c** linear
    d linear   **e** exponential   **f** reciprocal
    g exponential   **h** quadratic   **i** quadratic
    j linear   **k** linear   **l** quadratic

Page 211 Exercise 9.1

1 a $y = 3x$ **b** $y = \dfrac{12}{x}$
    c $y = 2^x$ **d** $y = x^2$
    e $y = -x + 3$ **f** $y = 4^x$

2 a

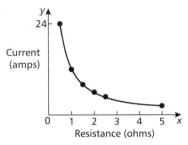

b Yes, $y = \dfrac{a}{x}$, $y = \dfrac{12}{x}$
c 3 amps
d 1·2 ohms

3 a

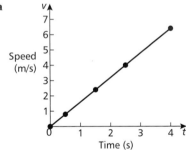

b $y = ax$
c $v = 1·6t$
d i 1·6 m/s
    ii 5·6 m/s
e 0·625 s

4 a

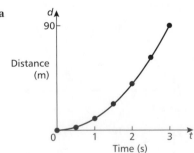

b $y = ax^2$
c $d = 10t^2$
d i 32·4 m
    ii 2·2 s, to 1 d.p.

5 a

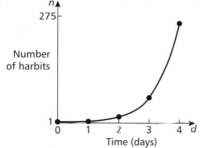

b $n = 4^d$ **c** 4096 harbits

6 a The base of the channel is $(36 - 2x)$ cm wide. The cross-section of the channel is rectangular, and the area of a rectangle is length times breadth. So, area of cross-section is $A(x) = x(36 - 2x)$ cm²

b

x	0	2	4	6	8	10	12	14	16	18
A(x)	0	64	112	144	160	160	144	112	64	0

From the symmetry of the parabola, the maximum value occurs at $x = 9$, i.e. at $A(x) = 162$. So the dimensions of the channel are breadth 18 cm, height 9 cm.

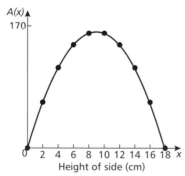

Page 214 Revise

1 a $\{-5, -2, 1, 4\}$ **b** $\{1, 7, 11\}$

2 a g means double
  b h means square and add 1

3 a $f: 0 \to -1$ **b** $g: 1 \to 4$
   $f: 1 \to 0$ $g: 2 \to 7$
   $f: 2 \to 3$ $g: 3 \to 10$
   $f: 3 \to 8$ $g: 4 \to 13$
   f means square $g: 5 \to 16$
   and subtract 1 g means multiply
                            by 3 and add 1

4 a i -2 **ii** 3
  b i -2 **ii** -5
  c i -3 **ii** -3
  d i 27
     ii 1
     iii 5·20, to 3 s.f.

5 a £400
  b the profit on 100 CDs sold
  c 25

6 $x = 9$

7 a $g(a) = 2a$
  b $g(2a) = 4a$
  c $g(a - 3) = 2a - 6$

8 $8t$

9 a 4   **b** $t = \frac{3}{8}$

10 a (-1, -9)   **b** -9 **c** $x = -1$
   d i -4 and 2   **ii** -2 and 0

11 a

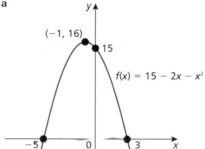

  b $-5 \leqslant x \leqslant 3$

12 a $t = 48, 40$
   b

Time (s)

Speed (m/s)

   c a hyperbola   **d** $vt = 4800$ **e** 60 s

13 a £45 or £1024   **b** $A(x) = 4^x$

14 a $y = -\frac{1}{2}x + 2$ **b** $y = 8 - 2x^2$

   c $y = 3^x$ **d** $y = \dfrac{12}{x}$

11 Trigonometric graphs and equations

Page 216 Exercise 1.1

1 a $\frac{5}{12}$ **b** $\frac{12}{13}$ **c** $\frac{24}{25}$
  d $\frac{24}{7}$ **e** $\frac{3}{5}$ **f** $\frac{3}{5}$

2 a $x = 6·10$ **b** $x = 5·60$ **c** $x = 4·80$
  d $x = 37·9°$ **e** $x = 37·1°$ **f** $x = 12·2$

3 a 8·87   **b** 13·1

4 216 m

5 a 40°
  b i AD = AB = 7·8 cm; DC = BC = 12 cm
     ii 15·4 cm

6 87·8 m

7 a 31 km   **b** 22 km

8 Q(17, 12)

Page 218 Exercise 2.1

1 a i $\frac{12}{13}$ **ii** $\frac{5}{13}$

  b i $\frac{4}{5}$ **ii** $\frac{4}{3}$

  c i $\frac{21}{20}$ **ii** $\frac{20}{29}$

2 a sin 0° = 0; cos 0° = 1; tan 0° = 0

  b sin 90° = 1; cos 90° = 0

  c tan 90° = BC or AC ÷ 0; division by 0 is not defined, so tan 90° is not defined

3 a i $\frac{1}{\sqrt{2}}$ **ii** $\frac{1}{\sqrt{2}}$ **iii** 1

  b i $\frac{1}{2}$ **ii** $\frac{\sqrt{3}}{2}$ **iii** $\frac{1}{\sqrt{3}}$

    iv $\frac{\sqrt{3}}{2}$ **v** $\frac{1}{2}$ **vi** $\sqrt{3}$

4

	0°	30°	45°	60°	90°
sin	0	$\frac{1}{2}$	$\frac{1}{\sqrt{2}}$	$\frac{\sqrt{3}}{2}$	1
cos	1	$\frac{\sqrt{3}}{2}$	$\frac{1}{\sqrt{2}}$	$\frac{1}{2}$	0
tan	0	$\frac{1}{\sqrt{3}}$	1	$\sqrt{3}$	—

5 a $x = \frac{5}{\sqrt{2}}$ **b** $x = 4\sqrt{3}$ **c** $x = 6\sqrt{3}$

6 $3 + \frac{5\sqrt{3}}{3}$ m

7 a 11·3 cm   **b** 35·3°

8 a AC = 170 cm   **b** AD = 188 cm

  c ∠CAD = 25·2°

9 0·15°

Page 222 Exercise 3.1

1 a i 0·80, 0·60, 1·3   **b i** 0·92, −0·38, −2·4

  c i −0·47, −0·88, 0·53   **d i** −0·96, 0·28, −3·4

  e i −0·92, −0·39, 2·4   **f i** 0·69, −0·72, −0·96

2 a

∠A	Quad	sin	cos	tan
65°	1	0·91	0·42	2·1
200°	3	−0·34	−0·94	0·36
285°	4	−0·97	0·26	−3·7
94°	2	1·0	−0·070	−14
190°	3	−0·17	−0·98	0·18
240°	3	−0·87	−0·5	1·7
160°	2	0·34	−0·94	−0·36
340°	4	−0·34	0·94	−0·36
275°	4	−1·0	0·087	−11
260°	3	−0·98	−0·17	5·7

  b i 1st and 2nd   **ii** 1st and 4th   **iii** 1st and 3rd

  c 1st

4 Positive: **a, f, h, k, l, n**;

  negative: **b, c, d, e, g, i, j, m, o**

5 a Positive   **b** negative   **c** positive

  d negative   **e** negative   **f** negative

  g negative   **h** negative   **i** negative

  j positive

6 a 0·087   **b** −0·268   **c** 0·139

  d −28·636   **e** −0·087   **f** −0·174

  g −0·017   **h** −1·540   **i** −0·966

  j 0·156

7 a Positive   **b** negative   **c** negative

Page 224 Exercise 4.1

1 a

$x°$	0°	30°	60°	90°	120°	150°	180°
$\sin x°$	0	0·5	0·9	1	0·9	0·5	0

$x°$	210°	240°	270°	300°	330°	360°
$\sin x°$	−0·5	−0·9	−1	−0·9	−0·5	0

b/c

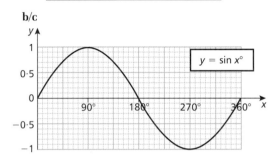

2 a i $x = 0°, 180°, 360°$

    ii $x = 90°$ **iii** $x = 270°$

  b (150, 0·5)   **c** (390, 0·5) and (510, 0·5)

  d (330, −0·5)   **e** (570, −0·5) and (690, -0·5)

3 a

$x°$	0°	30°	60°	90°	120°	150°	180°
$\cos x°$	1	0·9	0·5	0	−0·5	−0·9	−1

$x°$	210°	240°	270°	300°	330°	360°
$\cos x°$	−0·9	−0·5	0	0·5	0·9	1

b

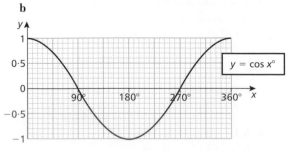

4 a i 90° and 270°   **ii** 0° and 360°   **iii** 180°

  b $x = 180°$

  c (300, 0·5)   **ii** (420, 0·5) and (660, 0·5)

  d i (240, -0·5)   **ii** (480, −0·5) and (600, −0·5)

5 a

$x°$	0°	30°	45°	60°	75°	105°	120°	135°	150°	180°
tan $x°$	0	0·6	1	1·7	3·7	−3·7	−1·7	−1	−0·6	0

$x°$	210°	225°	240°	255°	285°	300°	315°	330°	360°
tan $x°$	0·6	1	1·7	3·7	−3·7	−1·7	−1	−0·6	0

b/c

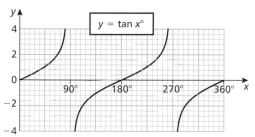

d the nearer x gets to 90 (from above 90 as well as from below 90), the larger tan $x°$ becomes

e as x approaches 270 from below, the greater tan $x°$ becomes; as x approaches 270 from above, the smaller tan $x°$ becomes (i.e. the greater its negative value)

6 a $x = 0, 180, 360, 540, 720$
  b i (225, 1), (405, 1), (585, 1)
    ii (315, −1), (495, −1), (675, −1)

Page 227 Exercise 5.1

1 a Max. 3, min. −3
  b i 1
    ii 3
  c

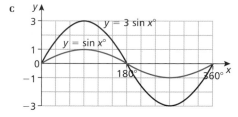

2 a Max. 2, min. −2
  b/c

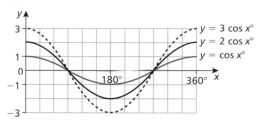

  d i period is 360°, amplitude 1
    ii period is 360°, amplitude 2
    iii period is 360°, amplitude 3

3 a/b

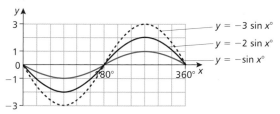

c 360° for each graph
d amplitudes are 1, 2 and 3
4 a Max. value n (positive),
    min. value $−n$ (negative)
  b 360°   **c** n (positive)
  d i

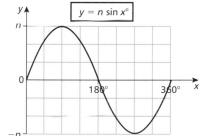

  ii

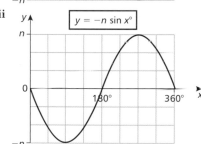

5 a Max. value n (positive),
    min. value $−n$ (negative)
  b 360°   **c** n (positive)
  d i

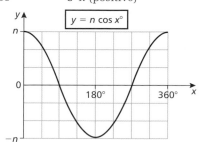

  ii

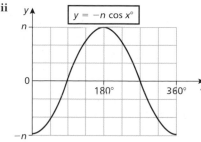

6 a

$x°$	0	45	90	135	180	225	270	315	360
$\cos 2x°$	1	0	−1	0	1	0	−1	0	1

b

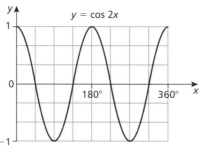

$y = \cos 2x$

c 180°

7 a 3

b

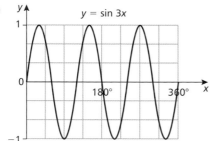

$y = \sin 3x$

8 3

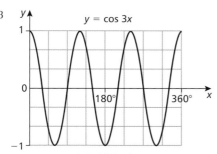

$y = \cos 3x$

9 a

$x°$	0	60	90	180	270	300	360
$\sin\frac{1}{2}x°$	0	0·5	0·7	1	0·7	0·5	0

$x°$	420	450	540	630	660	720
$\sin\frac{1}{2}x°$	−0·5	−0·7	−1	−0·7	−0·5	0

b

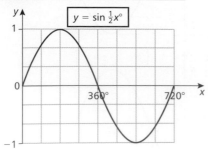

$y = \sin\frac{1}{2}x°$

c a half cycle

d 720°

10 a i Max. 1, min. −1; 1 cycle

ii

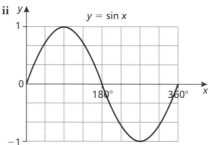

$y = \sin x$

b i Max. 1, min −1; 1

ii

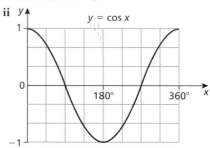

$y = \cos x$

c i Max. 4, min −4; 1 cycle

ii

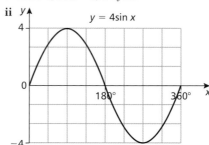

$y = 4\sin x$

d i Max. 4, min −4; 1 cycle

ii

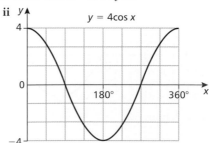

$y = 4\cos x$

e i Max. 1, min −1; 2 cycles

ii

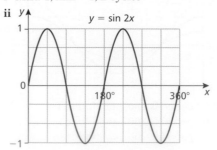

$y = \sin 2x$

f i Max. 4, min −4; 2 cycles

ii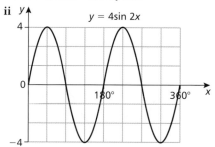

g i Max. 4, min. −4; 2 cycles

ii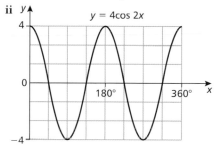

h i Max. 1, min −1; half a cycle

ii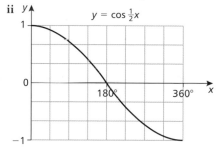

i i Max. $\frac{1}{2}$, min −$\frac{1}{2}$; half a cycle

ii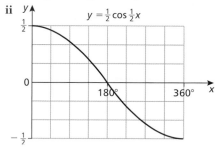

j i Max. 1, min −1; 2 cycles

ii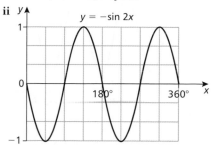

k i Max. 6, min. −6; 3 cycles

ii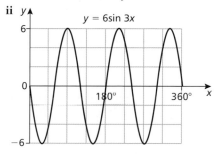

l i Max. 3, min. −3, 4 cycles

ii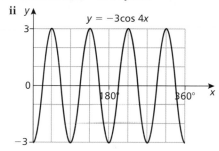

m i Max. 1, min. −1; one third of a cycle

ii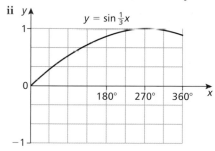

n i Max. 2, min. −2; half a cycle

ii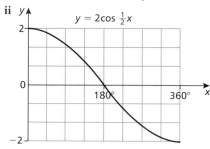

11 a i $y = \cos x°$
    ii $y = \cos 3x°$
 b i $y = \sin x°$
    ii $y = 3 \sin x°$
 c i $y = 2 \sin 3x°$
    ii $y = 4 \cos \frac{1}{2}x°$
 d i $y = -\cos x°$
    ii $y = -2 \cos x°$
 e i $y = -\sin \frac{1}{2}x°$
    ii $y = \frac{1}{2} \sin 2x°$

Page 231 Exercise 5.2

1 a i Max. 3, min. 1 **ii** 1 **iii** 360°

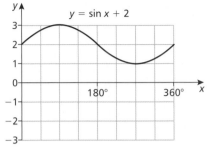

b i Max. 0, min. −2 **ii** 1 **iii** 360°

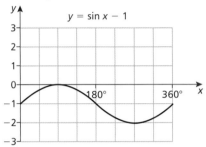

c i Max. 4, min. 2 **ii** 1 **iii** 360°

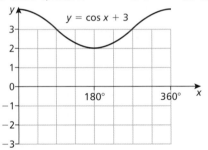

d i Max. −1, min. −3 **ii** 1 **iii** 360°

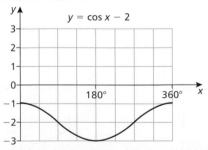

e i Max. 3, min. −1 **ii** 2 **iii** 360°

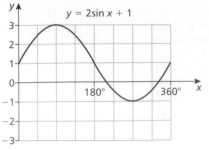

f i Max. 1, min. −3 **ii** 2 **iii** 360°

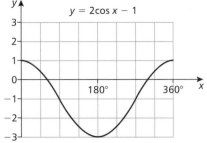

g i Max. 2, min. 0 **ii** 1 **iii** 180°

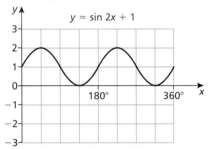

h i Max. 2, min. 0 **ii** 1 **iii** 360°

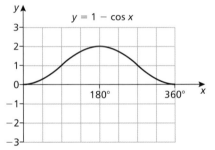

i i Max. 0, min. −1 **ii** 1 **iii** 720°

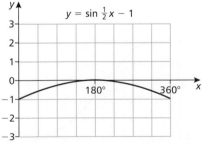

j i Max. 5, min. 1 **ii** 2 **iii** 120°

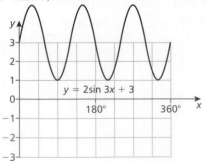

k i Max. 2, min. −4   **ii** 3   **iii** 180°

$$y = 3\cos 2x - 1$$

l i Max. 1·5, min. 0·5   **ii** 0·5   **iii** 720

$$y = \tfrac{1}{2}\cos \tfrac{1}{2}x + 1$$

2 a $y = 2\cos x° + 1$ **b** $y = 2\sin x° - 2$
c $y = 2.5\sin 2x° - 0.5$ **d** $y = 0.5\cos 2x° + 2$

Page 232 Challenge
a

x	0°	30°	45°	60°	90°	120°	135°	150°	180°
y	1	1·37	1·41	1·37	1	0·37	0	−0·37	−1

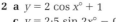

x	210°	225°	240°	270°	300°	315°	330°	360°
y	−1·37	−1·41	−1·37	−1	−0·37	0	0·37	1

i 1·41 to 3 s.f.   **ii** −1·41 to 3 s.f.
iii 1·41 to 3 s.f.   **iv** 360°

$$y = \sin x° + \cos x°$$

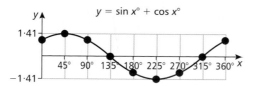

Page 233 Exercise 6.1
1 a 12°, 168°   **b** 78°, 282°   **c** 11°, 191°
d 24°, 156°   **e** 37°, 323°   **f** 37°, 217°
g 107°, 253°   **h** 107°, 287°   **i** 238°, 302°
j 120°, 300°   **k** 194°, 346°   **l** 118°, 242°
m 30°, 150°   **n** 120°, 240°   **o** 135°, 315°
p 270°   **q** 0°, 360°   **r** 0°, 180°, 360°
2 a 9°, 171°   **b** 123°   **c** 83°
3 a 30°, 150°   **b** 124°, 304°   **c** 71°, 289°
d 180°   **e** 104°, 284°   **f** 12°, 168°

4 a 60°, 120°   **b** 30°, 210°   **c** 150°, 210°
d 45°, 315°   **e** 225°, 315°   **f** 135°, 315°
5 a 38°, 142°, 218°, 322°
   b 37°, 143°, 217°, 323°
   c 35°, 145°, 215°, 325°
   d 60°, 120°, 240°

Page 234 Exercise 6.2
1 a A(17, 0·3), B(163, 0·3)
   b C(217, −0·6), D(323, −0·6)
   c 0·966
2 a P(90, 0), Q(270, 0)
   b R(60, 1), S(300, 1)
   c −0·347
3 a A(60, 2·5), B(300, 2·5), C(120, 1·5), D(240, 2·5)
   b i 3   **ii** 1   **c** 1·23
4 a i A(37, −0·4), B(143, −0·4)
   ii C(0, −1), D(180, −1)
   b max. (90, 0), min. (270, −2)
5 a i (6, 0·2) and (84, 0·2)
   ii (117, −0·8) and (154, −0·8)
   b i 0·866   **ii** −0·642

Page 235 Challenge
a $x = 45°, 225°, 135°, 315°$
b $x = 30°, 150°, 270°$
c $x = 90°, 270°, 60°, 300°$

Page 236 Exercise 7.1
1 a 1   **b** 1·7   **c** 1·9
d 1   **e** −2
2 a i 7·82   **ii** 5·52   **iii** 2·18
   b i 8   **ii** 0, 90, 180, …
3 a i 18·9 m   **ii** 8 m
   b 9 am   **c** 20 m   **d** 3 pm   **e** 4 m
4 a Max. 22 m, min. 2 m
   b max. after 4 seconds, min. after 8 seconds
   c 8 seconds
5 a Max. 192 cm, min. 168 cm
   b max. after 15 minutes, min. after 45 minutes
   c i 190 cm   **ii** 190 cm   **iii** 170 cm
   d 60 minutes, i.e. 1 hour!
   e

$$y = 180 + 12\sin 6t$$

6 a 5 cm   **b** 45 cm

  c i 10·9 cm   **ii** when $t = 7$, i.e. after 0·7 s

  d 25 cm   **e** 0·8 s (or 8°)

  f

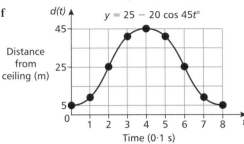

Distance from ceiling (m)

Time (0·1 s)

7 a At the start, i.e. at $n = 0$

  b $\frac{12}{100}$ of a second, i.e. at $n = 12$

  c 4 m; after $\frac{6}{100}$ of a second

  d i 3·73 m   **ii** 1 m

  e 12° ($\frac{12}{100}$ of a second)

Page 238 Exercise 8.1

1 a–c pupil's calculator check

2 a $\dfrac{\sin 100°}{\cos 100°}$

  b $\dfrac{\sin 200°}{\cos 200°}$

  c $\dfrac{\sin 340°}{\cos 340°}$

3 $\cos 90° = 0$; division by zero is not defined.

  Hence, $\dfrac{\sin 90°}{\cos 90°}$ is not defined.

  So $\tan 90°$ is not defined.

4 pupil's calculator check

5 a $\frac{1}{2} + \frac{1}{2} = 1$ **b** $\frac{3}{4} + \frac{1}{4} = 1$ **c** $1 + 0 = 1$

6 a $\sin x° = 0·8$

  b $\sin x° = 0·866$

  c $\sin x° = 0·436$

7 a $\cos x° = 0·954$

  b $\cos x° = 0·714$

  c $\cos x° = 0·6$

8 a $\pm 0·714$ **b** $\pm 0·866$ **c** $\pm 0·980$

9 a 0·51   **b** 0·961

10 $\sin^2 x° + \cos^2 x° \neq 1$

11 $\tan x° = -\frac{5}{4} \Rightarrow x° = 180° - 51·3°$

  or $360° - 51·3° = 129°$ or $309°$

12 a $x = 80·5°$ and $261°$ **b** $x = 26·6°$ and $207°$

  c $x = 146°$ and $326°$ **d** $x = 36·9°$ and $217°$

Page 239 Challenges

1 a $3(\sin^2 A + \cos^2 A) = 3 \times 1 = 3$

  b $\sin^2 x + 2 \sin x \cos x + \cos^2 x = 1 + 2 \sin x \cos x$

  c $\sin^2 x - 2 \sin x \cos x + \cos^2 x + \sin^2 x +$
    $2 \sin x \cos x + \cos^2 x = 1 + 1 = 2$

  d $\cos^2 y - \sin^2 y = \cos^2 y - (1 - \cos^2 y) =$
    $2 \cos^2 y - 1$

2 a LHS $= -\tan^2 x$;

  RHS $= 1 - \dfrac{1}{\cos^2 x} = \dfrac{\cos^2 x}{\cos^2 x} - \dfrac{1}{\cos^2 x}$

    $= \dfrac{\cos^2 x - 1}{\cos^2 x} = \dfrac{-\sin^2 x}{\cos^2 x} = -\tan^2 x =$ LHS

  b $(\cos A \tan A)^2 + \left(\dfrac{\sin A}{\tan A}\right)^2 = \left(\cos A \dfrac{\sin A}{\cos A}\right)^2$

    $+ \left(\sin A \dfrac{\cos A}{\sin A}\right)^2 = \sin^2 A + \cos^2 A = 1$

Page 241 Revise

1 a 8·0   **b** 37·9°   **c** 11·6

2 22·6°

3 a $\frac{15}{17}$ **b** $\frac{8}{15}$

4 a $-\sin 30°$ **b** $-\cos 50°$

  c $\tan 50°$ **d** $\cos 40°$

5 P(53·1, 0·6), Q(307, 0·6),
  R(143, −0·8), S(217, −0·8)

6 a Max. 7, min. −7

  b 360°   **c** 0 and 360

  d

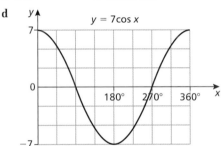

$y = 7\cos x$

7 a Max. 1, min. −1

  b 120°   **c** 0, 60, 120, 180, 240, 300, 360

  d

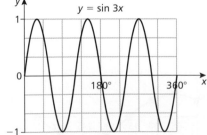

$y = \sin 3x$

8 a Max. 5, min. −1   **b** 720°

  c no values in $0 \leqslant x \leqslant 360$

  d

$y = 2 + 3 \sin \frac{1}{2} x$

9 **a** $y = 4 \cos 2x°$
   b $y = \sin \frac{1}{2}x°$
   c $y = 1 + \cos 3x°$
10 **a** Max. 18 m, min. 2 m
   b 2 m
   c 14 m
   d after 45 s
   e 180 s
   f

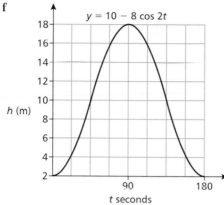

$y = 10 - 8 \cos 2t$

h (m) / t seconds

11 **a** $x = 58\cdot2°$ and $122°$
   b $x = 110°$ and $250°$
   c $x = 78\cdot2°$ and $258°$
   d $x = 75\cdot5°$ and $284°$
   e $x = 162°$ and $342°$
   f $x = 108°, 162°, 288°$ and $342°$
12 **a** $\sin x° = \pm 0\cdot917$
   b 2

12 Trigonometry and triangle calculations

Page 243 Exercise 1.1
1 **a** 7·39   **b** 38·9°   **c** 11·6
2 **a** 38·7 m   **b** 59·3°
3 **a** 66·5 m   **b** 103 m   **c** 58·5°
4 148°
5 **a** 312 m   **b** 628 m
6 **a** 135°   **b** 315°
7 **a** 20 cm²   **b** 39 cm²   **c** 54 cm²
8 **a** $\sin 10°$ **b** $-\cos 40°$ **c** $-\tan 35°$
   d $-\sin 75°$ **e** $\cos 15°$
9 **a** 11·5° and 168°   **b** 69·5° and 290°
   c 138° and 318°   **d** 106° and 254°
10 **a** 33·2 km   **b** 086·2°
11 **a** $x = ab$ **b** $x = \dfrac{ab}{c}$

   c $x = \dfrac{hn}{m}$ **d** $x = y \sin X$

   e $x = \dfrac{y \sin A}{\sin B}$ **f** $x = \dfrac{y \sin X}{\sin Y}$

Page 247 Exercise 2.1
1 **a** 7·96
   b i $\dfrac{5}{\sin 86°} = \dfrac{a}{\sin 37°}; a = 3\cdot02$

   ii $\dfrac{c}{\sin 34°} = \dfrac{12}{\sin 87°}; c = 6\cdot72$

   iii $\dfrac{b}{\sin 105°} = \dfrac{4\cdot2}{\sin 23°}; b = 10\cdot4$

   iv $\dfrac{x}{\sin 23°} = \dfrac{6}{\sin 27°}; x = 5\cdot16$

2 **a** $a = 4\cdot41$ cm   **b** $a = 6\cdot00$ cm   **c** $a = 9\cdot16$ cm
3 **a** 60°   **b** 7·25 cm
4 **a** 2·75 cm   **b** 131°   **c** 5·79 cm
5 **b** XY = 514 m
6 **a** 20·8 km   **b** 14·8 km
7 **a i** 1254 m   **ii** 854 m   **b** 754 m

Page 249 Exercise 3.1
1 **a** 31·5°   **b** 47·5°   **c** 41·5°
2 **a** $\angle S = 36\cdot7°, \angle R = 91\cdot3°$
   b $\angle F = 66\cdot1°, \angle E = 31\cdot9°$
   c $\angle Z = 36\cdot4°, \angle X = 25\cdot6°$
3 **a** $\angle P = 48\cdot3°, \angle Q = 66\cdot7°$
   b $\angle Y = 47\cdot7°, \angle Z = 68\cdot3°$
4 **a** $\angle LKM = 20\cdot5°$ **b** KM = 11·7 cm
5 **a** $\angle Y = 47\cdot2°, \angle Z = 27\cdot8°$
   b XY = 121 m
6 **a** $\angle P = 49\cdot8°, \angle Q = 67\cdot2°$ **b** 40·5 m
7 **a** 57·5°   **b** 208 m   **c** 46·0°   **d** 63·5°

Page 251 Exercise 3.2
1 **a** $\angle X = 54\cdot2°, \angle Y = 86\cdot8°$ or $\angle X = 126°$,
          $\angle Y = 15\cdot2°$
   b $\angle P = 69\cdot2°, \angle R = 74\cdot8°$ or $\angle P = 111°, \angle R = 33°$
2 $\angle FEG = 78\cdot3°$
3 14 m
4 **a i** 133 m   **ii** 162 m   **b** 86 m
   c Boat A – 102 m, boat B – 137 m
5 260 m, to 2 s.f.
6 **a i** 48°   **ii** 46°   **iii** 94°
   b 42·1°   **c** 090·1°   **d** 445 m

Page 252 Brainstormer
$\angle ACB = 59\cdot4°$ or $120\cdot6°$

Page 252 Challenges
1 **a** CD = 97·7 m
   b $\angle A = 115°, \angle B = 87°, \angle C = 141°, \angle D = 81°$,
          $\angle E = 116°$
2 Find an expression for PB in $\triangle$PBM and for PB
   in $\triangle$ABP. Equating the expressions leads to the
   formula:

   $$h = \frac{d \sin x° \sin y°}{\sin (x - y)°}; h = 197 \text{ m}$$

Page 254 Exercise 4.1

1 a 4·52 cm   **b** 7·02 cm   **c** 45·6 mm

2 a $b = 6{\cdot}16$ cm   **b** $c = 21{\cdot}6$ mm   **c** $b = 6{\cdot}12$ m

3 a $x^2 = y^2 + z^2 - 2yz \cos X$

  b $y^2 = x^2 + z^2 - 2xz \cos Y$

  c $z^2 = x^2 + y^2 - 2xy \cos Z$

4 a i $p^2 = q^2 + r^2 - 2qr \cos P$

  ii $q^2 = p^2 + r^2 - 2pr \cos Q$

  iii $r^2 = p^2 + q^2 - 2pq \cos R$

  b i 5·65   **ii** 10·9

  c $p^2 = q^2 + r^2$ i.e. Pythagoras' theorem

5 142 m

6 1350 m

7 a 7·54 m   **b** 7·10 m   **c** 6·52 m

8 8·92 cm

Page 256 Exercise 5.1

1 a $\cos B = \dfrac{a^2 + c^2 - b^2}{2ac}$ **b** $\cos C = \dfrac{a^2 + b^2 - c^2}{2ab}$

2 $\angle Q = 41{\cdot}4°$

3 a 38·2°   **b** 114°   **c** 73·5°

4 a $\cos P = \dfrac{q^2 + r^2 - p^2}{2qr}$, $\cos Q = \dfrac{p^2 + r^2 - q^2}{2pr}$,

    $\cos R = \dfrac{p^2 + q^2 - r^2}{2pq}$

  b i $\angle R = 95{\cdot}7°$ **ii** $\angle Q = 33{\cdot}6°$

5 a Apex angle is 53·6°; angles at base are 59·6°
    and 66·8°

  b apex angle becomes 49·5°; base angles 61·3°
    and 69·3°.
    Angle at apex is reduced; both base angles are
    increased

6 a 61·9°   **b** 100°

7 a 87·7°, 81·6°, 98·3°, 92·4°   **b** 56·5 m

Page 257 Exercise 5.2

1 $\angle K = 97{\cdot}9°$, $\angle L = 52{\cdot}4°$, $\angle M = 29{\cdot}7°$

2 13·4 cm

3 a 13 800 miles   **b** 19 500 miles

4 $\angle A = 38{\cdot}9°$, using cosine rule and right-angled
    triangle trig in 'half' of the isosceles triangle

5 b 15·7 m   **c** 44·9°

6 a 164°   **b** 210°

Page 258 Brainstormer

a 115°   **b** 13·7 cm

Page 258 Exercise 6.1

1 a 18·1 cm²   **b** 13·5 cm²   **c** 246 m²

  d 75·0 mm²   **e** 140 cm²   **f** 86·9 cm²

2 a 17·9 cm²   **b** 31·0 m²

3 a Sail D 7·42 m²

  b sail A 5·44 m²

    (Sail B is 6·10 m² and sail C is 5·64 m²)

4 a 14·7 cm²   **b** 26·2 cm²   **c** 388 m²

5 a 123 cm²   **b** 62·0 cm²

6 a 30° and 150°

7 a 84·3°   **b** 197 m²

8 3080 m² + 6800 m² = 9880 m²

9 53 600 cm²

Page 260 Challenges

1 3420 m²

2 426 m

Page 261 Investigation

a $A = 1{\cdot}3r^2,\ 2r^2,\ 2{\cdot}38r^2,\ 2{\cdot}60r^2$

b $A = \dfrac{nr^2}{2} \sin\left(\dfrac{360}{n}\right)^{\!\circ}$

c i $\pi r^2 \approx \dfrac{nr^2}{2} \sin\left(\dfrac{360}{n}\right)^{\!\circ}$ for large n so

    $\pi \approx \dfrac{n}{2} \sin\left(\dfrac{360}{n}\right)^{\!\circ}$

  ii 3·141 592 7

d by trials, $n = 5000 \to 3{\cdot}141\ 591\ 8$;
    $n = 10\ 000 \to 3{\cdot}141\ 592\ 4$;
    $n = 20\ 000 \to 3{\cdot}141\ 592\ 6$.
    This suggests $n > 20\ 000$

Page 262 Exercise 7.1

1 a Cosine rule

  b sine rule

  c cosine rule

2 a 17·5 cm   **b** 9·31 cm

3 a 56·2°

  b 57·8° or 122°

    (from the diagram, 122° seems more likely)

4 92·9 cm²

5 a 55·5°   **b** 27·6 m²

6 a 94·1°, 46·5°, 39·4°

  b 27·9 cm²

7 252 m

8 10·2 km

9 a 63·7 m   **b** 56·8 m

10 a 9·6 cm   **b** 53·4° or 127°   **c** 50·3°

Page 263 Exercise 7.2

1 a RT = 16·2 cm, SU = 5·98 cm

  b 41·1 cm²

2 43·8 cm²

3 a 62·2 km   **b** 108°

4 a AG = 39·1 m   **b** BG = 50·8 m   **c** AB = 77·3 m

5 Dom is given 1207 m², Tom is given 1220 m².
    Tom is given 13 m² more

6 a 82·3 cm   **b** $\angle BAC = 39{\cdot}0°$

Page 266 Revise

1 a $\dfrac{a}{\sin A} = \dfrac{b}{\sin B} = \dfrac{c}{\sin C}$

  b $\cos B = \dfrac{a^2 + c^2 - b^2}{2ac}$

  c $c^2 = a^2 + b^2 - 2ab \cos C$

  d $A = \tfrac{1}{2}ac \sin B$

2 a 11·1 cm   **b** 37·6°
3 a 11·5 m   **b** 47·2°
4 a i Sine rule   **ii** cosine rule
     iii cosine rule   **iv** Pythagoras' theorem
   b i 51·6°   **ii** 10·1 m
     iii 89·3°   **iv** 15 m
   c ii 34·6 m²   **iii** 12·4 cm²
5 a Every hour the hour hand moves 30°
     clockwise. 20 minutes is one third of an
     hour, so the hour hand moves 10° in 20
     minutes. Therefore, the angle between the
     hands is 30° + 30° + 30° + 30° + 10° = 130°
   b 27·3 cm
6 Area is 84 m²; $x = 9.48$ m
7 a 703 m   **b** 223 m
8 a i 83·9°   **ii** 53·5°   **b** 3·78 km

13 Chapter revision

Page 268 Revising Chapter 1

1 a 48·75 cm²   **b** 57·33 cm²   **c** 138 cm²
2 3876 cm²
3 11·5 m²
4 a 0·775 m³, 4·80 m²
   b 2·125 m³, 9·37 m² (or 11·9 including underneath)
5 11 500 cm³, 3000 cm² (3 s.f.)

Page 269 Revising Chapter 2

1 a £346   **b** £17 992
2 a £2468
   b i £2566·72   **ii** £30 800 ·64
3 a 38·5   **b** £278·01
4 a £638   **b** £2340
5 a £2432·23   **b** £807·78   **c** £1624·45
6 a £0   **b** £2287·78   **c** £3046·05
7 a £373·60   **b** £6025·40   **c** £17 517·60
8 a £3870·72   **b** £6903·60
9 £75 850·36
10 a 6·5%   **b** £12 596·77   **c** £18 000

Page 270 Revising Chapter 3

1 960 m
2 15 cm
3 b 1·04 km   **c** 040°
4 $21\frac{1}{3}$ m
5 a $\frac{1}{2}$ **b** 5·25 cm³   **c** 1·5 cm²
6 120 cm
7 a 42 cm   **b** $2\frac{2}{3}$ cm   **c** 12·5 cm
   d 15 cm   **e** 16·5 cm   **f** 2·4 cm

Page 271 Revising Chapter 4

1 a 138 km   **b** 154 mm²   **c** 27·1 cm
   d 23·6 m   **e** 15 cm
2 a 199·65
   b £250 gives £315·62 (£200 only gives £295·49)
3 a $V = \dfrac{h}{3}$ **b** $V = 13$

4 a $M = 5m + 5$ **b** 90   **c** design 48
5 a $k = Mw$ **b** $Q = \dfrac{L}{P}$ **c** $m = \dfrac{l}{T}$

   d $a = \dfrac{c}{5} + b$ or $a = \dfrac{c + 5b}{5}$

   e $Q = \sqrt{A + n}$ **f** $b = \dfrac{a}{3 - P}$
6 a $2y = \sqrt{4 - x^2} \Rightarrow 4y^2 = 4 - x^2 \Rightarrow x^2 = 4 - 4y^2$
     $= 4(1 - y^2) \Rightarrow x = 2\sqrt{1 - y^2}$
   b i 0·6   **ii** 1·6
7 a i halved   **ii** trebled   **iii** halved
   b i multiplied by 4
     ii doubled   **iii** divided by 3

Page 273 Revising Chapter 5

1 a $y = \frac{1}{2}$ **b** $x = 5$ **c** $k = 2$
   d $x = 5$ **e** $x = 1$ **f** $x = 2$
2 a $2(2x - 1) + 6(10 - x) = 48$
   b $x = 5$ **c** 9 m × 15 m
3 a $3y$ **b** $\dfrac{3m}{7}$ **c** $\dfrac{2y}{x}$ **d** $2k$
4 a $k = 2$ **b** $y = \frac{1}{2}$ **c** $n = 6$ **d** $x = \frac{15}{2}$
   e $y = 2$ **f** $y = \frac{3}{2}$ **g** $x = \frac{15}{14}$ **h** $x = \frac{4}{15}$
   i $x = 12$ **j** $x = 5$ **k** $m = 2$ **l** $x = 4$
   m $y = 5$ **n** $k = 5$ **o** $x = 24$ **p** $y = 10$
5 a sheep1: $x + 6 = \frac{6}{5}x$; sheep 2: $x - 8 = \frac{5}{7}x$;
     sheep 3: $x = \frac{5}{6}(x + 4)$
   b sheep 1: 30 kg; sheep 2: 28 kg; sheep 3: 20 kg

Page 274 Revising Chapter 6

1 a, b, e
2 a $\dfrac{3(x + 4)}{5(x + 4)} = \dfrac{3}{5}$ **b** $\dfrac{(x - 1)(x + 1)}{3(x + 1)} = \dfrac{x - 1}{3}$
   c $\dfrac{(x - 2)(x + 3)}{x(x + 3)} = \dfrac{x - 2}{x}$ **d** $\dfrac{(x + 2)(x - 1)}{(x + 2)(x + 3)} = \dfrac{x - 1}{x + 3}$
3 a $\dfrac{3a}{2b}$ **b** 6   **c** $\dfrac{x - 1}{x^2(x - 2)}$
4 a $\dfrac{1}{6}$ **b** $\dfrac{9xz}{2}$
   c $\dfrac{a^2}{2}$ **d** $\dfrac{3}{(x + 1)(2x - 3)}$
5 a $\dfrac{5a}{6}$ **b** $\dfrac{t^2 + 20}{5t}$ **c** $\dfrac{p + q}{pq}$
   d $\dfrac{4x + 1}{3(x + 1)}$ **e** $\dfrac{2h^2 + 3}{h}$
6 a $\dfrac{x}{20}$ **b** $\dfrac{2b - 3a}{ab}$ **c** $\dfrac{1 - d^2}{d}$
7 a i $\dfrac{10}{8 - x}$ **ii** $\dfrac{10}{8 + x}$
   iii $\dfrac{160}{(8 - x)(8 + x)} = \dfrac{160}{64 - x^2}$
   b $\sqrt{24} = 4.9$ km/h (1 d.p.)

Page 274 Revising Chapter 7

1 $x = 3$; $y = -2$

2 0.01

3 a $\frac{3}{4}$ **b** $\frac{4}{7}$

   c undefined   **d** 0

4 a

x	0	1	2	3	4
$y = 3x - 1$	-1	2	5	8	11

b

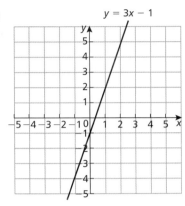

$y = 3x - 1$

   c 3

   d $(0, -1)$

5 a $y = -2x + 5$

   b $y = \frac{1}{5}x - 2$

6 a/b

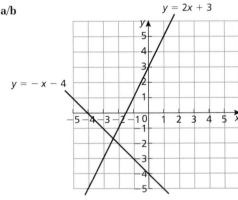

$y = 2x + 3$

$y = -x - 4$

c/d

$y = \frac{1}{3}x + 6$

$y = -\frac{1}{4}x$

7 a 9 cm

   b $h = -\frac{1}{20}W + 9$

8 a/b Pupils' lines of best fit may differ slightly.

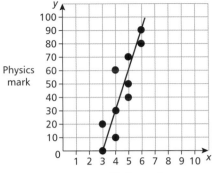

Physics mark / Maths mark

   c $y = 30x - 90$ **d** 45

Page 275 Revising Chapter 8

1 a $-\frac{1}{3}, -4$ **b** $-\frac{1}{5}, -2$ **c** ± 5 **d** ± 8

   e ± 5 **f** $\pm\frac{7}{4}$ **g** $\pm\frac{7}{6}$ **h** $-3, -7$

2 a $10.91, 0.09$ **b** $-0.21, -3.12$

   c $3.58, 0.42$ **d** $-8, 1$

3 a $x = 1$ gives $y = -3$ (negative);

     $x = 2$ gives $y = 1$ (positive);

     so $y = 0$ somewhere between $x = 1$ and $x = 2$

   b 1.88 **c** $-0.35, -1.53$

4 a $10.9, -0.9$ **b** $5, 1$

   c $7.6, 0.4$ **d** $5, -6$

5 a $16 - x$ **b** 14 cm, 2 cm

Page 276 Revising Chapter 9

1 a x^{14} **b** $12x^{16}$ **c** $3x^2$ **d** $\dfrac{3x}{4}$

   e $81x^8$ **f** $81x^8$ **g** $9x^4$ **h** $16x^3$

2 a $\dfrac{1}{x^{11}}$ **b** $\dfrac{1}{x^{20}}$ **c** $\dfrac{3}{x^{10}}$ **d** $\dfrac{1}{2x^7}$

   e $\dfrac{7}{6x^3}$ **f** $\dfrac{1}{x}$ **g** $\dfrac{12}{x^5}$ **h** $\dfrac{1}{x^8}$

   i $\dfrac{1}{x^{16}}$ **j** $\dfrac{3}{x^8}$ **k** $\dfrac{1}{6x^{20}}$ **l** $5x^6$

   m $\dfrac{x^7}{6}$ **n** $\dfrac{10x^{10}}{3}$ **o** $\dfrac{1}{x^2}$ **p** $\dfrac{1}{16x^4}$

3 a $x^7 - x^{15}$ **b** $x^2 - x^{-3}$

4 a 225   **b** 128   **c** $\frac{1}{243}$ **d** $\frac{1}{8}$

   e $\frac{1}{512}$ **f** 64   **g** 8000   **h** $\frac{9}{4}$

5 a $10\sqrt{5}$ **b** $7\sqrt{7}$ **c** $8\sqrt{6}$ **d** $9\sqrt{5}$

6 a $6\sqrt{2}$ **b** $2\sqrt{3} - 2\sqrt{2}$ **c** $\sqrt{6}$

   d $30\sqrt{2}$ **e** $\frac{9}{8}$ **f** $\frac{3}{4}$

7 a $\dfrac{\sqrt{22}}{22}$ **b** $\dfrac{3\sqrt{7}}{7}$ **c** $\dfrac{\sqrt{2}}{6}$ **d** $\dfrac{\sqrt{2}}{6}$

   e $\dfrac{\sqrt{6} + \sqrt{2}}{2}$ **f** $\dfrac{\sqrt{6} + 1}{5}$

   g $12 + 8\sqrt{2}$ **h** $\dfrac{10(\sqrt{6} + \sqrt{3})}{3}$

Page 277 Revising Chapter 10

1 a 10
  b 8
2 a $a = -8$
  b i 625
    ii $x = \frac{3}{2}$
3 -6
4 a $a + b = 1, -a + b = -5$
  b $a = 3, b = -2$
  c $p(x) = 3x - 2$
5 a

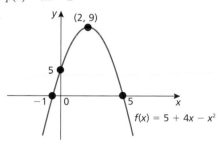

  b i 9   **ii** $(2, 9)$
    iii $x = 2$ **iv** $x = -1, x = 5$
6 a $y - 3x^2$ **b** $y - 2^x$ **c** $y = \dfrac{21}{x}$

7 a i 21 °C
    ii 13 °C
  b 8 hours
8 a £256 million
  b $P = 4^{t-1}$

Page 278 Revising Chapter 11

1 a

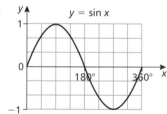

  b

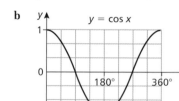

  c

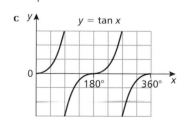

2 A(127, −0·6), B(233, −0·6)

3 a Max. 3, min. −3
  b 180°
  c

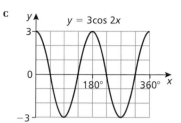

  d The graph of $y = 2 + 3 \cos 2x°$ is the graph of
    $y = 3 \cos 2x°$ translated 2 units up vertically.
    The max. value is 5, the min. value is −1.
4 a $y = 4 \sin 2x°$
  b $y = 2 \cos \frac{1}{2}x°$
5 $y = 1 + 2 \sin 3x°$
6 a Max. height 168 cm, min. height 152 cm
  b first max. after 3 hours, first min. after 9 hours
  c 8 cm   **d** 12 hours
  e i 164 cm   **ii** 167 cm
  f

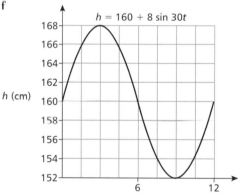

7 a $x = 26·7°$ and 153°
  b $x = 143°$ and 323°
  c $x = 39·8°$ and 220°
8 a $\cos x° = \pm 0·436$
  b 0·5

Page 279 Revising Chapter 12

1 a $x = 13·0$ m
  b $x = 10·4$ m
  c $x = 98·3°$
2 a 118°, $\angle PQR = 46°$
  b 40·5 m
3 $\cos A = \dfrac{9^2 + 8^2 - 7^2}{2 \times 9 \times 8} = \dfrac{2}{3}$
4 105 m
5 Area $\triangle ABC = 2428$ m², area $\triangle RST = 2076$ m²;
  so $\triangle ABC$ has the greater area by 352 m²
6 a 1360 km   **b** 125°   **c** 305°
7 a 106°
  b AE = 28·8 cm
  c 1960 cm²

14 Preparation for assessment

Page 281 Test A Part 1 (Non-calculator)

1 $2 \cdot 524 \times 10^5$ grams

2 £60 : £36

3 $\dfrac{\sqrt{10}}{5}$

4 $16\frac{2}{3}$ cm

5
```
1 | 5 5 6 7 7 7 7 8
2 | 1 4 4 5 5 9
3 | 2 4 8
4 | 0 1 3 4 6 7 8
5 | 0
```
   $n = 25 \qquad 1\,|\,5 = 15$

6 $\dfrac{x^2 + 12}{4x}$

7 $y = 2x + 3$

8 $x = 4, y = -2$

9 $x = -6$

10 a 0   b 6

11 12, 0, -8, -12, -12

12
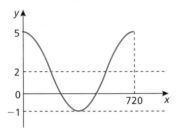

13 $x \geqslant 1$

Page 282 Test A Part 2 (Calculator)

1 a £1764
   b £1839·13

2 a $2p + q = 7, -3p + q = -13$
   b $p = 4, q = -1$
   c $f(x) = 4x - 1$

3 $x = 108 \cdot 4$ or $x = 288 \cdot 4$

4 a 1300 km, 1100 km
   b 6 hours
   c i 1250 km
      ii 1200 km
   d 24 hours
   e
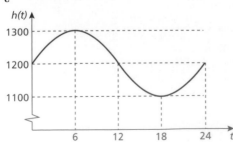

5 a

$\dfrac{1}{n}$	0·25	0·2	0·167	0·143	0·125
t	42	33·6	28	24	21

b $\dfrac{1}{n}$
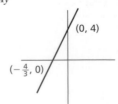

It is a straight line passing through the origin

c $t = \dfrac{168}{n}$ d 14 hours

6 a 4·8 m   b 106·3°   c 4·3 m²
   d 24·2 m²   e 12 100 m³ (3 s.f.)

7 Mean = 44·8, standard deviation = 27·4

8 a 18·4 m²   b £220·62

9 a 84   b £58·80

10 a $(-2)^3 - 2(-2) + 1 = -3$,
      $(-1)^3 - 2(-1) + 1 = 2$ positive when $x = -1$,
      negative when $x = -2$, somewhere between
      it must be zero.
    b $-1 \cdot 62$

11 The campaign has reduced the range, increased
    the mean and reduced the interquartile range.
    This means there is less variation in the
    number of magazines sold by the outlets and
    generally more magazines are sold.

12 a 2, 3, 4, 5; 3, 4, 5, 6; 4, 5, 6, 7; 5, 6, 7, 8; 6, 7,
       8, 9; 7, 8, 9, 10
    b $\frac{4}{24}$ or $\frac{1}{6}$

13 Harmonic 3·33, Geometric 3·53,
    Arithmetic 3·75

Page 284 Test B Part 1 (Non-calculator)

1 4×10^{-2}

2 $5\frac{11}{15}$

3 a $\dfrac{1}{x^{-3}}$ b $\dfrac{1}{x^{28}}$

4 $\dfrac{\sqrt{10}}{4}$

5 Lower 1·4, higher 4·9, median 3·3,
   LQ 2·3, UQ 4·25

6 $\dfrac{-x}{4y}$

7
(0, 4)

$\left(-\frac{4}{3}, 0\right)$

8 $x = 6, y = 12$

9 $x = 3$

10 a 3   b $\frac{1}{3}$

11

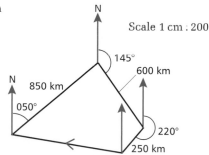

12 75

13 $x = \frac{4}{9}$

Page 285 Test B Part 2 (Calculator)

1 12% profit, £165, £82·50

2 a i £352·17   **ii** £18 312·84

   b i £364·50   **ii** £18 953·79

3 a 40 cm

   b 33·8°

4 48·2°, 311·8°

5 85·2 cm

6 32·2 km

7 a 298 cm

   b 7069 cm²

8 49 km

9 a

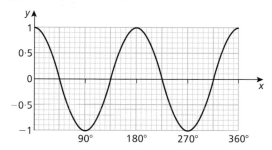

Scale 1 cm : 200

   b 279°   **c** 2 h 49 min

10 0·0172 or 1·72 × 10⁻²

11 4·9, 3·63

12 $\frac{11}{25}$

13 a 9, 14, 19, 24, 29

   b $M = 5n + 4$

   c 43

Page 287 Test C Part 1 (Non-calculator)

1 1·28 × 10⁻¹

2 $8\frac{1}{8}$

3 a $\dfrac{1}{x^{15}}$

   b $x^{2\frac{1}{2}} + \dfrac{1}{x}$

4 $8 - 4\sqrt{3}$

5 a $C = \frac{3}{4}t$

   b 18p

   c 120 minutes

6

Phone Use Campaign

Before		After
3 3 3 2 1 0		
8 8 8 7 7 6 5	6	
4 4 2 1 0	7	0 0 1 2 3 4 4
9 9 8 8 8 6 5	7	6 7 7 8
	8	2 2
	8	5 7 7 7 9
	9	2
	9	5 5 6 6
	10	0 0

$n = 25$ $n = 25$

7|6 represents 76 minutes

7 $\dfrac{3x - 2}{x}$

8 a $s = \frac{5}{2}t + 4$ **b** 29

9 $x = -3, y = 2$

10 $x = -3, x = 2$

11

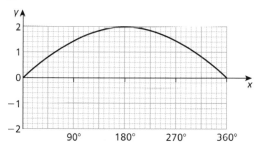

12 $x = 3, 2$

13 $y = 18$

Page 288 Test C Part 2 (Calculator)

1 a £29·08   **b** £11·90

2 a £2635   **b** £106 000

3 a 7·83 m²   **b** 2·70 m

4 $x = 194\cdot5°, 345\cdot5°$

5 a 16·97 miles   **b** 50·9 mph

6 50 cm

7 a 8 km/h   **b** 2 hours

8 3·46 × 10⁻¹³

9 $7\frac{1}{9}$ m²

10 a $x = 6\frac{2}{3}$ **b** $y = 5\frac{1}{4}$

11 a mean = 49·2, standard deviation = 2·42

   b mean is higher than stated average

12 $\frac{2}{5}$

13 a i $F = 80x$ **ii** $F = 50 + 75x$

   b 10 km

Page 290 Test D Part 1 (Non-calculator)

1 1·86 × 10⁸ miles

2 4

3 $7\sqrt{2}$

4 a $y = \dfrac{39}{\sqrt{x}}$ **b** $y = 4\cdot875$ **c** $x = 0\cdot25$

5

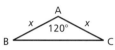

Exam results

23 34 45
17   55

10   20   30   40   50   60

Marks

6 $\dfrac{1-2x}{x+1}$

7 $y = -2x + 1$

8 $x = 4, y = -1$

9 a i

A
x 120° x
B   C

   ii BC = $12 - 2x$

  b Use Area $= \frac{1}{2}ac \sin B$ (angle B $= 30°$)

  b Area $= \frac{1}{2}ac \sin B = \frac{1}{2}(12 - 2x)x \sin 30°$

$$= \frac{1}{2}(12 - 2x)x.\frac{1}{2} = 3x - \frac{x^2}{2}$$

10 $x = 2, 3$

11 a $-\frac{2}{3}$ **b** -43

12 Mrs Jones: £34 000; Mr Jones: £26 000

13 $x = 1.59, -1.26$

Page 291 Test D Part 2 (Calculator)

1 a £1007·70   **b** 12·1%

2 a £365·60   **b** £2200

3 a 11·5 km   **b** 181°

4 P(44·4, 1·4), Q(135·6, 1·4)

5 2·66 metres

6 36·3 cm

7 63·5 miles per hour

8 $x = -0.25, -5$

9 a $\frac{2}{3}$ **b** $1\frac{7}{9}$ or 1·78 litres (to 3 s.f.)

  c 2250 cm²

10 Mean = 7·73, standard deviation = 0·79

11 3·8 and 4·2

12 a (male, 1), (male, 2), (male, 3), (male, 4)
      (female, 1), (female, 2), (female, 3), (female, 4)

  b $\frac{1}{8}$

13 a $\dfrac{x}{80}$ **b** $\dfrac{x+5}{40}$

  c $\dfrac{x+5}{40} - \dfrac{x}{80} = \dfrac{x+10}{80} = 5$; 390 km

Page 293 Test E Part 1 (Non-calculator)

1 1.08×10^{12}

2 $\frac{5}{3}(\sqrt{5} - \sqrt{2})$

3 a $C = \frac{5}{8}wt$ **b** 17·5p

4 a $Q_1 = 52$; $Q_2 = 75.5$; $Q_3 = 82$ **b** 15

5 $\dfrac{x+4}{2(x+3)}$

6 a $V = -1.9A + 20$ **b** £8600

7 Balance entries: 17·07; 125·01 DR;
    113·52; 82·91 DR

8 a $x = 90 - y$ **b** $x = 67.5$; $y = 22.5$

9 $-6y$

10 a $= 2$; $b = 4$; $c = 0$

11 12·5 cm

12 13 cm

13 $Q = \sqrt{A + 2a}$

Page 295 Test E Part 2 (Calculator)

1 a €1·48 = £1   **b** £25·68

2 a £860   **b** 15·6%

3 a 2·47 m   **b** 70·0°

4 a/b

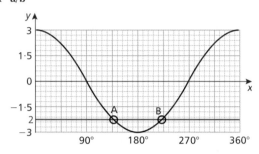

  c A(131·8, -2), B(228·2, -2)

5 $14^2 = 8.4^2 + 11.2^2 = 196$ so, by converse of Pythagoras' theorem, yes

6 499·4 cm³

7 a 1 h 35 min

  b new time will be $\frac{10}{11}$ of old time

8 -4 or $-\frac{2}{3}$

9 mean 62·2 W; standard deviation = 23·3 W

10 3·2 m²

11 a $x + 5y = 8.6$; $2x + 3y = 7.40$ **b** £4·60

12 a

Absences	2	3	4	5	6	7	8	9	10	11	12
Frequency	1	0	4	3	4	2	3	4	2	3	2

  b 0·28

13 a $h = \dfrac{2A}{a+b}$ **b** 4 cm

Page 297 Test F Part 1 (Non-calculator)

1 $14\sqrt{3}$

2 $\dfrac{2x-3}{1-3x}$

3 Lowest value = 11·3; highest value = 20; median = 15·25; lower quartile = 12·75; upper quartile = 16·5

4 $y = 3x - 17$

5 a $x + y = -5$; $x - y = 1$

  b $x = -2, y = -3$;
    the numbers are -2 and -3

6 $x = -\frac{4}{3}$

7 $\tan A = \frac{3}{4}$

8 $a = \frac{1}{2}, b = 1, c = 1$

9 $x = \pm 7$

10 18 cm

11 $x = 9.6$ cm

12 $13^2 = 5^2 + 12^2$ so the square on the hypotenuse is equal to the sum of the squares on the other two sides. Therefore, by the Converse of Pythagoras' theorem, $\triangle ABC$ is right-angled (at C)

13 $x = 3$ and $x = 6$

Page 298 Test F Part 2 (Calculator)

1 5.89×10^3

2 a £579   **b** £864

3 £68 300

4 a 7.31 m   **b** 26.7 m²

5 a max. 6, min. 2   **b** 120°
  c i 6   **ii** 5.78
  d

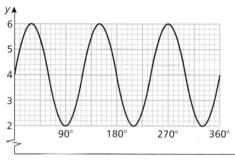

6 Each is 863 cm²

7 27.7 cm²

8 32 cm

9 $x = 1.90$ and -0.10

10 a Height of rebound (cm)

6	0 2 2 2 4 5 8 8
7	2 2 3 3 5 5 6 6 7 7 8 9
8	1 4 6 9
9	0

$n = 25$ 7|2 represents 72

b median = 75 cm

c mean is 73.76 cm, so median is slightly bigger

11 a $T_{total} = T_{down} + T_{up} = \dfrac{3}{x+3} + \dfrac{3}{x-3}$

b $\dfrac{3}{x+3} + \dfrac{3}{x-3} = 1 \Rightarrow x = 8$; 8 km/h

12 $\dfrac{a}{\sin x°} = \dfrac{2a}{\sin 30°} \Rightarrow a \sin 30° = 2a \sin x$

$\Rightarrow \sin x° = \dfrac{a \sin 30°}{2a} = \dfrac{\frac{1}{2}}{2} = \dfrac{1}{4}$

13 a $\dfrac{x^2}{36} + \dfrac{y^2}{9} = 1 \Rightarrow x^2 + 4y^2 = 36$

$\Rightarrow x^2 = 36 - 4y^2 = 4(9 - y^2)$

$\Rightarrow x = \sqrt{4(9 - y^2)} = 2\sqrt{(9 - y^2)}$

b $x = 3.6$